Dear Prof. Smith

Best wishes

[signature]
Jan 99

*Low-Voltage
CMOS VLSI Circuits*

Low-Voltage CMOS VLSI Circuits

James B. Kuo
Jea-Hong Lou
NTUEE
Taipei, Taiwan

A Wiley-Interscience Publication
JOHN WILEY & SONS, INC.
New York / Chichester / Weinheim / Brisbane / Singapore / Toronto

This text is printed on acid-free paper. ⊗

Copyright © 1999 by John Wiley & Sons, Inc. All rights reserved.

Published simultaneously in Canada.

No part of this publication may be reproduced, stored in a retrieval system or transmitted in any form or by any means, electronic, mechanical, photocopying, recording, scanning or otherwise, except as permitted under Section 107 or 108 of the 1976 United States Copyright Act, without either the prior written permission of the Publisher, or authorization through payment of the appropriate per-copy fee to the Copyright Clearance Center, 222 Rosewood Drive, Danvers, MA 01923, (978) 750-8400, fax (978) 750-4744. Requests to the Publisher for permission should be addressed to the Permissions Department, John Wiley & Sons, Inc., 605 Third Avenue, New York, NY 10158-0012, (212) 850-6011, fax (212) 850-6008, E-Mail: PERMREQ@WILEY.COM.

Library of Congress Cataloging-in-Publication Data:

Kuo, James B., 1956–
  Low-voltage CMOS VLSI circuits / James B. Kuo and Jea-Hong Lou.
    p.   cm.
  Includes bibliographical references and indexes.
  ISBN 0-471-32105-2 (cloth : alk. paper)
  1. Low voltage integrated circuits. 2. Integrated circuits—Very
large scale integration. 3. Metal oxide semiconductors,
Complementary.   I. Lou, Jea-Hong. II. Title
TK7874.66.K86   1999   98-37315
621.39'5—dc21

Printed in the United States of America

10 9 8 7 6 5 4 3 2 1

Acknowledgments

The authors would like to thank colleagues in Taiwan and professors at Stanford University for their support during the past eleven years. Without their encouragements this work would not have been possible. The authors would like to thank S. C. Lin, K. W. Su, C. K. Huang, C. Y. Huang, Y. M. Huang, J. C. Su, H. K. Sui, and K. H. Yuan, for their help on drawing the illustrations. The authors would also like to thank Ms. Cassie Craig, Mr. George Telecki, and Mr. Andrew Prince of John Wiley & Sons, Inc., for their help in publishing this book.

Preface

CMOS VLSI Technology has been progressing quickly for decades. As a result, information-oriented digital computer systems based on CMOS VLSI are changing our daily life at an unprecedented pace. Currently, CMOS VLSI is still advancing at a remarkable rate. In the future, VLSI digital chips having multi-billion transistors based on deep-submicron CMOS technology using low power supply voltages will be utilized to integrate low-power, high-speed, high-capacity information-oriented digital computer systems. For deep-submicron CMOS technology using a low power supply voltage, the techniques required for designing low-voltage VLSI CMOS circuits are challenging. Processing technology and devices have direct impacts on the circuit designs for low-voltage CMOS VLSI systems. Low-power VLSI system-oriented applications require innovative low-voltage CMOS VLSI circuits. In this book the authors approach this hot topic, from processing technology and devices to circuits and systems. Starting from fundamentals of deep-submicron CMOS technologies and devices, including SOI and BiCMOS, evolutions of CMOS static and dynamic logic circuit families are described. Diversified CMOS static and dynamic logic circuits based on various processing technologies, including BiCMOS and SOI, are analyzed. Specifically, dynamic threshold SOI logic circuits, multi-threshold standby/active techniques, and low-voltage BiCMOS logic circuits are included. Bootstrapped techniques for designing low-voltage CMOS VLSI circuits are presented. In addition, low-power circuit techniques, including bus architecture approaches and adiabatic logic for VLSI systems, are depicted. Fundamental difficulties in CMOS dynamic logic circuits for low-voltage operation, including race, noise, and charge sharing problems, and their solutions, are analyzed. True-single-phase-clocking dynamic logic circuits and high-speed dynamic flip-flops are analyzed. In the CMOS memory circuits, evolutions of volatile and non-volatile memory circuits are described. Fundamental operations of SRAM and DRAM circuits in terms of basic components and critical path analysis are included. Innovative low-voltage and low-power circuit techniques for realizing VLSI CMOS DRAM and SRAM, including embedded and application-specific memory circuits, are also analyzed. Special circuit techniques for SOI and BiCMOS technologies in realizing low-voltage DRAM and SRAM are also explained. Evolutions of non-volatile memory including ROM, EPROM, EEPROM, and flash are presented. Basic operations of recently emerged ferroelectric RAM (FRAM) suitable for use as low-voltage non-volatile memory are also included. In the last portion of the book, low-voltage CMOS VLSI systems are described. Starting from basic building blocks, including adders and multipliers, multi-port register files and cache memory are analyzed. In addition, analog and digital phase-locked loop circuits for generating high-frequency clocks in the VLSI systems are explained. Processing units including central processing unit (CPU), floating-point processing unit (FPU), and digital signal processing (DSP) unit are included. Low-voltage VLSI systems realized by other technologies including BiCMOS and SOI are also presented.

This book is written for undergraduate senior students and first-year graduate students interested in CMOS circuit designs. The arrangement of the book is designed for a three-unit course. This book is also suitable for engineering professionals interested in this field.

JAMES B. KUO

JEA-HONG LOU

Taipei, Taiwan
December 1998

About the Authors

Prof. James B. Kuo received a BSEE degree from National Taiwan University in 1977, an MSEE degree from Ohio State University in 1978, and a PhDEE degree from Stanford University in 1985. Since 1987, he has been with National Taiwan University, where he currently is a professor. He has published numerous international journal papers. He serves as an associate editor for the *IEEE Circuits and Devices Magazine* and an AdCom member for the IEEE Electron Devices Society.

Dr. Jea-Hong Lou received BSEE and PhDEE degrees from National Taiwan University in 1994 and 1998, respectively. His research specialty is circuit designs for CMOS VLSI.

Contents

Acknowledgments		v
Preface		vii
1	**Introduction**	**1**
	1.1 Why CMOS?	1
	1.2 Why Low-Voltage?	3
	1.2.1 Side Effects of Low-Voltage	6
	1.2.2 Supply Voltage Reduction Strategy	7
	1.3 Objectives	9
	References	11
2	**CMOS Technology and Devices**	**13**
	2.1 Evolution of CMOS Technology	13
	2.2 $0.25\mu m$ CMOS Technology	17
	2.3 Shallow Trench Isolation	18
	2.4 LDD	20
	2.5 Buried Channel	21
	2.6 BiCMOS Technology	24
	2.7 $0.1\mu m$ CMOS Technology	29
	2.8 SOI CMOS Technology	32

xii CONTENTS

2.9	Threshold Voltage	38
2.10	Body Effect	41
2.11	Short Channel Effects	42
2.12	Narrow Channel Effects	44
2.13	Mobility & Drain Current	44
2.14	Subthreshold Current	49
2.15	Electron Temperature	53
2.16	Velocity Overshoot	61
2.17	MOS Capacitances	61
2.18	Hot Carrier Effects	70
2.19	BSIM SPICE Models	74
2.20	Summary	79
	References	79
	Problems	83

3 CMOS Static Logic Circuits   85

3.1	Basic CMOS Circuits	85
3.2	CMOS Inverters	86
	3.2.1 CMOS Static Logic Circuits	96
	3.2.2 Difficulties for Low-Power and Low-Voltage Operations	98
3.3	CMOS Differential Static Logic	101
	3.3.1 Differential Cascode Voltage Switch (DCVS) Logic	101
	3.3.2 Differential Split-Level (DSL) Logic	103
	3.3.3 Differential Cascode Voltage Switch with Pass-Gate (DCVSPG) Logic	105
3.4	CMOS Pass-Transistor Logic	108
	3.4.1 Pass-Transistor Logic Fundamentals	108
	3.4.2 Other Pass-Transistor Logics	110
	3.4.3 CAD for Pass-Transistor Logic	113
3.5	BiCMOS Static Logic Circuits	115
	3.5.1 Standard BiCMOS	116
	3.5.2 Sub-3V BiCMOS	125
	3.5.3 1.5V BiCMOS Logic with Transient Feedback	126
	3.5.4 1.5V Bootstrapped BiCMOS Logic	127
3.6	SOI CMOS Static Logic	128
3.7	Low-Voltage CMOS Static Logic Circuit Techniques	132
	3.7.1 Bootstrapped CMOS Driver	132

	3.7.2	Multi-Threshold Standby/Active Techniques	135
3.8	Low-Power CMOS Circuit Techniques		140
	3.8.1	Bus Architecture Approach	145
	3.8.2	Adiabatic Logic	153
3.9	Summary		157
	References		157
	Problems		161

4 CMOS Dynamic Logic Circuits — 163

4.1	Basic Concepts of Dynamic Logic Circuits	163
4.2	Charge-Sharing Problems	167
4.3	Noise Problem	171
4.4	Race Problem	172
4.5	NORA	172
4.6	Zipper	177
4.7	Domino	178
	4.7.1 Latched Domino	182
	4.7.2 Skew-Tolerant Domino	185
	4.7.3 Multiple-Output Domino Logic (MODL)	188
4.8	Dynamic Differential	190
4.9	True-Single-Phase Clocking (TSPC)	196
4.10	BiCMOS Dynamic Logic Circuits	206
	4.10.1 1.5V BiCMOS Dynamic Logic Circuit	209
	4.10.2 1.5V BiCMOS Latch	213
	4.10.3 Charge-Sharing Problems	214
4.11	Low-Voltage Dynamic Logic Techniques	217
	4.11.1 Bootstrapped Dynamic Logic (BDL) Circuit	218
	4.11.2 Bootstrapped All-N-Logic TSP Bootstrapped Dynamic Logic	222
	4.11.3 Semi-Dynamic DCVSPG-Domino Logic Circuits	225
4.12	Summary	229
	References	229
	Problems	233

5 CMOS Memory — 235

5.1	SRAM	236
	5.1.1 Basics	236
	5.1.2 Memory Cells	239

		5.1.3	Decoders	243
		5.1.4	Bit-Line Related Architecture and Sense Amps	246
		5.1.5	SRAM Architecture	248
		5.1.6	Critical Path Analysis	250
		5.1.7	Application-Specific SRAM	257
		5.1.8	Low-Voltage SRAM Techniques	261
	5.2	DRAM		264
		5.2.1	Memory Cells	265
		5.2.2	Bit-Line Architecture	268
		5.2.3	Advanced Architecture	274
		5.2.4	Low-Power DRAM Techniques	276
		5.2.5	Low-Voltage DRAM Techniques	278
	5.3	BiCMOS Memories		286
		5.3.1	BiCMOS SRAM	287
		5.3.2	BiCMOS DRAM	295
	5.4	SOI Memory		299
		5.4.1	SOI SRAM	299
		5.4.2	SOI DRAM	302
	5.5	Nonvolatile Memory		307
		5.5.1	ROM	308
		5.5.2	EPROM	311
		5.5.3	EEPROM	314
		5.5.4	Flash Memory	317
	5.6	Ferroelectric RAM (FRAM)		328
	5.7	Summary		337
		References		337
		Problems		343
6	**CMOS VLSI Systems**			**345**
	6.1	Adders		345
		6.1.1	Basic Principles	345
		6.1.2	Carry Look-Ahead Adder	347
		6.1.3	Manchester CLA Adder	349
		6.1.4	PT-Based CLA Adder	351
		6.1.5	PT-Based Conditional Carry-Select CLA Adder	352
		6.1.6	Enhanced MODL Adder	355
		6.1.7	Carry-Skip CLA Circuit	357
		6.1.8	Parallel and Pipelined Adders	359

6.2	Multipliers		363
	6.2.1	Modified Booth Algorithm	366
	6.2.2	Advanced Structure	368
6.3	Register File and Cache Memory		370
6.4	Programmable Logic Array		380
6.5	Phase-Locked Loop		386
6.6	Processing Unit		394
	6.6.1	Central Processing Unit (CPU)	394
	6.6.2	Floating-Point Processing Unit	408
	6.6.3	Digital Signal Processing Unit	412
6.7	Other Technologies		418
	6.7.1	BiCMOS Systems	418
	6.7.2	SOI Systems	423
6.8	Summary		427
	References		427
	Problems		431

Index   **433**

1
Introduction

Low-voltage operation has become a general trend for CMOS VLSI circuits and systems. In this chapter, evolution of the low-voltage CMOS VLSI systems is briefly described, followed by the objectives of this book.

1.1 WHY CMOS?

Since the invention of the transistors, semiconductor technology has been progressing at a rapid pace. Various semiconductor fabrication processing technologies have been developed. As shown in Fig. 1.1, digital IC technologies can be divided into two groups: active substrate and inert substrate. In the active substrate group, it is divided into two categories: silicon and III-V. In the inert substrate group, there are thin-film and thick-film technologies. Depending on the property of the thin-film, they can be categorized as silicon-on-insulator (SOI), polysilicon thin-film transistors (TFT), and amorphous TFT. In the active substrate, for the III-V group, there are GaAs MESFET, heterojunction bipolar transistors (HBT), and modulation doped FETs (MODFET). Until now, silicon technologies have been the most widely used in the semiconductor industry. In the silicon group, there are bipolar and MOS technologies. The bipolar technologies can be further divided into three parts: transistor-triggered logic (TTL), integrated injection logic (I^2L), and emitter-coupled logic (ECL). In MOS, there are PMOS, NMOS, and CMOS technologies. Combining CMOS with bipolar, BiCMOS technology has also been developed before. Among so many digital IC technologies, CMOS technology has become the dominant technology for VLSI. In the future, CMOS technology will still be a leading technology for VLSI for several reasons. Silicon material is easy to obtain—wafer cost is cheap. In addition, CMOS

2 INTRODUCTION

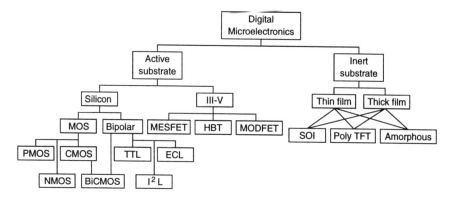

Fig. 1.1 Family of digital IC.

circuits are less difficult to design. Compared to other devices, the physics of CMOS devices is easier to understand—CAD of the CMOS devices is easier to implement. Along with the progress in the CMOS processing technology, performance of CMOS circuits has been improved consistently. Among all technologies, CMOS technology offers the lowest power delay product. Under an identical speed performance, the circuits realized by CMOS technology consume the least power. Therefore, under a certain power consumption limit, using CMOS technology a chip can have the most transistor count—using CMOS technology the most functions can be integrated in a chip. This is why CMOS is the dominant technology for VLSI.

Fig. 1.2 shows gate delay versus power consumption of various logic families[1]. At a designated power delay product value, CMOS offers the smallest power-delay product among all technologies. CMOS technology is the most suitable for realizing VLSI systems. One important reason for CMOS technology becoming the dominant technology for VLSI is that CMOS devices are continuously scaled down. A down-scaled CMOS device brings in an increased driving current, reduced parasitics, reduced power dissipation per gate, and an increased packing density. Thus, the cost per function can be reduced.

Fig. 1.3 shows the evolution of CMOS technology based on ISSCC[2] and the 1997 National Technology Roadmap for Semiconductors (NTRS) report [3]. As shown in the figure, the number of transistors in a microprocessor chip has been increasing quickly and the power consumption per transistor has been decreasing accordingly. Channel length of CMOS devices has been scaled down exponentially with time. Along with the progress in the processing technology, power supply voltage (V_{DD}) is also shrunk accordingly. In the next century, for the sub-$0.1\mu m$ CMOS technology, the power supply voltage is expected to be 1.5V or below.

WHY LOW-VOLTAGE? 3

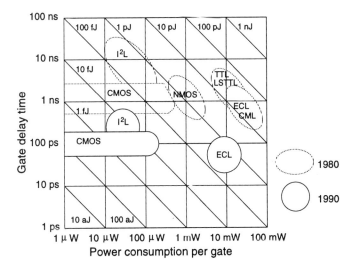

Fig. 1.2 Gate delay versus power consumption of various logic families. (Adapted from Kohyama[1].)

1.2 WHY LOW-VOLTAGE?

Why is low-voltage used for CMOS VLSI? As shown in Fig. 1.4[6], based on the constant voltage (CV) scaling, when the channel length is scaled down, its lateral electric field in a device increases to a large extent. Therefore, device reliability is degraded. Due to the decrease in the gate oxide thickness, the vertical electric field also increases, which leads to reduced oxide reliability. In order to reduce vertical and lateral electric fields, reduction of power supply voltage is necessary—low voltage. Along with down-scale of the CMOS devices, device count of a CMOS VLSI chip increases quickly. Although CMOS has the lowest DC power dissipation, its dynamic power dissipation increases quickly along with the progress in the CMOS processing

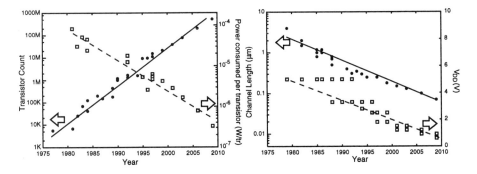

Fig. 1.3 Evolution of the CMOS technology based on the 1997 NTRS report.

4 INTRODUCTION

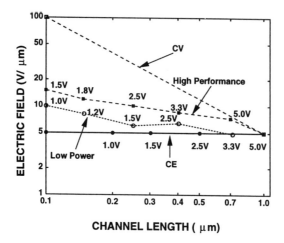

Fig. 1.4 Lateral electric field versus channel length in the CMOS devices. (Adapted from Davari et al.[6].)

technology. As shown in Fig. 1.5, power dissipation of microprocessors and digital signal processors increases along with the progress in CMOS VLSI technology[7]. Therefore, the increased power consumption results in a raised ambient temperature. Thus the device performance is degraded and the circuit performance is less stable. Thus good packaging is required for dissipating heat of a VLSI chip. For a general plastic package, it can handle up to around 1W of power dissipation. To handle higher power dissipation up to 10W, expensive ceramic packaging is required. For present-day high-performance microprocessors, a fan or other cooling device may be required. Based on trends as shown in the figure, power dissipation capability for future CMOS VLSI circuits may still be not enough. As shown in Fig. 1.6, power consumption of the future CMOS VLSI chips is still going up quickly[7]. Efficient power dissipation of CMOS VLSI chips is a challenge for engineering future CMOS VLSI systems using down-scaled CMOS devices.

Lowering power supply voltage is the most efficient method in reducing power dissipation of a CMOS VLSI chip. The dynamic power dissipation of a CMOS circuit is given by[11]:

$$P = \alpha_{0 \to 1} \cdot C_L \cdot V_{DD}^2 \cdot f_{CLK},$$

where $\alpha_{0 \to 1}$ represents the probability of the logic gate output to change from 0 to 1, which is the switching activity, C_L is the load capacitance, and f_{CLK} is the clock frequency. From the above equation, dynamic power dissipation depends on the switching activity, the output capacitive load, the supply voltage, and the clock frequency. Due to the increased demands on the system performance, the clock frequency increases. Power supply voltage can be reduced to shrink dynamic power dissipation. Via proper circuit and system designs, switching activity can be reduced to decrease power dissipation. Output load capacitance can be reduced by

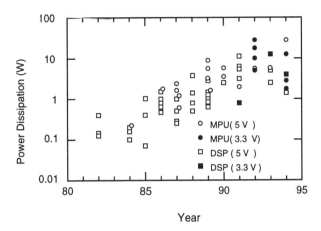

Fig. 1.5 Power dissipation of microprocessors and digital signal processors. (From Kuroda and Sakurai [7], ©1995 IEICE.)

an advanced CMOS technology or by reducing device dimensions to reduce power dissipation. Among these three approaches to reduce power dissipation, reduction of the power supply voltage is the most effective way since when V_{DD} is lowered, power consumption is reduced in square law: $\propto V_{DD}^2$. Thus, to reach low-power goal, low-voltage is also the most straightforward approach.

Lowering power dissipation is also important for a high-performance CMOS VLSI system. When power dissipation of a CMOS VLSI chip increases, the increased ambient temperature worsens the electromigration reliability problems. For a DRAM chip, a rise in the junction temperature results in an increase in the leakage current. As a result, the data retention time (refresh time) is lowered. Thus, lowering power dissipation is important for high-performance VLSI circuits. Recently, due to rapid progress in mobile computing and communication, demands on low power are on lengthening battery life. The progress in battery technology has not yet matched the speed of increase in power dissipation of a VLSI chip. Therefore, lowering power supply voltage is a good approach in accomplishing high performance or meeting low-power (low-energy) requirements.

Table 1.1 shows the strategies in converting a high-performance chip to a low-power chip using various power reduction methods[8]. As shown in the table, the DEC Alpha 21064 chip operating at a supply voltage of 3.45V and power dissipation of 26W has been used as a starting point. As shown in the table, among all power reduction approaches, the supply voltage reduction approach is the most effective. When the supply voltage is scaled down from 3.45V to 1.5V, power dissipation is reduced by 5.3×. Other approaches cannot match the low-voltage approach.

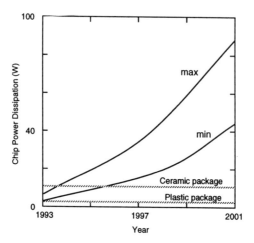

Fig. 1.6 Prediction of CMOS power consumption for a VLSI chip. (Adapted from Kuroda and Sakurai [7].)

1.2.1 Side Effects of Low-Voltage

As described before, lowering power supply voltage can offer increased device reliability and reduced power dissipation. Lowering power supply voltage also brings in disadvantages. Fig. 1.7 shows the bit-line sensing delay versus the operating voltage and the threshold voltage of an SRAM circuit[9]. As shown in the figure, when the supply voltage is scaled down, the speed of the CMOS SRAM is degraded since the sensing delay is inversely proportional to the driving current of the device: $T_d \propto \frac{C_L \Delta V}{I_{Dmax}}$, where the maximum drain current is expressed as $I_{Dmax} = \frac{1}{2}\mu_n C_{ox} \frac{W}{L}(V_{DD} - V_{TH})^2$. When V_{DD} is lowered, the sensing time increases. As shown in the figure, when the threshold voltage of the sense amp is decreased, its speed increases. In order to lessen the degradation of the circuit performance, when V_{DD} is lowered, threshold voltage (V_{TH}) should be reduced too. In addition to shrinking the delay time, reduction of threshold voltage also provides a smaller sensitivity to the variation in the supply voltage.

Reduction of threshold voltage also brings in disadvantages—an increased subthreshold current results in a larger DC leakage current. As a result, the DC power consumption is increased. As shown in Fig. 1.8, when the DRAM size increases, DC current due to the subthreshold leakage current increases[10]. For a 1G-bit DRAM, the DC subthreshold leakage current is comparable to the AC switching current. At a larger DRAM capacity, the DC current is even larger than the AC switching current— DC current dominates power consumption of the whole DRAM chip. Although reducing threshold voltage offers an enhancement in the circuit performance, the increased subthreshold leakage current raises the DC power dissipation, which determines the battery life of a mobile system since most of the time it is in the sleep mode.

Alpha 21064: 26W @ 200MHz and 3.45V

Methods	Power Reduction	Expected Power
Reduce V_{DD} 3.45V → 1.5V	5.3×	4.9W
Reduce function	3×	1.6W
Scale process	2×	0.8W
Reduce clock load	1.3×	0.6W
Reduce clock rate	1.25×	0.5W

Table 1.1 Strategies in converting a high-performance chip to a low-power chip. (Adapted from Montanaro et al.[8].)

1.2.2 Supply Voltage Reduction Strategy

As described in the previous section, lowering supply voltage offers many tradeoffs. Depending on applications, there are two strategies in the reduction of supply voltage—high performance and low power approaches. For the high-performance approach, lowering supply voltage is targeted to raise system reliability—the electromigration reliability, the hot-carrier reliability, the oxide stress reliability, and other reliabilities related to high electric field and temperature. For the high-performance approach, power supply voltage is not scaled down aggressively. Instead, under the scaled supply voltage, circuit performance is optimized. For the low-power approach, it is for mobile systems, which emphasize lengthening battery life. The degraded performance of the circuits at a reduced supply voltage can be compensated by using advanced technology. In addition, the degraded performance of the circuits due to the reduced supply voltage can also be made up by adopting system parallelism and pipelining.

When gate length is shrunk, threshold voltage cannot be scaled down accordingly due to subthreshold leakage consideration. At a reduced supply voltage, downscaling of devices may lead to degradation of the device performance. In the industry, a relationship between minimum supply voltage and minimum threshold voltage has been found:

$$V_{DD,min} = 3V_{TH,min},$$

where the minimum supply voltage is set to about three times the minimum threshold voltage. Considering the practical low-voltage limit[11], the minimum supply voltage can be determined by the minimum threshold voltage. The minimum threshold voltage is determined by three factors: (1) threshold voltage variation due to process fluctuation, (2) threshold voltage variation due to temperature effect, and (3) on-off current ratio of the device. Therefore, the minimum threshold voltage is expressed

8 INTRODUCTION

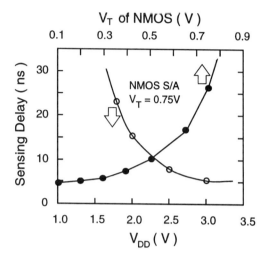

Fig. 1.7 Sensing delay versus operating voltage and threshold voltage of an SRAM. (Adapted from Matsuzawa [9].)

as:

$$V_{TH,min} = \Delta V_{T,proc} + \Delta V_{T,temp} + \left(n \cdot \frac{kT}{q} \cdot ln10\right) \cdot log\frac{I_{on}}{I_{off}},$$

where $(n \cdot \frac{kT}{q} \cdot ln10)$ is called subthreshold slope (mV/dec), which means the required change in the gate voltage when the subthreshold current shrinks 10 times. n represents $n = 1 + \frac{C_D}{C_{ox}} = 1 + \frac{1}{C_{ox}}\sqrt{\frac{\epsilon_{si}qN_A}{2(2\phi_f+V_{SB})}}$, which will be described in Chapter 2, where C_D is the depletion capacitance in the silicon substrate, and C_{ox} is the unit area gate oxide capacitance. For a CMOS technology, assume its subthreshold slope is 76mV/dec and the fluctuations due to process and temperature variations are $0.05V$. At 2-decade and 5-decade on/off current ratios, the minimum threshold voltages are 0.25V and $0.48V$, respectively. Therefore, from the above equation, the minimum supply voltages are $V_{DD,min} = 0.75V$ and $1.43V$, respectively. For lowering the minimum threshold voltage ($V_{TH,min}$), the threshold voltage fluctuation due to process variation and temperature coefficient should be reduced as much as possible. In addition, lowering the subthreshold slope to close to its ideal value—$\frac{kT}{q}ln10 \cong 60mV/dec$—should also be the goal. Generally speaking, a current on-off ratio of 10^2 is acceptable for VLSI logic circuits. The goal with a current on-off ratio of 10^5 is for reaching the goal of low DC standby leakage for VLSI memory circuits.

Fig. 1.9 shows the threshold voltage versus supply voltage at various goals—active power, standby power, and clock frequency. The arrow in the figure represents that if a certain goal is to be enhanced, which direction for threshold voltage and supply voltage should go. If standby power is to be reduced, threshold voltage should be raised and supply voltage should be reduced with an emphasis on threshold

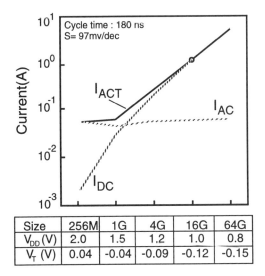

Fig. 1.8 Trend in active current of DRAM chips. (From Itoh et al.[10], ©1995 IEEE.)

voltage. If active power is to be reduced, supply voltage should be reduced and threshold voltage should be increased with an emphasis on supply voltage. To increase performance, supply voltage should be increased and threshold voltage should be decreased, which is in conflict with the last two trends. Considering the approach in realizing three goals—optimization of standby power, active power, and performance, an optimized situation with the triangle bounded by these three curves becoming a point can be found.

1.3 OBJECTIVES

Fig. 1.10 shows the key factors in engineering low-voltage CMOS VLSI. As shown in the figure, from low-voltage CMOS technology and low-voltage CMOS devices, low-voltage CMOS static and dynamic logic circuits are designed. From low-voltage CMOS static and dynamic logic circuits, low-voltage CMOS memory circuits and low-voltage CMOS VLSI systems are integrated. In this book, from circuit point of view, effects of low-voltage operation are analyzed. The evolution of the digital circuit techniques for implementing CMOS VLSI systems are described in the book. This book emphasizes the fundamental techniques to realize low-voltage CMOS digital circuits.

There are six chapters in this book. In Chapter 2, basics and evolution of the CMOS technology and CMOS devices are described. Starting from $0.8\mu m$ CMOS technology, $0.25\mu m$ CMOS technology is described. Key processing techniques including LDD (lightly doped drain) and STI (shallow trench isolation) and challenges for realizing deep-submicron CMOS technology such as $0.1\mu m$ technology are re-

10 INTRODUCTION

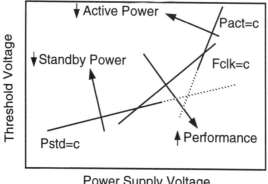

Fig. 1.9 Threshold voltage versus supply voltage. (From Stork [12], ©1995 IEEE.)

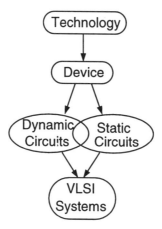

Fig. 1.10 Engineering low-voltage CMOS VLSI.

viewed. In addition, BiCMOS and silicon-on-insulator (SOI) CMOS technologies are also described. In CMOS devices, starting from threshold voltage, drain current, subthreshold conduction, electron-temperature dependent mobility, capacitance, and hot-electron effects are described. Finally in Chapter 2, fundamental device models used in the BSIM1 SPICE circuit simulation program are reviewed.

In Chapter 3, the evolution of innovative static logic circuits is described. Various static logic techniques including differential logic, pass-transistor logic, BiCMOS static logic, and SOI logic are described. Low-voltage static logic circuit techniques are also included. In the final portion of Chapter 3, low-power static logic circuit techniques are presented. In Chapter 4, CMOS dynamic logic circuits are described. Starting from the fundamental techniques, various logic families including NORA, Zipper, domino, latched domino, multiple output domino, skew-tolerant domino, and true-single-phase clocking (TSPC) dynamic logic are introduced. In the final portion of Chapter 4, BiCMOS dynamic logic circuits are analyzed, followed by the low-voltage dynamic logic circuit techniques.

In Chapter 5, various memory families including volatile memory—SRAM and DRAM, and non-volatile memory—ROM, EPROM, EEPROM, flash memory, and the ferroelectric memory (FRAM) are described. Various circuit technologies for low-voltage environment are included. In Chapter 6, basic circuits in CMOS VLSI systems are described. Basic circuits include adder, multiplier, register file, content addressable memory (CAM), cache memory, programmable logic array (PLA), and phase-locked loop (PLL). In addition, fundamental structures and related circuit techniques for VLSI systems such as floating-point unit (FPU), central processing unit (CPU), and digital signal processor (DSP) are described. In the final portion of Chapter 6, circuit techniques for BiCMOS and SOI systems are presented.

REFERENCES

1. S. Kohyama, *Very High Speed MOS Devices*, Clarendon Press: Oxford, 1990.

2. *International Solid-State Circuits Conference (ISSCC)* Digests.

3. 1997 National Technology Roadmap for Semiconductors, *http://notes.sematch.org/97pelec.htm*.

4. J. E. Brewer, "A New and Improved Roadmap: The NTRS Provides the High-Level Vision Necessary to Facilitate Future Advancements in Semiconductor Technology," *IEEE Circuits and Devices Magazine*, **14**(3), 13–18 (1998).

5. P. K. Vasudev and P. M. Zeitzoff, "Si-ULSI with a Scaled-Down Future: Trends and Challenges for ULSI Semiconductor Technology in the Coming Decade," *IEEE Circuits and Devices Magazine*, **14**(3), 19–29 (1998).

6. B. Davari, R. H. Dennard, and G. G. Shahidi, "CMOS Scaling for High Performance and Low Power—The Next Ten Years," *IEEE Proc.*, **83**(4), 595–606 (1995).

7. T. Kuroda and T. Sakurai, "Overview of Low-Power ULSI Circuit Techniques," *IEICE Trans. Electron.*, **E78-C**(4), 334–343 (1995).

8. J. Montanaro, R. T. Witek, K. Anne, A. J. Black, E. M. Cooper, D. W. Dobberpuhl, P. M. Donahue, J. Eno, G. W. Hoeppner, D. Kruckemyer, T. H. Lee, P. C. M. Lin, L. Madden, D. Murray, M. H. Pearce, S. Santhanam, K. J. Snyder, R. Stephany, and S. C. Thierauf, "A 160-MHz, 32-b, 0.5-W CMOS RISC Microprocessor," *IEEE J. Sol. St. Ckts.*, **31**(11), 1703–1714 (1996).

9. A. Matsuzawa, "Low-Voltage and Low-Power Circuit Design for Mixed Analog/Digital Systems in Portable Equipment," *IEEE J. Sol. St. Ckts.*, **29**(4), 470–480 (1994).

10. K. Itoh, K. Sasaki, and Y. Nakagome, "Trends in Low-Power RAM Circuit Technologies," *IEEE Proc.*, **83**(4), 524–543 (1995).

11. K. Shimohigashi and K. Seki, "Low-Voltage ULSI Design," *IEEE J. Sol. St. Ckts.*, **28**(4), 408–413 (1993).

12. J. M. C. Stork, "Technology Leverage for Ultra-Low Power Information Systems," *IEEE Proc.*, **83**(4), 607–618 (1995).

2
CMOS Technology and Devices

In this chapter, CMOS technology and devices are described. First, the evolution of the CMOS technology is described, followed by a 0.25μm CMOS technology. Shallow trench isolation (STI) and the lightly-doped drain (LDD) structure used in the deep-submicron CMOS technology are also described. Then BiCMOS technology is depicted, followed by 0.1μm CMOS technology and SOI CMOS technology. Then CMOS device behaviors are analyzed. From threshold voltage, body effect, short channel effect, and narrow channel effect are analyzed. Then, from the mobility model in terms of the velocity overshoot phenomenon, the drain current model including the electron temperature effect for deep-submicron CMOS devices is presented. Finally, the capacitance model and hot-carrier effects of the CMOS devices are described. In addition, the BSIM SPICE models for deep-submicron CMOS devices are summarized.

2.1 EVOLUTION OF CMOS TECHNOLOGY

Two major products of the digital VLSI circuits are DRAMs and microprocessors. The evolution of DRAMs is emphasized on the miniaturization of CMOS devices. The evolution of CMOS technology has a direct impact on the performance of DRAMs—state-of-the-art DRAMs are always based on leading CMOS technology. Fig. 2.1 shows the evolution of DRAMs based on papers published in ISSCC[1]. As shown in Fig. 2.1, the size of DRAMs increased from 1K based on 8μm MOS technology in 1970 to 64M based on 0.35μm CMOS technology in 1995. For the past 25 years, the size of DRAMs became four times bigger for every two years and the size of the MOS devices used was scaled down accordingly. According to the

14   CMOS TECHNOLOGY AND DEVICES

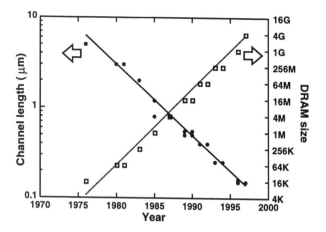

Fig. 2.1 Evolution of DRAMs.

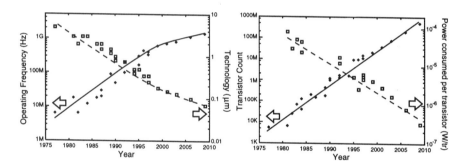

Fig. 2.2 Evolution of CMOS microprocessors.

trend, in year 2003, 1G-bit DRAMs, which are based on $0.18\mu m$ CMOS technology, will be in mass production. In year 2010, 128G-bit DRAMs may be in the market.

In addition to DRAMS, the evolution of CMOS technology also has had a strong impact on the progress of microprocessors. Fig. 2.2 shows the evolution of the microprocessors based on the papers published in ISSCC digest[1] and 1997 NTRS report[2]. From the figure, the evolution of the CMOS microprocessor chips is impressive. In 1980, there were only 30,000 transistors in a microprocessor chip based on $3\mu m$ CMOS technology. In 1992, in a microprocessor chip there were two million transistors based on $0.6\mu m$ CMOS technology. As shown in the figure, the number of transistors in a microprocessor chip increased exponentially and the size of transistors decreased accordingly. As a result, the number of functions in a microprocessor chip increased drastically. Based on this trend, in 2010, a microprocessor chip will have over 100 million transistors based on sub-$0.1\mu m$ CMOS technology. In 1995, for a multi-million microprocessor chip based on a $0.4\mu m$ CMOS technology, its power supply voltage was 3.3V and its power dissipation was 10W, operating at a

Parameter	CE	CV
W, L, t_{ox}	$\frac{1}{s}$	$\frac{1}{s}$
V_{DD}, V_{TO}	$\frac{1}{s}$	1
C_{ox}	$\frac{1}{s}$	$\frac{1}{s}$
I_D	$\frac{1}{s}$	s
Delay	$\frac{1}{s}$	$\frac{1}{s^2}$
DC Power	$\frac{1}{s^2}$	s
Power-Delay Product	$\frac{1}{s^3}$	$\frac{1}{s}$
Parasitic Capacitance ($\frac{c_{ox}}{t_{ox}}WL$)	$\frac{1}{s}$	$\frac{1}{s}$
Substrate Doping Density	s	s

Table 2.1 Scaling laws of MOS devices.

clock frequency of 200MHz. For the past twenty years, the number of transistors in a microprocessor chip increased 100,000 times and the clock frequency increased 250 times. Its speed in terms of million instructions per second (MIPS) increased 10,000 times. In addition to the progress in CMOS technology, the rapid progress of the microprocessor chip was due to the increase in the clock frequency and the advances in the circuit and the system designs. The speed performance of a microprocessor chip is related to the consumed power. At larger power consumption, speed is faster. While the speed of the microprocessor rose in the past, power consumption also increased. Currently, speed performance of an advanced microprocessor is over 300 MHz at power consumption of 30W. Considering two million transistors on the chip implemented by 0.4μm technology, the power consumption per transistor is 5μW. In the past 25 years, although the number of transistors in a microprocessor chip increased from 4K to over 2 million, the power consumed per chip stayed about the same. As a result, the consumed power per transistor decreased from 8×10^{-4}W/tr to near 2×10^{-6}W/tr. In the future microprocessor systems, low power is the basic trend. How to achieve high speed for a system with an increased number of transistors at decreased power consumption per transistor is a challenge.

In the past, there were two scaling laws for MOS devices[3][4]. One is the constant electric field (CE) scaling law. The other one is the constant voltage (CV) scaling law. As shown in the table, based on the CE scaling law, the lateral and the vertical dimensions of an MOS device are scaled down by a factor of s ($s > 1$). In order to maintain the internal electric field, the supply voltage and the threshold voltage need to be scaled down by a factor of s. Therefore, the substrate doping density is increased by s times and the drain current is decreased by s times. The delay is shrunk by s times and the power consumption becomes smaller by s^2 times, hence the power delay product is decreased by s^3 times. For a deep-submicron CMOS device,

16 CMOS TECHNOLOGY AND DEVICES

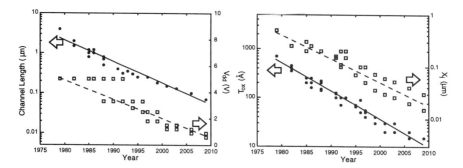

Fig. 2.3 Evolution of CMOS technology.

down-scaling of the supply voltage and the threshold voltage is difficult. Therefore, CE scaling is not practical for deep-submicron CMOS devices. Based on the CV scaling law, the voltage is maintained, however, the current increases s times and the power consumption increases s times. The delay decreases s^2 times and the power delay product shrinks s times.

Fig. 2.3 shows the down-scaling of the MOS devices based on the information from the published papers in ISSCC[1] and the 1997 report from the National Technology Roadmap for Semiconductors[2]. As shown in Fig. 2.3, the channel length of a CMOS device was scaled down from $4\mu m$ in 1979 to $0.08\mu m$ in 1995. The junction depth of the source/drain region was scaled down from $0.5\mu m$ to $0.1\mu m$. The gate oxide was scaled down from 70nm to 30nm. The substrate doping density increased from $10^{16} cm^{-3}$ to $10^{18} cm^{-3}$. The supply voltage shrunk from 5V to 1.5V. The threshold voltage shrunk from 0.8V to 0.4V. For the past 20 years, the channel length was shrunk by 50 times, the junction depth by 5 times, and the gate oxide by 23 times. However, the substrate doping density increased 100 times. The supply voltage shrunk 3 times and the threshold voltage 2 times. In 1995, a deep-submicron MOS device with a channel length of less than 50nm was reported[5]. As shown in the figure, down-scaling of the MOS devices has been slowed down. What are the limits in scaling down of MOS devices? The limits in down-scaling of an MOS device are on the processing technology, device physics, and circuits. The limit in the processing technology is on photolithography and etching. In addition, further down-scaling of the $0.4\mu m$ wide interconnect metal lines may be difficult. The limits in the device physics include the tunneling current problems for an MOS device with a very thin gate oxide[6]. When the source/drain junction is very shallow, the increased source/drain resistance may worsen the transconductance. In addition, when the junction depth of the LDD N^- region is shrunk to below 10nm, the large source/drain resistance may degrade the device performance. Low-voltage is the trend in future VLSI CMOS technology. For future sub-$0.1\mu m$ CMOS technology beyond 2005, its power supply voltage may be below 1V, the gate oxide thickness maybe close to 2nm and the junction depth close to 20nm. As shown in the figure, the supply voltage can't be scaled down according to the shrinkage of the dimensions.

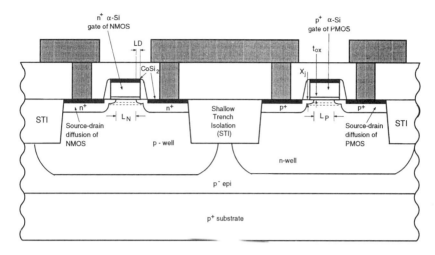

Fig. 2.4 Cross-section of a 1.8V 0.25μm CMOS technology. (Adapted from Luo et al.[13].)

As a result, the electric field in the device is increased substantially. The high electric field of the MOS device, which cannot be overlooked, may limit further down-scaling of the device. From circuit point of view, the requirements on future VLSI circuits are low voltage, low power dissipation, and high speed. Threshold voltage should be scaled down accordingly for the future VLSI circuits. Considering subthreshold conduction, threshold voltage cannot be scaled accordingly with shrinkage of power supply voltage. As a result, a drop in the V_{DD}/V_{TH} ratio leads to degraded speed performance.

2.2 0.25μM CMOS TECHNOLOGY

Fig. 2.4 shows a cross-section of a 1.8V 0.25μm CMOS technology[13]. As shown in the figure, in the 1.8V 0.25μm CMOS technology, high energy/dose retrograde well implants have been used to lower the n-well sheet resistance and to reduce the source/drain junction capacitances. Super-steep retrograde channels have been used for good subthreshold leakage control at a low threshold voltage. NO-nitrided gate oxide and amorphous silicon gate have been used to suppress boron penetration in the PMOS device. Shallow source/drain extensions coupled with halo implants have been used to improve the short-channel and the narrow-channel threshold voltage rolloffs. Deep source/drain regions with an offset spacer have been used to eliminate junction leakage problems. Co salicide for reduction of the gate and the source/drain resistances has been applied.

The $0.25\mu m$ CMOS process starts with the thin-epi on p^+-type substrate. Critical dimensions are defined by deep UV (248nm) photolithography using an undyed resist. After completing a shallow trench process for device isolation, a phosphorus

18 CMOS TECHNOLOGY AND DEVICES

No.	Mask
M1	n-well
M2	Field oxide
M3	Polysilicon
M4	n^+ S/D implant
M5	Contact
M6	Metal1
M7	Via
M8	Metal2

Table 2.2 Masks for a 0.25μm CMOS technology.

implant at 850KeV and a boron implant at 450KeV are used to define n and p wells, respectively. Either boron or indium is used for adjusting the NMOS threshold voltage and phosphorus or antimony for PMOS. After removal of the thin sacrificial oxide layer, 4.5nm gate oxidation and a deposition of 200nm amorphous silicon follow. For the NO-nitrided gate oxide, a nitric oxide anneal is performed after the initial thermal oxidation. A 20nm Si-rich nitride layer is deposited as an anti-reflective coating to improve notching and critical-dimensional control for gate patterning. A highly selective RIE process is utilized to minimize oxide consumption during gate etch. Shallow extension ($\cong$50nm deep) is made after gate re-oxidation. Halo implantation is also applied to help reduce the threshold voltage rolloff and to improve punchthrough immunity. The deep source/drain implants are performed after nitride spacer formation with a junction depth between 130nm and 160nm. A self-aligned silicide layer with Ti or Co is formed by a conventional two-step rapid thermal annealing (RTA) process. For TiN-capped Co salicide, a Co deposition is followed by an additional 20nm TiN deposition before the first RTA step.

As shown in Table 2.2, eight layout masks are needed for the 0.25μm n-well CMOS process. In this 0.25μm CMOS technology, two layers of metal are used. For an advanced CMOS technology, in order to facilitate interconnects, more than two interconnects may be necessary—multi-layer interconnects. If a third metal line is used in the 0.25μm CMOS technology, another via mask is required. Fig. 2.5 shows the layout of a CMOS inverter circuit using the design rules for the 0.25μm CMOS technology.

2.3 SHALLOW TRENCH ISOLATION

When the dimension of a CMOS device is above 0.25μm, LOCOS (local oxide) can be used to provide isolation between devices as shown in Fig. 2.6. In the LOCOS

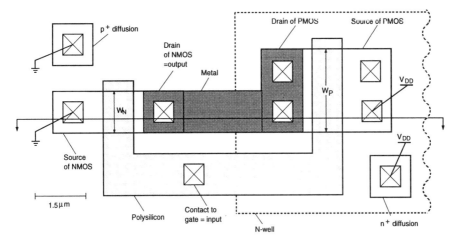

Fig. 2.5 Layout of a CMOS inverter using a $0.25\mu m$ CMOS technology.

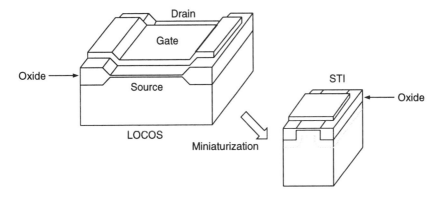

Fig. 2.6 LOCOS versus STI in scaling a MOS device.

structure, bird's beak may limit device density. Shallow trench isolation (STI) technology as shown in Fig. 2.6 is used to increase device density of a CMOS technology. Fig. 2.7 shows the fabrication processing sequence of the shallow-trench isolation technology[14]. A gate oxide of 5nm is grown first, followed by thin polysilicon nitride deposition. The isolation pattern over the nitride/thin polysilicon/gate oxide/substrate stack is defined. These layers are anisotropically etched in sequence to form $0.4\mu m$-deep shallow trenches in the substrate. After photoresist strip, the trench sidewalls are re-oxidized to form a thin thermal oxide liner. The trench is refilled with CVD oxide, followed by a densification anneal. CMP—chemical mechanical polishing—is subsequently carried out to planarize the CVD oxide and the nitride, which serves as the polishing stop layer. After the remaining nitride is removed, channel implants and high-energy retrograde well implants are performed sequentially through polysilicon and gate oxide. Second polysilicon and tungsten

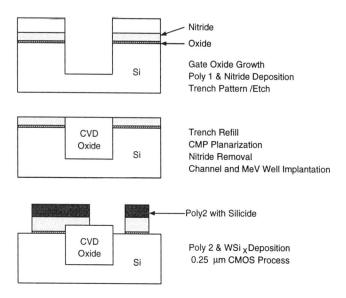

Fig. 2.7 Fabrication processing sequence of shallow-trench isolation technology. (Adapted from Chen et al.[14].)

silicide layers are deposited to form a stacked polycide gate. Then gate patterning, source/drain formation, and backend metallization are carried out for the 0.25μm CMOS process.

2.4 LDD

For deep-submicron CMOS devices, the internal lateral electric field is high. Under the high electric field, the hot carrier effects cause aging and avalanche breakdown, which are not advantageous for operation of a related circuit. The hot carrier effects should be reduced. The best way to reduce hot carrier effects is by reducing the lateral electric field in the channel of the device. The lightly-doped drain (LDD) structure is used to reduce the lateral electric field.

Fig. 2.8 shows the key processing sequence of the LDD NMOS technology [15][16]. Until the polysilicon gate is formed, the processing sequence is identical to that for standard NMOS technology. After the polysilicon gate is formed (b), before the source/drain N^+ implant (f), an N^- implant is carried out to form the N^- LDD region (c). Then a layer of low-temperature oxide (LTO), which is about the same thickness as the polysilicon gate, is deposited by CVD (d). At the edge of the polysilicon gate, this LTO layer is thicker. Then, the LTO layer is etched away by an anisotropic LTO plasma etch (e). When the LTO layer in most area is gone, at the edge of the polysilicon gate, there is some LTO left—the parabolic-shape sidewall

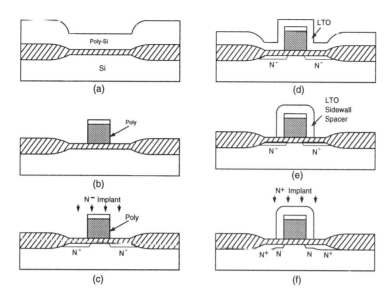

Fig. 2.8 Processing sequence of an LDD NMOS technology (Adapted from Tsang et al.[15].)

spacer. Then an N^+ implant is carried out to form the source/drain region (f). Due to passivation by the sidewall spacer, the N^+ source/drain region is at some distance away from the channel region under the gate. Therefore, the N^- LDD region can be regarded as the parasitic resistance region. When the device is biased at a large current, there is a substantial voltage drop over this parasitic resistance region—the effective voltage drop over the channel region is reduced. Therefore, the peak of the lateral electric field can be reduced substantially. As shown in Fig. 2.9, by using the LDD structure, the peak value of the lateral electric field profile is reduced by 25% as compared to the structure without using it. Since the hot electron effects are exponentially proportional to the electric field, a 25% cut in the peak electric field can bring down the hot electron effects substantially. The LDD structure exists at both source and drain. The LDD structure near the drain reduces the hot electron effects. The LDD structure near the source also influences the performance of the device— the LDD structure near the source lowers the transconductance of the device, which is the side effect of the LDD structure. Although the transconductance and the drain current are reduced slightly by adding the LDD structure, the hot-electron effects can be reduced exponentially. Therefore, the use of of LDD has been justified.

2.5 BURIED CHANNEL

At a light substrate doping density, using an N^+ polysilicon gate, without any channel implant, the PMOS device is enhancement-mode ($V_T < 0$). However, its threshold voltage may be very negative. For the NMOS device, it tends to be depletion type ($V_T < 0$). In a CMOS circuit, enhancement-mode devices are needed. In order

22 CMOS TECHNOLOGY AND DEVICES

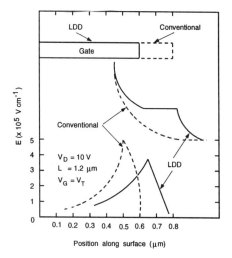

Fig. 2.9 Electric field distributions in the lateral channel of LDD and conventional NMOS devices. (From Ogura et al.[16], ©1980 IEEE.)

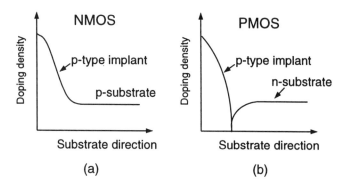

Fig. 2.10 Doping profile in the substrate direction of (a) NMOS and (b) PMOS devices.

to maximize the current driving capability for CMOS devices, the magnitude of the threshold voltage should be reduced. For an enhancement-mode CMOS device, the magnitude of the threshold voltage is between 0.4V and 0.8V. The magnitude of the threshold voltage of the enhancement-mode CMOS device cannot be set to close to zero due to the subthreshold slope consideration, which will be discussed in detail later. In order to produce CMOS enhancement-mode devices using the N^+ polysilicon gate with appropriate threshold voltages, ion implantation is a necessary procedure. For the CMOS devices using N^+ polysilicon gate without ion implants, the threshold voltage of the PMOS device is too negative and the threshold voltage of the NMOS device is not positive enough. Both PMOS and NMOS devices need a p-type ion implant (boron) to adjust their threshold voltages toward the positive direction. For an NMOS device, the substrate is p-type. By implanting p-type dopants into the substrate, a non-uniform doping profile with a peak at the surface in the

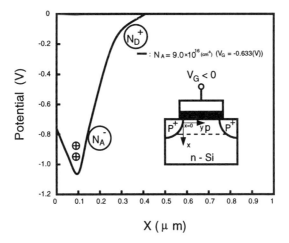

Fig. 2.11 Potential distribution in the substrate direction of a buried PMOS device.

substrate direction as shown in Fig. 2.10(a) can be seen.

For a PMOS device, the situation is different. For a PMOS device, its substrate is n-type. By implanting the p-type dopants into the substrate, the doping profile in the substrate direction is counter-doped—a pn junction exists in the surface[18] as shown in Fig. 2.10(b)—a buried channel. The buried channel structure in the PMOS device is due to the counter-doped channel. In contrast, the NMOS device using the N^+ polysilicon gate does not have a buried channel since no pn junction exists in the substrate. Fig. 2.11 shows the electrostatic potential profile in the substrate direction of a buried-channel PMOS device[17]. In this figure, the origin of the x-axis is at the silicon/oxide interface. As shown in the figure, in the region near the surface, a non-zero slope in the potential profile can be seen, which implies the existence of non-zero electric field in this region. Due to the non-zero electric field, the region near the $Si - SiO_2$ interface is fully depleted. At some distance from the interface, the slope of the potential and hence the electric field are zero. At this location with a zero electric field, holes exist—the buried channel. Further deeper into the substrate, a region with a non-zero slope can be identified—the depletion region of the pn junction. Between the buried channel and the oxide interface, due to the counter-doped channel, the region is depleted with the negative space charge—N_A^-. Between the buried channel and the substrate, the region is also depleted with the negative space charge N_A^-. At a negative gate voltage, the holes in the buried channel are attracted by the electric fields generated by the negative gate voltage and the negative space charge. As a result, holes stay at a location at some distance from the oxide interface with the lowest potential—the buried channel, where the electric field is zero. In an NMOS device, since there is no pn junction in the substrate, the surface of the substrate is more heavily doped with p-type dopants. The carrier is the negatively charged electron supplied by the source and the drain. When a channel exists in the NMOS device, a positive gate voltage is imposed. Under this situation, the surface area is fully depleted with negative space charge N_A^-. In addition, electrons are attracted toward the gate by the electric field due to the positive gate voltage. Therefore, the existence of the buried channel in the PMOS device is due to the p-type region under the gate—holes tend to be attracted by the electric field produced by both the N_A^- space charge region and the negative gate voltage. For a buried-channel PMOS device, when the imposed gate voltage is very negative, the buried channel disappears. Instead, the surface channel dominates. In the surface channel, the transport of the carriers is at the oxide interface, which may affect the carrier mobility. In the buried channel, the transport of the carriers is not via the silicon surface. The surface scattering effect has influence in the carrier transport. Therefore, the carrier mobility in the buried channel is higher. However, for an MOS device with a buried channel, it is more difficult to turn off—the subthreshold slope is less steep. Therefore, a buried-channel MOS device may have a larger carrier mobility, however its subthreshold slope is poorer. Due to the subthreshold slope consideration. The buried-channel MOS device is not suitable for use in the memory cell in DRAM, which is composed of an access transistor and a storage capacitor. When the memory cell is not accessed, the access transistor is turned off. Under this situation, the leakage current of the access transistor determines the time that the charge in the storage capacitor can be sustained before the memory cell is refreshed. A small leakage current lengthens the refresh time of a memory cell, which is helpful for the performance of a DRAM circuit. Therefore, buried-channel MOS devices may not be helpful for use in the DRAM memory cell.

2.6 BiCMOS TECHNOLOGY

There have been spin-offs from the CMOS technology. One successful spin-off from CMOS is BiCMOS technology[19][20]. In the 1980s, BiCMOS technology was a popular technology. Fig. 2.12 shows the cross-section of a $0.8\mu m$ BiCMOS technology[19]. This BiCMOS technology is based on a $0.8\mu m$ twin-tub CMOS technology. The bipolar device is placed in the n-well. In this BiCMOS structure, the bipolar device is formed by adding base and deep-collector implants to the CMOS technology. The bipolar device in this BiCMOS technology has a unity gain frequency of 8GHz. In the 1980s, by adding 8GHz bipolar devices, the access time of an SRAM could be shortened by two times as compared to the CMOS ones. For a medium-size semiconductor company in the 1980s, in order to enhance the speed performance of an SRAM, there were two choices. One was to upgrade the $0.8\mu m$ CMOS technology to $0.6\mu m$. Upgrading the CMOS technology needed a large investment in processing equipment. It might take up to two years in upgrading a CMOS technology from $0.8\mu m$ to $0.6\mu m$. The other choice to enhance the speed performance of an SRAM was via the BiCMOS technology. By incorporating the bipolar device in the $0.8\mu m$ CMOS technology, no large investment on the processing equipments was required. The $0.8\mu m$ CMOS process was only needed to be slightly modified to include two extra masks and implants. Only a few extra engineers were

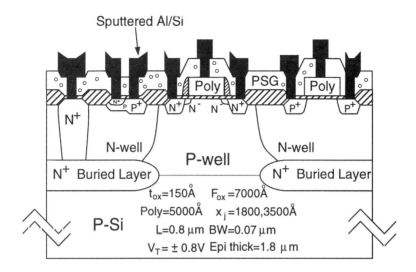

Fig. 2.12 Cross-section of a 0.8μm BiCMOS technology. (Adapted from Iwai et al.[19].)

required to modify the CMOS process and to analyze the bipolar device. In addition, less than a year was required. The upgrade of the speed performance of the SRAM was accomplished by the BiCMOS circuits used in the word line decoder and bipolar sense amps. Therefore, the development of BiCMOS technology for SRAM was a good choice for a semiconductor company in the 1980s. In the 1980s, several world-class semiconductor companies were developing BiCMOS technology. Nowadays, many semiconductor companies have stopped developing BiCMOS technology. Instead, they are fully dedicated to deep-submicron CMOS technology. The saturation of the BiCMOS technology nowadays reflects the bottlenecks.

As shown in Fig. 2.12, in the $0.8\mu m$ BiCMOS technology[19], only NMOS devices have the LDD structure. No LDD structure exists in the PMOS devices. This is because the hot hole effect of the $0.8\mu m$ PMOS device is much smaller than the hot electron effect of the $0.8\mu m$ NMOS device. In addition to the unity gain frequency, other key parameters for a bipolar device are current gain, base-emitter/base-collector junction capacitances, collector-substrate parasitic capacitance, and base resistance. The overall speed performance of a BiCMOS circuit is mainly determined by the bipolar device. If the bipolar device is not fast enough, circuit designers had better use the pure CMOS devices. In addition, the advantages of the bipolar device can be expedited only when the BiCMOS circuit is innovatively designed for a circuit with a large output load.

In the $0.8\mu m$ BiCMOS fabrication process, from the p-type silicon wafers, a thermally grown oxide of 100nm is formed for passivation. After the buried layer photolithography, a high-dose arsenic implant is used to form the buried layer. A p-type thin-film of $1.8\mu m$ is deposited by epitaxy. After the n-well photolithography,

No.	Mask
M1	Buried Layer
M2	N-well
M3	Field Oxide
M4	Deep Collector
M5	Polysilicon
M6	Base
M7	N^+ S/D Implant
M8	Contact
M9	Metal1
M10	Via
M11	Metal2

Table 2.3 Masks for a 0.8μm BiCMOS technology.

the n-well region is counter-doped by an n-type implant. Then a pad oxide layer of 15nm is thermally grown, followed by a silicon nitride layer of 30nm deposited by CVD. After the field region photolithography, the nitride layer is selectively etched—only the nitride on the active region remains. A field oxide layer of 600nm is thermally grown. After stripping the nitride, a sacrificial oxide of 15nm is grown. After deep-collector photolithography, a high-energy high-dose phosphorus implant is used to form the N^+ deep-collector region. Then, two blanket boron implants are used to adjust the threshold voltages of the PMOS and NMOS devices. After stripping the sacrificial oxide, a gate oxide of 15nm is thermally grown. After this step, the LDD structure using the sidewall spacer technique described before is formed. Then a high-dose boron implant is used to form the source/drain region of the PMOS device and the extrinsic base region of the bipolar device. After the base photolithography, a boron implant is used to generate the intrinsic base region. A high-dose arsenic implant is used to produce the source/drain region of the NMOS device and the emitter region of the bipolar device. After this step, passivation, contact hole formation, and multi-layer interconnects are identical to that in the CMOS technology described before.

Table 2.3 shows the masks in the 0.8μm BiCMOS technology. In this 0.8μm BiCMOS technology, 11 masks are required in defining the layout. Compared with the 0.8μm CMOS, the 0.8μm BiCMOS technology has only two extra masks: the deep collector mask and the base mask. The deep collector mask is used for accomplishing the N^+ deep collector for the bipolar device. The base mask is used for realizing the doping profile of the intrinsic base region. The base contact for the bipolar device is formed when the PMOS source/drain region for the CMOS process is implanted. The emitter of the bipolar device is formed by the N^+ source/drain as

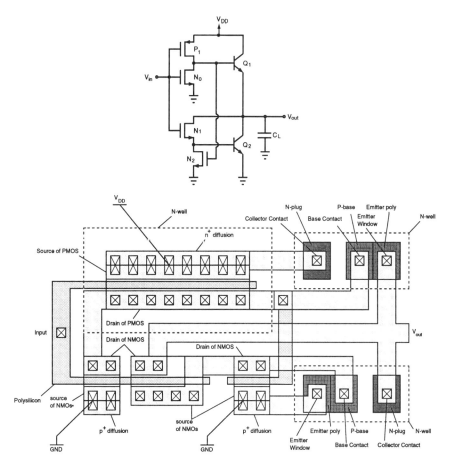

Fig. 2.13 Layout of a 0.8μm BiCMOS inverter.

in the CMOS process. For all complexity of whole processing sequence, the 0.8μm BiCMOS process is only 10 to 15% more complicated than the 0.8μm CMOS process.

Fig. 2.13 shows the layout of a 0.8μm BiCMOS inverter. As shown in the figure, the NMOS device N_1 and the bipolar device Q_2 constitute the BiNMOS device to enhance the driving capability during the pull-down transient. The PMOS device P_1 and the bipolar device Q_1 form a BiPMOS device, which is used to strengthen the driving capability during the pull-up transient. The NMOS device N_0 is used to help evacuate the mobile minority carriers from the base of Q_1 during the turn-off process. The NMOS device N_2 is used to help the removal of minority carriers from the base of Q_2 during the turn-off process. As shown in the figure, this BiCMOS inverter is based on a design using a supply voltage of 5V.

28 CMOS TECHNOLOGY AND DEVICES

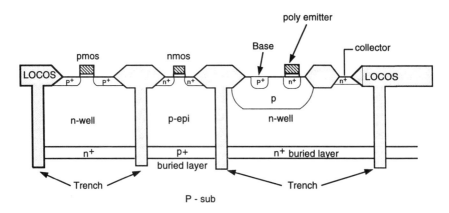

Fig. 2.14 Cross-section of a 0.5μm BiCMOS technology (Adapted from Liu et al.[22].)

In a typical VLSI logic circuit, if fan-out is large, the parasitic capacitance associated with the bus lines can be large. The driving capability of the conventional CMOS driver may not be sufficient. As a result, the switching speed may be degraded. By using the npn bipolar device to enhance the driving capability of the PMOS device, the BiPMOS device can be used to shorten the switching speed during the pull-up transient[21]. Similarly, by using the npn bipolar device to enhance the driving capability of the NMOS, the BiNMOS device can shorten the switching speed during the pull-down transient. Using the BiCMOS driver, the switching speed can be improved by 2 to 3 times. The output voltage of the BiCMOS driver cannot be full swing as the power supplies. The output high voltage cannot reach V_{DD} and the output low voltage cannot reach V_{SS}. The non-full-swing of the BiCMOS driver is a bottleneck for applications in low-voltage VLSI circuits.

Fig. 2.14 shows cross-section of a 0.5μm BiCMOS technology[22]. This BiCMOS technology uses trench isolation. In addition, the LDD structure is used for both NMOS and PMOS devices. Polyemitter structure is used for the bipolar device. The polycide technology has been used to lower the sheet resistance of the polysilicon layer. By adopting the III-V compound technology, an advanced SiGe-base heterojunction bipolar device has been included in the BiCMOS technology[23]. By incorporating the germanium, which has a smaller bandgap in the base, the basewidth can be shrunk to increase the unity gain frequency without affecting the current gain of the bipolar device. In an advanced SiGe-base HBT, the base width can be as narrow as 30nm. The development of an advanced deep-submicron BiCMOS technology can be justified if the bipolar device has a higher speed. In the 1990s, BiCMOS technology became saturated since the advanced CMOS technology provides fast CMOS devices. In addition, the non-full-swing of the output voltage further limits the development of BiCMOS technology. For CMOS technology, device performance and packing density can be enhanced by scaling down the dimensions. In BiCMOS technology, the intrinsic portion of the bipolar devices is vertical, which is

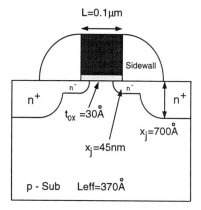

Fig. 2.15 Cross-section of a 0.1μm MOS device (Adapted from Yan et al.[24].)

not easily shrunk with the lateral scaling in the CMOS devices. This is another weak point of the BiCMOS technology.

2.7 0.1μM CMOS TECHNOLOGY

The channel length of an advanced deep-submicron CMOS devices is less than 0.1μm. In another fifteen years, the channel length of the CMOS devices can be shrunk to 10nm. The 1Gbit DRAM announced before was based on a 0.18μm CMOS technology. With a 0.1μm CMOS technology, 4Gbit DRAM can be produced. At 0.01μm(10nm), 256Gbit is possible. In 1998, the impact of the 64Mbit DRAM on the performance of a computer is already unprecedented. In 1998, the circuit speed of a CPU is over 300MIPS (million instructions per second). In 2001, what kind of applications can a personal computer with 4Gbit DRAM chips and a 10 GIPS(giga instructions per second) CPU be used for? This is probably beyond our current imagination. Both the 4Gbit DRAM and 10 GIPS CPU will be built by deep-submicron CMOS technology.

Fig. 2.15 shows the cross-section of a 0.1μm MOS device[24]. In this 0.1μm MOS device, the LDD structure with the sidewall spacer is used. The effective channel length of the device is 37nm. In order to reduce the device second order effect, the junction depth of the N^- LDD region is 3nm. The gate oxide thickness is 3nm. This device is designed for 1V operation. In order to avoid punchthrough of the source and the drain depletion regions, the substrate doping density has been raised to $10^{18} cm^{-3}$. Therefore, the body effect of this device is higher than the MOS devices based on a CMOS technology greater than 0.1μm. The key technology for realizing this 0.1μm CMOS device is the photolithography and etching technology for realizing the very small polysilicon gate.

30   CMOS TECHNOLOGY AND DEVICES

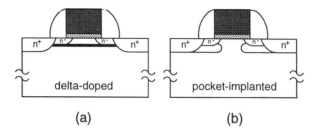

Fig. 2.16 Deep submicron MOS structures: (a) the delta-doped structure; and (b) the pocket-implanted structure. (Adapted from Asai and Wada[25].)

When the channel length of a CMOS device is scaled down to 0.1 μm, in addition to the LDD structure described before, two other device structures as shown in Fig. 2.16 can be used[25]. In the delta-doped MOS device, the delta-doped region is used to avoid channel punchthrough. Above the delta-doped region, the low doping region is used to reduce the short channel effect. In the pocket-implanted MOS device, below the LDD region the pocket-implanted highly doped region has been used to reduce the channel punchthrough. Using the pocket-implanted structure, the drain induced barrier lowering (DIBL) effect can also be suppressed.

For CMOS devices scaled down to below 0.1 μm, there are bottlenecks in the device technology. When the gate oxide is shrunk to its limit at 3nm, a further shrinkage may lead to direct tunneling, which may degrade the device performance. Therefore, other insulating materials with a higher permittivity (than silicon dioxide) are required to reduce the internal electric field for overcoming direct tunneling problems. For a deep-submicron MOS device, a shallow source/drain junction may cause a high source/drain resistance, which is not helpful for low-voltage operation. Therefore, how to reduce the sheet resistance of the source/drain region for a very shallow source/drain junction is a challenge for designing deep-submicron MOS devices. When the CMOS VLSI technology is further shrunk to the sub-0.1 μm regime, the metal interconnect lines also need to be scaled down. For a thinner and narrower metal line, its sheet resistance and conducting current density increase. As a result, the increased electromigration may burn out the metal line. How to reduce the electromigration and the sheet resistance in the thin metal interconnect lines is another challenge for designing the deep-submicron CMOS technology. For low-voltage operation, low-threshold MOS devices are required. For the MOS devices designed with a low threshold voltage, its subthreshold conduction current is more noticeable when the gate voltage is V_G=0V. How to reduce the threshold voltage for low-voltage operation without increasing the subthreshold current is also a challenge. Throughput of the photolithography steps for sub-0.1 μm CMOS technology such as E-beam lithography decreases when the dimensions of the devices are shrunk. How to maintain the throughput of photolithography for the sub-0.1 μm CMOS technology is also a challenge.

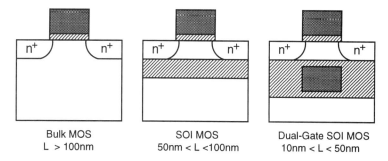

Fig. 2.17 Evolution of device structures for deep-submicron CMOS technology.

When CMOS technology is scaled down to below $0.1\mu m$, conventional device structures may encounter difficulties. As shown in Fig. 2.17, some device structures may be important for sub-$0.1\mu m$ CMOS technology[26]. As shown in the figure, due to the device structure, in the SOI MOS devices junction capacitances are reduced. Owing to the ultra-shallow junctions, short-channel and narrow-channel effects are much smaller. Therefore, SOI MOS devices have potentials for use in $0.1\mu m$–$0.05\mu m$ CMOS technology. Owing to the superior control capability, dual-gate SOI CMOS devices possess advantages of the fully-depleted SOI CMOS devices without floating body effects. Thus, dual-gate SOI CMOS devices have potentials to be used for $0.05\mu m$–$0.01\mu m$ CMOS technology.

When CMOS technology is shrunk to below $0.25\mu m$, the influence of the local interconnects is important, as circuit performance is affected by the interconnects. As shown in Fig. 2.18[2], conventional interconnect metal lines using Al and oxide as the dielectric may have a high sheet resistance and a high dielectric constant. Interconnect metal lines using copper may have a lower sheet resistance and a better electromigration reliability as compared to the Al line. If a low-k dielectric is used, the parasitic capacitance between the interconnect lines can be lowered. As shown in the figure, a better speed performance can be obtained if Cu and low-dielectric are used for the interconnect metal lines. In the future, interconnect metal lines using Cu and low-k dielectric are the trend.

Fig. 2.19 shows the cross-section of a CMOS technology scaled from $0.35\mu m$ to $0.1\mu m$ for integrating microprocessor chips[26]. As shown in the figure, in the $0.1\mu m$ CMOS technology, Cu and low-k interconnect lines are used to integrate VLSI systems. In addition, the aspect ratio (height/width) of the interconnect metal lines tends to increase such that the sheet resistance of the interconnect metal lines will not increase while the metal line width is scaled down. In $0.35\mu m$ CMOS technology, tungsten via plugs have been replaced by Cu for the purpose of having a low resistance. The use of the low low-k dielectric reduces the parasitic capacitances—when the device and interconnect metal lines are scaled down, the shrinkage in the separation distance may increase the inter-metal line coupling capacitance. Using a low-k dielectric, the inter-metal line coupling capacitance can be lowered such that

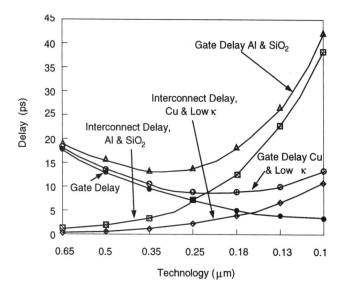

Fig. 2.18 Gate and interconnect delays versus technology (Adapted from 1997 NTRS report[2].)

the coupling noise and the speed performance can be improved.

Based on the predictions from the 1997 report of National Technology Roadmap for Semiconductors[2], CMOS VLSI technology will evolve continuously. Via scaling threshold voltage and using multiple threshold voltage, switched threshold voltage, dynamic threshold voltage, and SOI technologies, the difficulty in shrinking the threshold voltage while down-scaling the power supply voltage can be overcome. Via advanced transistor structures, the difficulties in scaling down the devices to improve device performance can be overcome. Improvement of the interconnect technology can be done via the Cu and low-k dielectric approach. Other evolution of technologies includes intensive use of CAD tools to shrink the development time for VLSI.

2.8 SOI CMOS TECHNOLOGY

Recently, SOI (silicon-on-insulator) CMOS technology has been regarded as another major CMOS technology for VLSI[27]. Due to the buried oxide structure for providing an excellent isolation capability, there is no latchup in SOI CMOS. In addition, immunity to radiation effect is high. Transconductance of the SOI CMOS devices is better than that of bulk ones. Subthreshold slope of the SOI MOS devices is better[28]. Due to the silicon thin-film structure, the electric field effect in the SOI CMOS devices is smaller, as compared to the bulk CMOS devices. In a bulk CMOS device, the depletion region is irregularly influenced by the lateral and vertical electric fields in the substrate. In an SOI CMOS device, the depletion

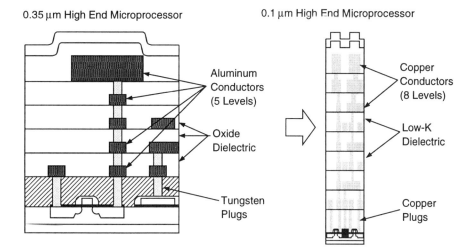

Fig. 2.19 Challenge in the deep-submicron VLSI CMOS technology (From Vasudev and Zeitzoff[26], ©1998 IEEE.)

region is limited to the thin-film region. As a result, to some extent, the electric field effect in the SOI CMOS devices is more regular. Thus, the short channel effect of the SOI CMOS devices is smaller as compared to the bulk devices. Due to oxide isolation, the parasitic capacitances are smaller. However, inferior heat dissipation and larger source/drain resistance are the disadvantages of the SOI CMOS technology.

In this section, a $0.25\mu m$ SOI CMOS technology based on SIMOX wafers using a two-step LOCOS isolation technique[29] as shown in Fig. 2.20, is described. Fig. 2.21 shows the basic steps of implant and anneal in fabricating the SIMOX (separation by implantation of oxygen) wafers[30]. As shown in the figure, first, oxygen is implanted into the silicon wafer. The profile of the implanted oxygen dopants is Gaussian-shape with its peak at some distance below the surface. With a high-temperature anneal, oxygen reacts with silicon to form a buried oxide layer around the peak of the oxygen profile. During the high-temperature anneal, the oxygen dopants above or below the buried oxide move to gather at the peak region, and thus react with silicon to form silicon dioxide. As a result, a single-crystal silicon thin-film above the buried oxide layer can be formed. If necessary, the thickness of the single-crystal silicon thin-film can be increased by epitaxy. As shown in Fig. 2.20, in the $0.25\mu m$ SIMOX SOI CMOS technology[29], SIMOX wafers with p-type silicon substrate and a 90nm buried oxide are used. Atop the buried oxide is a silicon thin-film of 320nm. The thickness of the top silicon thin-film is reduced to 56nm by a sacrificial oxidation and a subsequent oxide etch. A pad oxide layer of 70nm is grown, followed by a silicon nitride layer of 50nm deposited by LPCVD. An RIE etch is used to remove the nitride layer over the field oxide region. A field oxidation followed by a wet etch is used to form silicon islands with rounded corners. A pad oxide layer of 70nm and another CVD silicon nitride layer of 10nm are formed. Only the nitride

34 CMOS TECHNOLOGY AND DEVICES

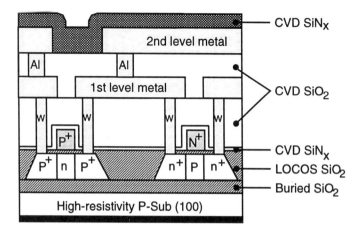

Fig. 2.20 Cross-section of a 0.25μm SIMOX SOI CMOS technology (Adapted from Ohno et al.[29].)

layer over the active device region and the sidewall of the silicon island is left due to a larger thickness before the nitride etch. A field oxide is produced by the second LOCOS step at 1100°C. Due to the existence of the nitride layer at the sidewall of the silicon island, bird's beaks have been substantially minimized. After the second LOCOS step, the thickness of the field oxide is about identical to the thickness of the silicon thin-film. Therefore, a planar surface can be expected. The remaining nitride layer is removed by a wet etch. The PMOS thin-film region is implanted by phosphorus to a density of 10^{17}cm^{-3}, while the NMOS thin-film region is masked by a patterned photoresist layer. The NMOS thin-film region is implanted with the BF$_2$ dopants to a density of 10^{17}cm^{-3}. Then the pad oxide is etched off. A 7nm gate oxide is grown by thermal oxidation. A 0.3μm phosphorus-doped polysilicon layer is deposited by LPCVD. An ECR (electron cyclotron resonance) etch is used to remove the polysilicon layer except over the n-channel gate region, which is masked by photoresist. A thin nitrogen-doped CVD polysilicon layer is deposited, followed by a 0.3μm P^+ polysilicon layer. The purpose of the nitrogen-doped polysilicon layer is used to prevent the boron in the P^+ polysilicon from penetrating through the gate oxide into the silicon thin-film during the subsequent high-temperature cycles. An ECR etch is used to remove the P^+ layer except over the p-channel gate region, which is masked by the patterned photoresist. Thus, the N^+ poly-gate for the NMOS device and the P^+ poly-gate for the PMOS device have been done. Then, the wafer surface is covered by an silicon nitride layer, which is deposited by CVD, and an N^+ polysilicon layer. The top N^+ polysilicon layer is removed an ECR etch except some remains at the sidewalls of the poly-gates, which becomes the polysilicon sidewall spacer. A boron implant into the PMOS region generates its source/drain region, while the NMOS region is masked by a patterned photoresist. The use of the polysilicon sidewall spacer is used to prevent influence of the implant lateral straggle in the channel region during the source/drain implant. Similarly, the NMOS

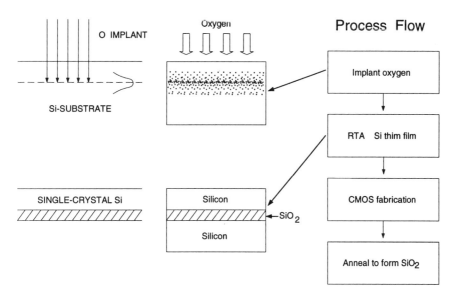

Fig. 2.21 Basic steps of implant and anneal in fabricating the SIMOX wafers.

source/drain regions are formed by the phosphorus implant while the PMOS region is masked by the patterned photoresist. The polysilicon sidewall spacers are etched off by a wet etch. Then the wafer is annealed by RTA (rapid thermal anneal) at 1000°C such that the implanted dopants can be activated. Low-temperature oxide (LTO) is deposited over the wafer by CVD. After planarization, contact areas are defined by the patterned photoresist. An RIE (reactive ion etching) etch removes the LTO and the silicon nitride in the contact areas. Then the wafer is deposited with a layer of Ti and TiN film by sputtering, followed by a tungsten film by CVD. The contact areas are masked by a patterned photoresist. The W/TiN/Ti film outside the contact area is removed. Then the photoresist is stripped, followed by the deposition of the AlSiCu/Ti/TiN film. The interconnect metal lines are defined by patterning the AlSiCu/Ti/TiN layer by photoresist. The AlSiCu/Ti/TiN layer is etched off except the interconnect metal line areas. Then the wafer is deposited with CVD LTO, followed by planarization of the surface. The via areas between interconnects are defined by the patterned photoresist. After the LTO in the via areas is etched off, the via areas are filled with the Al. Then the wafer is covered with an Al/Ti layer, which is patterned by photolithography to define the second metal interconnect area. The unwanted Al/Ti is removed by an Al etch. After stripping photoresist, the wafer is covered with a layer of silicon nitride for passivation, followed by the electrode pad opening and the post-metallization anneal. This completes the processing sequence of the $0.25\mu m$ SOI CMOS technology.

Due to the buried oxide isolation, there is no latchup problem in the SOI CMOS structure. Therefore, NMOS and PMOS devices can be placed against each other sharing the same contact. Therefore, as shown in Fig. 2.22, the density of the SOI

36 CMOS TECHNOLOGY AND DEVICES

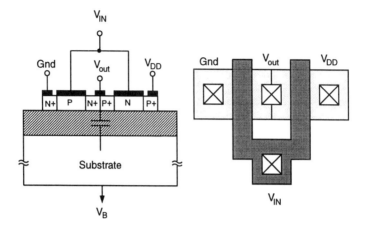

Fig. 2.22 Layout of a 0.5μm SOI CMOS inverter.

CMOS circuits can be high. However, due to the floating body structure, the SOI CMOS inverter suffers from the floating body effect, which is different from the bulk CMOS case.

Depending on the thickness of the thin-film, the SOI CMOS devices can be divided into thick-film and thin-film. As shown in Fig. 2.23, if the thin-film is thick, only the top portion of the thin-film is depleted and the bottom portion is neutral. This type of SOI CMOS devices is called partially-depleted SOI CMOS devices. If the thin-film is fully depleted, this type of the SOI CMOS devices is called fully-depleted SOI CMOS devices. In a partially-depleted SOI CMOS device, the threshold voltage is insensitive to the thin-film thickness since the depletion region is independent of the film thickness. Therefore, for the partially-depleted thin-film SOI CMOS devices, the threshold voltage is more stable. However, due to the neutral region in the thin-film in a partially-depleted SOI CMOS device, the floating body effect is more serious. In addition, the depletion region is more susceptible to the influence from the source and the drain regions, therefore the device second order effect is more serious. Furthermore, in some partially-depleted thin-film SOI CMOS devices, due to the neutral region, an unsmooth transition in the drain current characteristics—kink effect[31]—can be identified. The kink effect in partially-depleted SOI NMOS devices is due to the impact ionization in the high electric field region in the lateral surface channel and the activation of the parasitic npn bipolar device—the floating body effect.

In a fully-depleted thin-film SOI CMOS device, regardless of the biasing condition, the thin-film is fully depleted. In addition, a larger transconductance can be expected due to the larger gate control capability over the thin-film. Therefore, the device second order effects including the short channel effect and the narrow channel effect are smaller. In addition, the subthreshold slope is improved. However, in a fully-

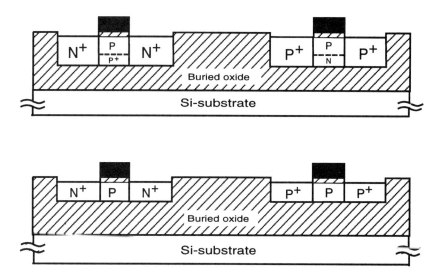

Fig. 2.23 SOI CMOS devices: partially-depleted versus fully-depleted.

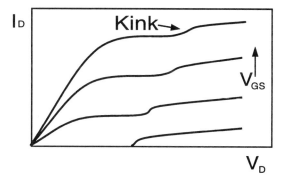

Fig. 2.24 The kink effect in the partially-depleted SOI NMOS device.

depleted device, the threshold voltage may be sensitive to the thickness of the thin-film, which may vary during the fabrication process. Therefore, the threshold voltage stability of the fully-depleted SOI CMOS devices may be poorer.

Until now in this chapter, CMOS technologies have been described. From here, fundamental CMOS device behaviors are described. Starting from the threshold voltage, the body effect, the short channel effect, and the narrow channel effect are analyzed. Then, from the mobility phenomenon including the prepinchoff saturation, and velocity overshoot, the drain current for the deep-submicron CMOS devices including the electron temperature effects is reported. Finally, capacitances of MOS devices and hot carriers effects are reported.

2.9 THRESHOLD VOLTAGE

For a PMOS device, the threshold voltage is defined as the gate voltage when hole inversion with a density comparable to that of the substrate concentration starts to exist at the Si/SiO_2 interface. When the gate voltage is not negative enough to reach the threshold voltage, the width of the depletion region under the gate is a function of the gate voltage. Here, the threshold voltage formula is derived. By solving the Poisson's equation in the depletion region under the channel, the electrostatic potential at the oxide interface is: $\Psi_S = -\frac{qN_D}{2\epsilon_{si}}x_d^2$, where N_D is the donor density in the substrate, ϵ_{si} is the silicon permittivity, and x_d is the depth of the depletion region. Considering the voltage drop in the gate oxide between the gate electrode and the oxide interface, one obtains: $V_G = E_{ox}t_{ox} + \Psi_S$, where E_{ox} is the electric field in the oxide, and t_{ox} is the thickness of the gate oxide. Applying Gauss law at the interface, since the displacement is continuous, one obtains: $\epsilon_{ox}E_{ox} = \epsilon_{si}E_{si} = qN_D x_d$, where ϵ_{si} is the permittivity of the silicon, ϵ_{ox} is the oxide permittivity, and E_{si} is the electric field at the oxide interface in the silicon. The displacement at the oxide interface is equal to the total space charge in the depletion region in the substrate. The width of the depletion region in the substrate can be expressed as a function of the gate voltage. From the depletion width (x_d), the total charge in the substrate is $Q = qN_D x_d$. At the oxide interface, when the hole inversion exists, the surface potential is $\Psi_s = -2\phi_f$, where ϕ_f is: $\phi_f = \frac{kT}{q}\ln\frac{N_D}{n_i}$ is Fermi voltage, where $x_d = \sqrt{\frac{2\epsilon_{si}2\phi_f}{qN_D}}$ is the depletion width. At this moment, the related gate voltage is the threshold voltage:

$$V_T = -2\phi_f - \frac{Q_B}{C_{ox}}. \tag{2.1}$$

The second term in the above equation can be regarded as the voltage drop over the gate oxide. By considering the substrate as the bottom plate of the capacitor and the gate electrode as the top plate, the voltage drop over the gate oxide is expressed as the total space charge on the bottom plate divided by the oxide capacitance. Considering the work function difference between the gate and the silicon (ϕ_{MS}) and the surface state density (Q_{SS}) at the Si/SiO_2 interface, the threshold voltage becomes:

$$V_T = \phi_{MS} - 2\phi_f - \frac{Q_{SS}}{C_{ox}} - \frac{Q_B}{C_{ox}}. \tag{2.2}$$

If the polysilicon gate is used, the work function difference is simplified as:

$$\phi_{MS} = \phi_f - \phi_{f(M)}. \tag{2.3}$$

When computing the work function difference, the polysilicon gate can be regarded as single-crystalline silicon. Thus $\phi_{f(M)}$ is the work function of the polysilicon gate. Therefore the threshold voltage formula is further simplified as:

$$V_T = -\phi_{f(M)} - \phi_f - \frac{Q_{SS}}{C_{ox}} - \frac{Q_B}{C_{ox}}. \tag{2.4}$$

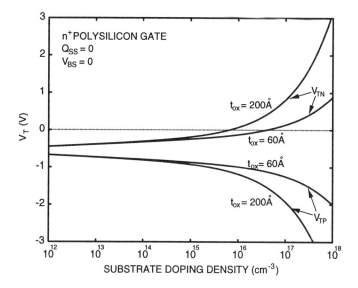

Fig. 2.25 Threshold voltage versus substrate doping density of CMOS devices.

The dopant of the polysilicon gate and the substrate can be n-type or p-type. $\phi_{f(sub)}$ and $\phi_{f(M)}$ can be computed using the following equations:

$$\phi_f = \frac{kT}{q}\ln\frac{N_D}{n_i} \quad for \quad n-type, \quad (2.5)$$

$$\phi_f = \frac{kT}{q}\ln\frac{n_i}{N_A} \quad for \quad p-type.$$

Note that the flat-band voltage define as $V_{FB} = \phi_{MS} - \frac{Q_{ss}}{C_{ox}}$ is used. From the above equations, for n-type, ϕ_f is positive. For p-type, ϕ_f is negative. Based on Eq. (2.4), Fig. 2.25 shows the threshold voltage versus the substrate doping density of CMOS devices. An N^+ polysilicon gate is assumed. In addition, Q_{SS} is assumed to be zero. As shown in the figure, when the substrate doping density is low, the threshold voltage is less sensitive to the substrate doping density. When the gate oxide is thin, the threshold voltage is not sensitive to the substrate doping density. When the substrate doping density is low, the threshold voltage of the PMOS device is negative—the enhancement-type and the threshold voltage of the NMOS device is negative—the depletion-type. When the substrate doping density is large, the threshold voltage of the NMOS device may be greater than zero—the enhancement-type. When the threshold voltage of a CMOS device is not appropriate, ion implantation can be used to adjust the value. By implanting p-type dopants, the threshold voltage is shifted toward the positive direction. Implanting n-type dopants shifts the threshold voltage toward the negative direction. The amount of the shift in the threshold voltage can be determined by the dose of the implant: $\Delta Q = C_{ox}\Delta V_T$. For example, by implanting boron with a dose of $10^{12} cm^{-2}$ to the substrate of the MOS device with a gate oxide

Threshold Voltage Formula				
poly Gate	MOS	$-\phi_{f(M)}$	$-\phi_{f(sub)}$	$-\frac{Q_B}{C_{ox}}$
N^+	n-ch	−	+	+
N^+	p-ch	−	−	−
P^+	n-ch	+	+	+
P^+	p-ch	+	−	−

Table 2.4 Signs in the threshold voltage formula.

of 6nm, the threshold voltage is shifted by:

$$\Delta V_T = \frac{10^{12} \times 1.6 \times 10^{-19}}{\frac{3.9 \times 8.85 \times 10^{-14}}{60 \times 10^{-8}}} = 0.28V.$$

Table 2.4 shows the signs in the threshold voltage formula. As shown in Table 2.4, if N^+ polysilicon gate is used, $-\phi_{f(M)}$ is negative. If P^+ polysilicon gate is used, $-\phi_{f(M)}$ is positive. The substrate of the NMOS device is p-type, therefore $-\phi_{f(sub)}$ is positive. The substrate of the PMOS device is n-type, thus $-\phi_{f(sub)}$ is negative. Due to the ionized acceptors in the substrate of the NMOS device (N_A^-), therefore $-\frac{Q_B}{C_{ox}}$ is positive. Due to the ionized donor in the substrate of the PMOS device (N_D^+), $-\frac{Q_B}{C_{ox}}$ is negative. Depending on the circuit applications, in the CMOS technology both enhancement-type and depletion-type devices may be required.

Fig. 2.26 shows the definition of the enhancement-type and the depletion-type CMOS devices. For enhancement-type devices, the threshold voltage of the NMOS device is positive and the threshold voltage of the PMOS device is negative. For depletion-type devices, the threshold voltage of the NMOS device is negative and the threshold voltage of the PMOS device is positive. From Table 2.4, using an N^+ polysilicon gate without any implant, the PMOS device is already enhancement-type. However, its threshold voltage can be very negative. If a less negative threshold voltage is needed for the enhancement-type PMOS device, an implant of p-type dopant is required to shift the threshold voltage toward the positive direction. Since the substrate of the PMOS device is n-type, implanting p-type dopants counter-dopes the n-type substrate. As a result, a pn junction near the silicon surface is produced [18] as shown in Fig. 2.27. Similarly, using P^+ polysilicon gate, without any implant, the NMOS device is already enhancement-type—its threshold voltage is positive. However, its value may be very positive. In order to lower the threshold voltage, an n-type implant is required. As a result, an np junction exists near the silicon surface of the NMOS device using a P^+ polysilicon gate.

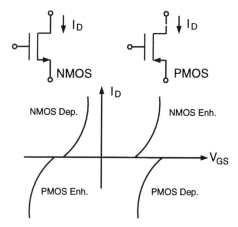

Fig. 2.26 Definition of enhancement-type and depletion-type CMOS devices.

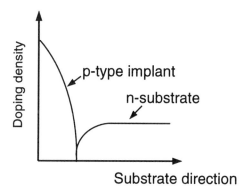

Fig. 2.27 Doping profile in the PMOS device.

2.10 BODY EFFECT

For an NMOS device as shown in Fig. 2.28, when the back gate bias changes from ground to negative, the threshold voltage shifts toward the positive direction. This is called the body effect, or called the back gate bias effect. Considering the back gate bias effect, the previous threshold voltage formula for the zero source-body biased ($V_{SB} = 0$) needs to be modified to include the $V_{SB} \neq 0$ cases. The body effect can be regarded as the change in the depletion width of the reverse-biased pn junction between the source and the substrate due to the change in the source-body voltage. Therefore, the threshold voltage considering the body effect becomes:

$$V_T = -\phi_{f(M)} - \frac{Q_{ss}}{C_{ox}} - \phi_{f(sub)} - \frac{Q_B}{C_{ox}},$$
$$Q_B = -\sqrt{2\epsilon_{si}qN_A(V_{SB} - 2\phi_{f(sub)})}. \tag{2.6}$$

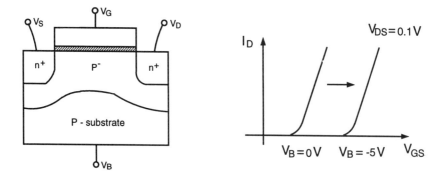

Fig. 2.28 Influence of the back gate bias in the threshold voltage.

From Eq. (2.6), the threshold voltage formula becomes:

$$V_T = V_{T0} + \frac{Q_{B0} - Q_B}{C_{ox}}, \quad (2.7)$$

$$= V_{T0} + \gamma\left(\sqrt{V_{SB} - 2\phi_{f(sub)}} - \sqrt{-2\phi_{f(sub)}}\right),$$

where γ is the body effect coefficient ($\gamma = \frac{\sqrt{2\epsilon_{si}qN_A}}{C_{ox}}$), V_{T0} is the zero-biased threshold voltage, and Q_{B0} is the zero-biased depletion charge ($Q_{B0} = -\sqrt{2\epsilon_{si}qN_A} \cdot \sqrt{(-2\phi_{f(sub)})}$). From the body effect formula, when the substrate doping density becomes higher, the body effect is more serious. This implies that at a higher substrate doping density, the threshold voltage is more influenced by the source-body voltage (V_{SB}). From circuit design point of view, the body effect coefficient should be as small as possible. Lowering of the body effect coefficient can be accomplished by lowering the substrate doping density and by decreasing the thickness of the gate oxide. In the conventional n-well CMOS technology, since the n-well doping density is higher than the substrate, the body effect of the PMOS device is higher than that of the NMOS device. When scaling down CMOS technology, the substrate doping density is increased and the gate oxide becomes thinner. As a result, their influences in the body effect coefficient when scaling down the CMOS technology are mixed.

2.11 SHORT CHANNEL EFFECTS

As shown in Fig. 2.29, at a channel length of smaller than a micron, when the channel length of an MOS device is scaled down, the threshold voltage becomes smaller—short channel effect [32].

For an MOS device with a long channel, the depletion region under the gate is near rectangular—the influence from the source and the drain can be neglected. Therefore, the 1D threshold voltage formula described in the previous section is applicable. However, when the channel length is short as shown in Fig. 2.29, the

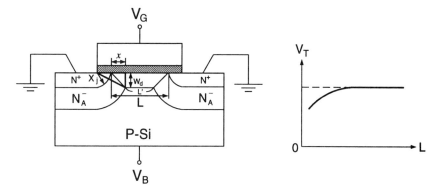

Fig. 2.29 Short channel effect of an NMOS device (Adapted from Yau[32].)

depletion region under the gate is not rectangular. Instead, it is trapezoidal-shape. Under this situation, the influence of the source and the drain cannot be neglected. The threshold voltage needs to be modified to be:

$$V_T = \phi_{MS} - 2\phi_f - \frac{Q_{SS}}{C_{ox}} - \frac{Q'_B}{C_{ox}}, \tag{2.8}$$

$$Q'_B = -qN_AW_d\left(\frac{L+L'}{2}\right)\frac{1}{L},$$

where Q'_B is the total space charge region in the trapezoidal-shape depletion region. From Fig. 2.29, based on the basic geometry theorem, one obtains the following formulas:

$$L' = L - 2x = L - 2x_j\left(\sqrt{1 + \frac{2W_d}{x_j}} - 1\right), \tag{2.9}$$

$$V_T = \phi_{MS} - 2\phi_f - \frac{Q_{SS}}{C_{ox}} - \frac{Q_B}{C_{ox}}\left[1 - \left(\sqrt{1 + \frac{2W_d}{x_j}} - 1\right)\frac{x_j}{L}\right], \tag{2.10}$$

where x_j is the junction depth of the source/drain region. As shown in Fig. 2.29, from the short-channel threshold voltage formula, when the channel length is shortened, the threshold voltage becomes smaller. When the channel length is long, the short-channel threshold voltage formula is simplified to the 1D threshold voltage formula. From Eq. (2.10), with a smaller junction depth (x_j), the threshold voltage is less sensitive to the short channel effect. In addition, a thinner gate oxide leads to a smaller short channel effect. Therefore when scaling down MOS devices for deep-submicron VLSI technology, the source/drain junction depth and the gate oxide thickness are also scaled down accordingly to reduce the short-channel effect.

44 CMOS TECHNOLOGY AND DEVICES

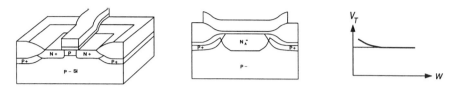

Fig. 2.30 Narrow channel effect of a bulk device using LOCOS.

2.12 NARROW CHANNEL EFFECTS

For a bulk MOS device using LOCOS for device isolation, when its channel width is scaled down to submicron, its threshold voltage increases at a decreased channel width as shown in Fig. 2.30. This is the narrow channel effect[33][34]. Narrow channel effect is due to the influence of the depletion region under the edge of the field oxide surrounding the active region. Using a similar approach as for the short channel effect to consider the effect of the surrounding field oxide on the total space charge in the depletion region, the narrow channel threshold voltage formula is:

$$V_T = \phi_{MS} - 2\phi_f - \frac{Q_{SS}}{C_{ox}} - \frac{Q_B}{C_{ox}}\left(1 + \frac{\alpha W_d}{W} - \frac{x_j}{L}\left(1 + \frac{2}{3}\frac{\alpha W_d}{W}\right)\right. \\ \left. \left(\sqrt{1 + \frac{2W_d}{x_j}} - 1\right)\right), \tag{2.11}$$

where α, which is between 0 and 1, is determined by the doping profile and topography of the field oxide at the Si-SiO$_2$ interface surrounding the active region. Specifically, the bird's beak shape of the field oxide region around the active region affects the narrow channel effect. In addition, the channel stop implant under the field oxide surrounding the active region also influences the narrow channel effect—with the channel stop implant, α is larger.

In order to increase the device density of a VLSI technology, deep-submicron bulk CMOS devices using shallow trench isolation have been used. Different from using LOCOS technology, deep-submicron CMOS devices using STI have a much smaller narrow channel effect when the channel with of the device is scaled down. Fig. 2.31 shows the narrow channel effect of NMOS devices using shallow trench isolation and LOCOS[35]. Due to the elimination of the bird's beak and the channel stop implant, the narrow channel effect of the STI NMOS device is much smaller.

2.13 MOBILITY & DRAIN CURRENT

In an MOS device, carrier mobility is an important factor in the drain current formula. Carrier mobility is related to the electric field[36]. As shown in Fig. 2.32, at a low electric field, electron mobility is larger than hole mobility. Therefore, in CMOS

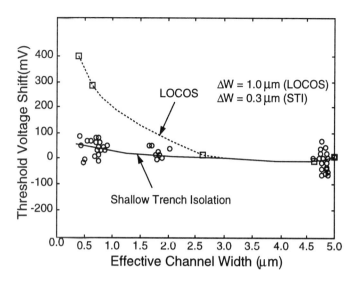

Fig. 2.31 Narrow channel effect of NMOS devices using STI and LOCOS. (From Davari et al.[35], ©1988 IEEE.)

technology, under an identical situation the drain current of an NMOS device is larger than that of a PMOS device. Under a high electric field, the difference between electron and hole mobilities shrinks. The electron mobility in an NMOS device is influenced by the lateral electric field and the vertical electric field simultaneously.

(a) Vertical electric field

Consider an NMOS device as shown in Fig. 2.33. At location y in the channel, the voltage drop over a region dy is:

$$I_D \frac{dy}{\mu_n Q_I(y) W} = dV(y), \qquad (2.12)$$

where $Q_I(y) = C_{ox}(V_G - V(y) - V_T)$. From Eq. (2.12), from source to drain, the drain current is:

$$I_D = \mu_n(y) Q_I(y) W \frac{dV(y)}{dy}. \qquad (2.13)$$

The electron mobility is inversely proportional to the vertical electric field in the channel:

$$\mu_n(y) = \frac{\mu_{n0}}{1 + \alpha_\theta \overline{E_x(y)}}, \qquad (2.14)$$

where $\overline{E_x}$ is the average vertical electric field in the inversion layer. Considering the vertical electric field above the inversion layer at the oxide interface (E_{xs}) and the

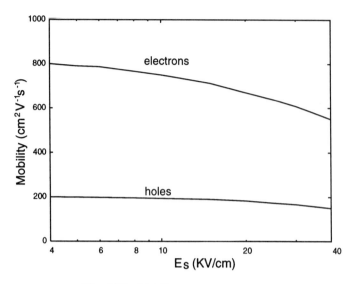

Fig. 2.32 Mobility versus electric field.

vertical electric field below the inversion layer (E_{xb}), the average vertical electric field is:

$$\overline{E_x(y)} = \frac{E_{xs}(y) + E_{xb}(y)}{2}, \qquad (2.15)$$

where E_{xs} and E_{xb} can be computed by considering the inversion charge (Q_I) and the depletion charge (Q_B) as:

$$E_{xs}(y) = -\frac{Q_I + Q_B}{\epsilon_{si}},$$
$$E_{xb}(y) = -\frac{Q_B}{\epsilon_{si}}. \qquad (2.16)$$

From Eqs. (2.14)–(2.16), the vertical electric field electron mobility model is:

$$\mu_n = \frac{\mu_{n0}}{1 - \frac{\alpha_\theta}{2\epsilon_{si}}(Q_I + 2Q_B)}. \qquad (2.17)$$

Therefore, the drain current is:

$$I_D = WQ_I(y)\frac{\mu_{n0}}{1 - \frac{\alpha_\theta}{2\epsilon_{si}}(Q_I + 2Q_B)}\frac{dV(y)}{dy}. \qquad (2.18)$$

Integrating Eq. (2.18), the drain current of the NMOS device biased in the triode region ($V_D < V_G - V_T$) is:

$$I_D = \mu_{n(eff)}\frac{W}{L}\int_0^{V_D} Q_I(y)dV(y),$$

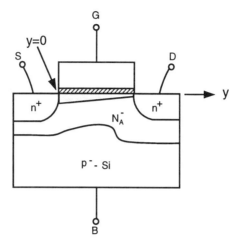

Fig. 2.33 Cross-section of an NMOS device under study.

$$\mu_{n(eff)} = \frac{\frac{1}{2}\mu_{n(eff)}C_{ox}\frac{W}{L}(2(V_G - V_T)V_D - V_D^2)}{\frac{\mu_{n0}}{\frac{1}{L}\int_0^L(1 - \frac{\alpha_\theta}{2\epsilon_{si}}(Q_I + 2Q_B))dx}}, \quad (2.19)$$

where $\mu_{n(eff)}$ is the effective electron mobility. Using an approximation approach: $dx = (L/V_D)dV$, the effective electron mobility model is:

$$\mu_{n(eff)} = \frac{\mu_{n0}}{\frac{1}{V_D}\int_0^{V_D}(1 - \frac{\alpha_\theta}{2\epsilon_{si}}(Q_I + 2Q_B))dV}$$

$$= \frac{\mu_{n0}}{1 + \theta(V_G - V_T) + \theta_B V_{SB}}. \quad (2.20)$$

From Eq. (2.20), the effective electron mobility is inversely proportional to the front gate voltage and the back gate voltage.

(b) Lateral electric field

When the drain voltage further increases to greater than $(V_D > V_G - V_T)$, the channel may pinch off near the drain. The channel potential at the pinchoff point is $(V_G - V_T)$. When the drain voltage is large $(V_D > V_G - V_T)$, in the depletion region between the pinchoff point and the drain, there is a large lateral electric field. As a result, the electrons travel at a saturated velocity over this region. After the channel pinches off, a further increase in the drain voltage moves the pinchoff point toward the source. Therefore, the effective channel length becomes shorter. Thus, the drain current is slightly increased when the drain voltage is increased. This is called channel length modulation. Considering channel length modulation, the drain

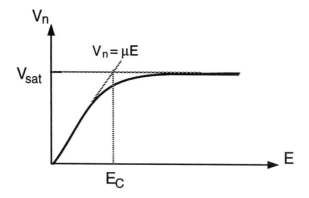

Fig. 2.34 Electron velocity versus electric field.

current formula in the saturation region is:

$$I_D = \frac{1}{2}\mu_{n(eff)}C_{ox}\frac{W}{L-\Delta L}(V_G - V_T)^2, \qquad (2.21)$$

where ΔL is the distance between the pinchoff point and the drain:

$$\Delta L = \sqrt{\frac{2\epsilon_{si}(V_D - (V_G - V_T))}{qN_A}}. \qquad (2.22)$$

The above equation is based on the depletion width formula for a reverse-biased one-sided pn junction, which is made of the N^+ drain region and P^- substrate region, biased at the drain voltage of V_D and the voltage of $V_G - V_T$ at the pinchoff point. The derivation of the drain current formula in this section is based on an assumption—the gradual channel approximation, which assumes that the change in the lateral potential in the channel is gradual such that the analysis of the lateral direction and the vertical direction can be separated. The interaction between the lateral direction and the vertical direction is negligible. The vertical direction determines the quantity of the electrons in the channel and the lateral direction determines the drift current. Based on the gradual channel approximation, both the lateral and the vertical direction behaviors can be analyzed in terms of one-dimension equations. Therefore, the analysis is simplified. The gradual channel approximation is applicable only when the channel length of the device is long. When channel length is short, 2D analysis is required.

Fig. 2.34 shows the electron traveling velocity versus the electric field. As shown in the figure, at a low electric field the electron traveling velocity is linearly proportional to the electric field. When the electric field exceeds a critical electric field (E_C), the electrons are traveling at the saturation velocity. As described before, for a long-channel NMOS device biased in the saturation region, between the pinchoff point and the source, the lateral electric field is small and hence the velocity of the traveling electrons is linearly proportional to the lateral electric field. When the lateral electric field increases, the velocity of the traveling electrons increases.

Between the pinchoff point and the drain, the lateral electric field is large. Therefore, the electrons travel at the saturated velocity. In a short-channel NMOS device biased in the saturation region, before the pinchoff point in the channel, due to a large lateral electric field, the electrons may already travel at the saturated velocity—prepinchoff velocity saturation[37]. Under this situation, the drain current can be expressed as the product of the saturated velocity and the electron quantity at the onset of the prepinchoff velocity saturation:

$$I_D = Q_I W v_{sat} = C_{ox} W (V_G - V_T - V'_{DSAT}) \mu_{n(eff)} E_C, \qquad (2.23)$$

where V'_{DSAT} is the channel potential at the onset of the prepinchoff velocity saturation. Considering the region between source and the location with onset of the prepinchoff velocity saturation as a new device, the drain current can also be expressed as:

$$I_D = \mu_{n(eff)} C_{ox} \frac{W}{L} \left(V_G - V_T - \frac{V'_{DSAT}}{2} \right) V'_{DSAT}, \qquad (2.24)$$

where the drain saturation voltage of the new device is V'_{DSAT}. From Eqs. (2.23) and (2.24), the channel potential at the onset of the prepinchoff velocity saturation is:

$$V'_{DSAT} = V_G - V_T + E_C L - \sqrt{(V_G - V_T)^2 + (E_C L)^2}. \qquad (2.25)$$

In the above equation, a two-segment piecewise-linear approach for the electron velocity has been assumed: When $E < E_C$, $v_n = \mu_n E$; when $E > E_c$, $v_n = v_{sat}$. From the above equation, when the channel length is long—$V'_{DSAT} = V_G - V_T$—the prepinchoff velocity saturation does not exist in the device. When the channel length is small—$V'_{DSAT} < V_G - V_T$—which means that the onset of velocity saturation occurs before the pinchoff point—prepinchoff velocity saturation. Without the prepinchoff velocity saturation, the transconductance is linearly proportional to the gate voltage—when the gate voltage increases, the transconductance increases as shown in Fig. 2.35. At a short channel, since the influence of the prepinchoff velocity saturation is important, the transconductance is independent of the gate voltage as shown in the figure.

2.14 SUBTHRESHOLD CURRENT

For a bulk NMOS device biased at $V_G < V_T$, its drain current is not zero. Instead, the drain current is exponentially proportional to the gate voltage—the subthreshold characteristics[46] as shown in Fig. 2.36.

Consider a bulk NMOS device as shown in Fig. 2.33. When the surface potential in the channel reaches $\Psi_s = -2\phi_f$, strong inversion occurs. In the channel, the electron density is greater than the p-type acceptor density in the substrate. At strong inversion, the current in the channel is caused by the drift current. When the surface potential is $-\phi_f < \Psi_s < -2\phi_f$, the electron density at the Si/SiO$_2$ interface is not

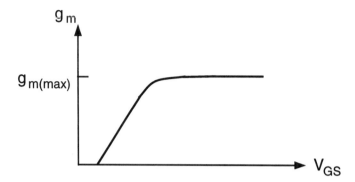

Fig. 2.35 Transconductance versus the gate voltage of an NMOS device biased at a V_{DS}.

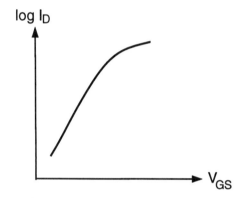

Fig. 2.36 Subthreshold characteristics of a bulk NMOS device.

greater than the acceptor doping density in the substrate. Under this situation, it is called weak inversion. At weak inversion, the current in the channel is caused by diffusion, which can be expressed as:

$$I_D = -qW\Delta x D_n \frac{dn}{dx} = -qW\Delta x D_n \frac{n(0) - n(L)}{L}, \qquad (2.26)$$

where Δx is the channel depth, which is usually 5nm–10nm. $n(0)$ is the electron density in the channel near the source and $n(L)$ is the electron density in the channel near the drain. $n(0)$ and $n(L)$ can be expressed as:

$$\begin{aligned} n(0) &= n_i exp\left(\frac{\Psi_S - V_{SB} - (-\phi_f)}{kT/q}\right), \\ n(L) &= n_i exp\left(\frac{\Psi_S - V_{DB} - (-\phi_f)}{kT/q}\right). \end{aligned} \qquad (2.27)$$

From Eqs. (2.26) and (2.27), the drain current at weak inversion is:

$$I_D = \frac{W}{L}\frac{kT}{q}D_n n_i \sqrt{\frac{\epsilon_{si}}{2qN_D\Psi_S}} exp\left(\frac{\Psi_S - V_{SB} - (-\phi_f)}{kT/q}\right)(1 - e^{-\frac{V_{DS}}{kT/q}}). \qquad (2.28)$$

Since the gate-source voltage determines the surface potential, the gate-source voltage can be expressed as:

$$V_{GS} = V_{GB} - V_{SB}$$
$$= \phi_{MS} - \frac{Q_{SS}}{C_{ox}} + \Psi_S - V_{SB} + \frac{\sqrt{2q\epsilon_{si}N_A(\Psi_S - V_{SB})}}{C_{ox}}. \quad (2.29)$$

From Newton's approximation method, V_{GS} can be expressed as a linear function at the reference point ($V_{GS}|_{\Psi_s=-1.5\phi_f}$):

$$V_{GS} = (V_{GS}|_{\Psi_S=-1.5\phi_f}) + \frac{\partial V_{GS}}{\partial \Psi_S}(\Psi_S - (-1.5\phi_f)). \quad (2.30)$$

where the reference point at $\Psi_s = -1.5\phi_f$ is chosen in the middle of the weak inversion region ($-\phi_f < \Psi_s < -2\phi_f$). From Eq. (2.30), one obtains:

$$\frac{\partial V_{GS}}{\partial \Psi_S} = 1 + \frac{C_d}{C_{ox}} = n, \quad (2.31)$$

$$C_d = \frac{q\epsilon_{si}N_A}{\sqrt{2q\epsilon_{si}N_A(\Psi_S + V_{SB})}}.$$

Therefore, the drain current becomes:

$$I_D = \mu_n \left(\frac{kT}{q}\right)^2 C_d \frac{W}{L} \left(e^{\frac{q}{nkT}(V_{GS}-V_{GS}|_{\Psi=-1.5\phi_f})-\frac{q(-\phi_f)}{2kT}}\right)\left(1 - e^{-\frac{qV_{DS}}{kT}}\right). (2.32)$$

From Eq. (2.32), at weak inversion, the drain current is exponentially proportional to the gate-source voltage, which implies that at weak inversion the transport of electrons in the channel is via diffusion. At weak inversion, the MOS device functions like a bipolar device, except the 'n' in Eq. (2.32). In the bipolar device, $n = 1$. At weak inversion of an MOS device, $n > 1$. From Eq. (2.30), for a thinner gate oxide and/or a more lightly doped substrate, n is closer to 1. When n is closer 1, the subthreshold slope is steeper—the MOS device turns off faster. From Eq. (2.32), a more heavily doped substrate leads to a larger n, hence a worse subthreshold slope. From Eq. (2.32), one obtains:

$$log(I_d) \propto (loge)\frac{V_{GS}}{nkT/q}. \quad (2.33)$$

The subthreshold slope is usually expressed in terms of mV/decade, which indicates that when the drain current drops 10 times, how many volts of the gate voltage should change. From Eq. (2.33), when the gate voltage changes $(1/loge)nkT/q$ volts, the drain current changes 10 times. For a $0.25\mu m$ CMOS technology with a gate oxide of 7.5nm and the substrate doping density of $6 \times 10^{17} cm^{-3}$, from Eqs. (2.32) and (2.33), the subthreshold slope is 94.8 mV/decade. If a steeper subthreshold slope is needed, the substrate should be more lightly doped. When the MOS device is used in DRAMs, the subthreshold slope should be as steep as possible—the value in terms

52 CMOS TECHNOLOGY AND DEVICES

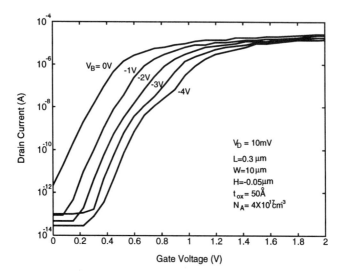

Fig. 2.37 Subthreshold characteristics of an STI NMOS device biased at a drain voltage of $V_D = 10mV$ and a back gate bias from 0V to $-3V$. (From Chen et al.[14], ©1996 IEEE.)

of mV/decade should be as small as possible. A steeper subthreshold slope leads to a shorter transition region between on and off. The leakage current of an MOS device determines the hold time of the memory cell in DRAM, which implies that the time that the charge in the storage capacitor of the memory cell can be retained. The hold time determines the refresh time of a DRAM. In order to increase the hold time of the DRAM memory cell, the threshold voltage of an NMOS device is usually designed to be a large value such that the leakage current can be small when $V_{GS} = 0$. However, when the threshold voltage is large, the drain current may be degraded. In fact, due to subthreshold leakage current consideration, the threshold voltage of a deep-submicron NMOS device almost remains at a fixed value in spite of the down-scaling of the CMOS technology. For example, assume that the subthreshold slope of an NMOS device is 80mV/decade. If the off current is needed to be 7 orders of magnitude smaller than the on current for the NMOS device, the threshold voltage of the device needs to be designed to be 0.56V. Due to the subthreshold slope consideration, the threshold voltage of an MOS device cannot be easily scaled down when designing a deep-submicron CMOS technology. The threshold voltage of most deep-submicron enhancement-mode MOS devices is usually between 0.4V and 0.8V.

For deep-submicron CMOS VLSI technology, shallow trench isolation (STI) technology has been replacing conventional LOCOS techniques to become the mainstream device isolation technology as described before. For CMOS devices using shallow trench isolation, due to the trench oxide sidewall, field crowding at the edge of the sidewall results in anomalous subthreshold conduction—the subthreshold current hump phenomenon. Fig. 2.37 shows the subthreshold characteristics of an STI NMOS device biased at a drain voltage of $V_D = 10mV$ and a back gate bias from

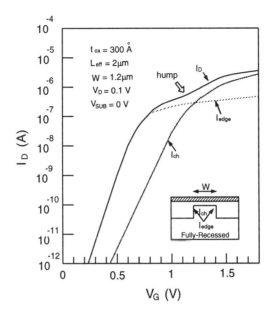

Fig. 2.38 Subthreshold characteristics of an NMOS device with a fully recessed field oxide and a channel width of 1.2μm. (From Shigyo and Dang [38], ©1985 IEEE.)

0V to -3V[14]. The subthreshold current hump phenomenon is especially noticeable at a negative back gate bias, which is due to the different back-gate-bias sensitivities of the center and the edge channels. As shown in the figure, when the back gate bias becomes more negative, the current hump phenomenon becomes more noticeable. The current hump phenomenon can be understood by considering the subthreshold characteristics of an NMOS device with a fully recessed field oxide and a channel width of 1.2μm[38]. As shown in the figure, the conducting drain current is composed of the center channel current and the edge channel current. Due to the fringing field, the edge channel has a smaller threshold voltage as compared to the center channel. Therefore, the edge channel turns on earlier. Thus, the subthreshold current hump exists. From this reasoning, the body effect as shown in Fig. 2.37, can be explained. At a more negative back gate bias, the current hump phenomenon of the STI NMOS device is more noticeable due to the edge channel effect.

2.15 ELECTRON TEMPERATURE

For a deep-submicron CMOS device with a channel length of smaller than 0.2μm, the conventional drift-diffusion device model may not be sufficient. The high electric field effects cannot be overlooked[39]. Under a high electric field, the temperature of the traveling electron may be much higher than the lattice temperature. Fig. 2.39[40] shows the electron temperature profiles in the lateral channel of a 0.1μm NMOS device. Note that only the portion before the pinchoff point is shown. The electron

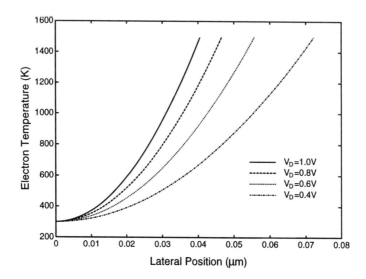

Fig. 2.39 Electron temperature profiles in the 0.1μm NMOS device (Adapted from Ma and Kuo[40].)

temperature can be as high as 1400K. At such a high electron temperature, the degradation of the electron mobility cannot be neglected. Therefore, energy transport— the energy balance equation[40] becomes an important factor to consider when analyzing the current conduction in the channel. Inclusion of the electron temperature in the deep-submicron device model becomes a necessary step.

Consider a deep-submicron NMOS device as shown in Fig. 2.40[41]. From the energy balance equation, in a unit volume in the channel in the device, in a unit time, under the lateral electric field, the acquired energy of a moving electron (JE) is equal to the energy transferred from the electron to the lattice and the gradient of the energy flux:

$$JE = \frac{3}{2}nk\frac{T_n - T_0}{\tau_\epsilon} + \frac{dS}{dy}, \qquad (2.34)$$

where J is the current density, E is the electric field, T_n is the electron temperature, n is the electron density, τ_ϵ is the energy relaxation time, and S is the electron energy flux. Note that y is the lateral channel direction. Energy relaxation time is a time unit used to describe the energy absorption capability of the lattice. Under a certain amount of JE, a longer energy relaxation time represents that the energy absorption capability of the lattice is not good, thus the electron temperature is high. A small energy relaxation time implies that the energy absorption capability of the lattice is strong, hence the electron temperature is low. The energy relaxation time can be expressed as the effective RC time constant of a unit volume silicon material: $\tau_\epsilon = (1/nq)(\epsilon_{si}/\mu_n)$, which is related to the permittivity and the mobility in the silicon. As for the channel transit time, the energy relaxation time is also a reference

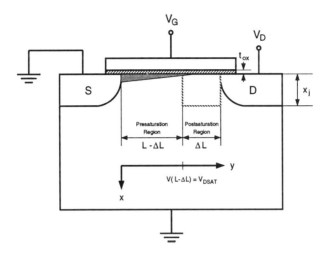

Fig. 2.40 Cross-section of a deep-submicron NMOS device.

to evaluate the heat dissipation capability of the lattice. When the electric field in the channel is not high such that the energy of the moving electron is not high, the temperature of the moving electron is identical to the lattice temperature. However, when the electric field in the channel is high, the energy of the moving electron is high. Hence, the bottleneck of the heat dissipation capability of the lattice becomes noticeable. From Eq. (2.34), since there is not enough time to dissipate heat, the electron temperature rises. S in Eq. (2.34) is the electron energy flux, which can be expressed as:

$$S = -\kappa \frac{dT_n}{dy} - \frac{5}{2} J \frac{kT_n}{q}, \qquad (2.35)$$

where κ is the thermal conducting coefficient. In an NMOS device, the drain current is basically via the drift current:

$$J = nq\mu_n E. \qquad (2.36)$$

From Eqs. (2.34)–(2.36), one obtains:

$$q\mu_n E^2 = \frac{3}{2} k \frac{T_n - T_0}{\tau_\epsilon} - \frac{5}{2} k\mu_n E \frac{dT_n}{dy}. \qquad (2.37)$$

In Eq. (2.37), the heat conduction term has been neglected. The electron temperature affects the electron mobility:

$$\mu_n = \frac{\mu_s}{1 + \alpha \frac{k}{q}(T_n - T_0)},$$
$$\alpha = \frac{3\mu_s}{2v_{sat}^2 \tau_\epsilon}, \qquad (2.38)$$

where μ_s is the surface mobility due to surface scattering: $\mu_s = \frac{\mu_0}{1+\theta_1(V_{GS}-V_T)}$. From Eqs. (2.37) and (2.38), one obtains a differential equation in terms of the electron temperature:

$$E^2 = \frac{3k}{2q}\frac{1+\alpha\frac{k}{q}(T_n-T_0)}{\mu_s}\frac{T_n-T_0}{T_\epsilon} - \frac{5kE}{2q}\frac{dT_n}{dy}. \quad (2.39)$$

From Eq. (2.39) the electron temperature distribution can be obtained as long as the electric field is known.

For a deep-submicron NMOS device, the IV characteristics can be divided into two regions: triode ($V_D < V_{DSAT}$) and saturation ($V_D > V_{DSAT}$).

(1) Triode region

For a deep-submicron NMOS device biased in the triode region, the lateral electric field can be regarded as linear:

$$E(y) = E_0 + ay, \quad (2.40)$$

where $E_0 = -\frac{\eta_1 V_D}{L_{eff}}$, $a = -\frac{2(1-\eta_1)V_D}{L_{eff}^2}$, and E_0 is the electric field at source, which is related to the average electric field in the channel. L_{eff} is the effective channel length. η_1 is a fitting parameter. a is the slope of the electric field. Therefore, the differential equation in terms of the electron temperature can be simplified as:

$$(E_0+ay)^2 = \frac{3k}{2q}\frac{1+\alpha\frac{k}{q}(T_n-T_0)}{\mu_s}\frac{T_n-T_0}{T_\epsilon} - \frac{5k(E_0+ay)}{2q}\frac{dT_n}{dy}. \quad (2.41)$$

In order to solve Eq. (2.41), two boundary conditions are required. At the source end, the electron temperature is the lattice temperature. Using a linear approximation formula, the electron temperature is expressed as:

$$T_n = Ay + B. \quad (2.42)$$

From Eqs. (2.41) and (2.42), one obtains:

$$(E_0+ay)^2 = \frac{3k}{2q}\frac{1+\alpha\frac{k}{q}(Ay+B-T_0)}{\mu_s}\frac{Ay+B-T_0}{T_\epsilon} - \frac{5k(E_0+ay)}{2q}A. \quad (2.43)$$

From the coefficient of the second order term, one obtains: $A = -\frac{2qav_{sat}T_\epsilon}{3k}$. Considering at the drain end ($y = L_{eff}$), from Eq. (2.43), one obtains: $B = T_0 - AL_{eff} - \frac{q}{2\alpha k} + \frac{q\sqrt{1-\frac{20\mu_s^2 T_\epsilon}{3v_{sat}}(E_0+aL_{eff})+\frac{4\mu_s^2}{v_{sat}^2}(E_0+aL_{eff})^2}}{2\alpha k}$. The drain current is expressed as:

$$I_D = -WQ_m(y)\mu_n\frac{dV(y)}{dy}, \quad (2.44)$$

$$Q_m(y) = -C_{ox}(V_G - V_{ON} - 2a_0V(y)),$$

where C_{ox} is the unit area oxide capacitance, V_{ON} and a_o are used to consider the drain induced barrier lowering (DIBL), and the body effect: $V_{ON} = V_T - \epsilon V_D$, $a_0 = \frac{1}{2} + \frac{1}{4}\frac{\gamma}{\sqrt{2\phi_f}}$. Therefore, the drain current becomes:

$$I_D = \mu_{eff} C_{ox} \frac{W}{L_{eff}}[(V_G - V_{ON})V_D - a_0 V_D^2], \quad (2.45)$$

where μ_{eff} is the effective electron mobility, which can be approximated as:

$$\mu_{eff} = \frac{L_{eff}}{\int_0^{L_{eff}} \frac{1}{\mu_n} dy} = \frac{2}{\frac{1}{\mu_n(0)} + \frac{1}{\mu_n(L_{eff})}}. \quad (2.46)$$

At the source end, the electron mobility is $\mu_n(0) = \mu_s$. Therefore, the effective mobility becomes:

$$\mu_{eff} = \frac{4\mu_s}{3 + \sqrt{1 - \frac{20a\mu_s^2 \tau_\epsilon}{3v_{sat}}(E_0 + L_{eff}) + \frac{4\mu_s^2}{v_{sat}^2}(E_0 + aL_{eff})^2}},$$

$$= \frac{4\mu_s}{3 + \sqrt{1 + KV_{DS}^2}}, \quad (2.47)$$

where $K = \frac{4(2-\eta_1)^2 \mu_s^2}{v_{sat}^2 L_{eff}^2} - \frac{40(1-\eta_1)(2-\eta_1)\mu_s^2 \tau_\epsilon}{3v_{sat} L_{eff}^3}$. From the above equation, the effective mobility is dependent on the electric field and its gradient. The term with the negative sign in the denominator of Eq. (2.47) is a function of τ_ϵ. For a larger thermal relaxation time, the effective mobility is larger. When the surface mobility (μ_s) is large, due to the thermal relaxation time effect, the effective mobility may become greater than the saturated velocity—the velocity overshoot phenomenon[42].

For a deep-submicron NMOS device with a shallow source/drain junction, its source/drain resistance may be very large. Therefore, its effect cannot be overlooked. Considering the source/drain resistance, the effective V_{GS} and V_{DS} becomes:

$$V_{GS} = V_{gs} - I_D R_S, \quad (2.48)$$
$$V_{DS} = V_{ds} - I_D(R_S + R_D),$$

where V_{gs} and V_{ds} are the terminal gate-source and drain-source voltages, respectively. The effective threshold voltage becomes: $V_{ON} = V_T - \xi V_{ds} + \xi I_D(R_S + R_D)$. Therefore, considering the effective source/drain resistance, the drain current becomes:

$$\frac{I_D}{\mu_{eff}} = C_{ox}\frac{W}{L_{eff}}([V_{gs} - I_D R_S - V_T + \xi V_{ds} - \xi I_D(R_S + R_D)]$$
$$\cdot [V_{ds} - I_D(R_S + R_D)] - a_0[V_{ds} - I_D(R_S + R_D)]^2), \quad (2.49)$$

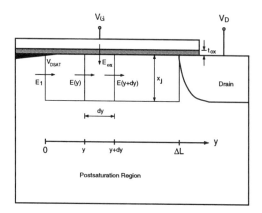

Fig. 2.41 The post-saturation region in the NMOS device.

where $\dfrac{1}{\mu_{eff}} = \dfrac{3+\sqrt{1+KV_{ds}^2} - \dfrac{KV_{ds}}{\sqrt{1+KV_{ds}^2}} I_D(R_S+R_D)}{4\mu_s}$, $I_D = \dfrac{A_2 - \sqrt{(A_2^2 - 4A_1A_3)}}{2A_1}$, $A_1 = (R_S+R_D)[R_S + (\xi - a_0)(R_S+R_D) + \dfrac{L_{eff}KV_{ds}}{4\mu_s C_{ox} W\sqrt{1+KV_{ds}^2}}]$, $A_2 = \dfrac{L_{eff}(3+\sqrt{1+KV_{ds}^2})}{4\mu_s C_{ox} W} + R_S V_{ds} + (R_S+R_D)[V_{gs} - V_T + 2(\xi - a_0)V_{ds}]$, and $A_3 = [V_{gs} - V_T + (\xi - a_0)V_{ds}]V_{ds}$.

(2) Saturation region

When the NMOS device is biased in the saturation region as shown in Fig. 2.40, the channel region can be divided two regions: (a) the pre-saturation region and (b) the post-saturation region. These two regions are separated by the saturation point, where the channel potential is V_{dsat}:

$$V_{dsat} = V_{dsat1} + V_P - \sqrt{V_{dsat1}^2 + V_P^2}, \qquad (2.50)$$

where V_p is the V_{ds} when $(dI_D/dV_{ds}) = 0$. V_{dsat1} is the drain voltage when velocity saturation occurs at drain:

$$V_{dsat1} = V_{DSAT1} + I_{DSAT1}(R_S + R_D), \qquad (2.51)$$

At the same time, $-\mu_n(L_{eff})(E_0 + aL_{eff}) = \eta_2 v_{sat}$, $E_0 = -\eta_1 \dfrac{V_{DSAT1}}{L_{eff}}$, $a = -\dfrac{2(1-\eta_1)V_{DSAT1}}{L_{eff}^2}$, and $\mu_n(L_{eff})$ is the electron mobility at drain, η_2 is used to consider the difference between the actual electron traveling velocity and the saturated velocity at the saturation point. Therefore, V_{DSAT1} is

$$V_{DSAT1} = \dfrac{\eta_2 v_{sat} L_{eff}^2}{(2-\eta_1)(1-\eta_2^2)\mu_s L_{eff} + \tfrac{10}{3}(1-\eta_1)\eta_2^2 \mu_s \tau_e v_{sat}}. \qquad (2.52)$$

When $V_{DS} = V_{DSAT1}$, the drain current is:

$$I_{DSAT1} = \dfrac{\mu_{eff1} C_{ox} \tfrac{W}{L_{eff}}((V_{gs} - V_T + \xi V_{DSAT1})V_{DSAT1} - a_0 V_{DSAT1}^2)}{1 + \mu_{eff1} C_{ox} \tfrac{W}{L_{eff}} V_{DSAT1} R_S}, \qquad (2.53)$$

where $\mu_{eff1} = \frac{4\mu_s}{3+\sqrt{1+KV_{DSAT1}^2}}$.

(a) Pre-saturation region

As shown in Fig. 2.40, in the pre-saturation region, the drain current is similar to the drain current for the triode region except that L_{eff} is replaced by $L_{eff} - \Delta L$ and V_D by V_{DSAT}.

(b) Post-saturation region

As shown in Fig. 2.41, in the post-saturation region, at location y, applying 2D Gauss law in the Gauss box with a width of dy in the lateral direction and from the oxide interface to the junction depth in the vertical direction, one obtains:

$$-\frac{\epsilon_{ox}}{\epsilon_{si}}\frac{V_G - V(y)}{t_{ox}}dy - x_j E(y) + x_j E(y + dy) = -\frac{q(N_A + n)}{\epsilon_{si}}x_j dy. \quad (2.54)$$

Differentiating Eq. (2.54), one obtains:

$$\frac{d^2V(y)}{dy^2} = \frac{1}{\lambda^2}(V(y) - V_{DSAT}) - a, \quad \lambda = \sqrt{\frac{\epsilon_{si} x_j t_{ox}}{\epsilon_{ox}}}. \quad (2.55)$$

The boundary condition for this differential equation is that at the saturation point ($y = 0$) is: $-\frac{\epsilon_{ox}}{\epsilon_{si}}\frac{V_G - V_{DSAT}}{t_{ox}} + x_j a = -qx_j \frac{N_A+n}{\epsilon_{si}}$. Solving the differential equation (Eq. (2.55)) using the boundary condition, the potential in the post-saturation region is:

$$V(y) = V_{DSAT} + a\lambda^2 + \frac{\lambda(E_1 - a\lambda)}{2}e^{-\frac{y}{\lambda}} - \frac{\lambda(E_1 + a\lambda)}{2}e^{\frac{y}{\lambda}}, \quad (2.56)$$

where $E_1 = -\frac{(2-\eta_1)V_{DSAT}}{L_{eff}-\Delta L}$. At the saturation point, the channel potential is V_{DSAT} and the electric field is E_1, from Eq. (2.56), one obtains:

$$V_D = V_{DSAT} + a\lambda^2 + \frac{\lambda(E_1 - a\lambda)}{2}e^{-\frac{\Delta L}{\lambda}} - \frac{\lambda(E_1 + a\lambda)}{2}e^{\frac{\Delta L}{\lambda}}. \quad (2.57)$$

Therefore, considering that the drain voltage is V_D, one obtains: $a = -\frac{2(1-\eta_1)V_{DSAT}}{L_{eff}^2}$, and $E_1 = -\frac{(2-\eta_1)V_{DSAT}}{L_{eff}}$. Solving Eq. (2.57), the width of the post-saturation region is:

$$\Delta L' = \lambda \ln\left(\frac{-(V_D - V_{DSAT} - a\lambda^2) - \sqrt{(V_D - V_{DSAT} - a\lambda^2)^2 + \lambda^2(E_1^2 - a^2\lambda^2)}}{\lambda(E_1 + a\lambda)}\right),$$

$$\Delta L = s\Delta L'. \quad (2.58)$$

Considering the effect of the source resistance, the above equation is still applicable as long as V_{DS} is replaced by V_{ds} and V_{DSAT} by V_{dsat}. The above equations are

Parameter	Value
L_{eff}	$0.2\mu m$
R_S	300Ω
T_o	$300K$
V_T	$0.6V$
t_{ox}	$7.5nm$
v_{sat}	$1 \times 10^7 cm/sec$
x_j	$180nm$
η_1	0.95
η_2	0.9
μ_0	$350 cm^2/V \cdot sec$
θ_1	0.8
θ_2	0.01
τ_ϵ	$0.2ps$
ξ	0.07

Table 2.5 Parameters of the $0.2\mu m$ NMOS device under test.

the analytical drain current model considering the electron temperature.

Consider a deep-submicron NMOS device[43] with its parameters as listed in Table 2.5. The channel length of the device is $0.2\mu m$. The gate oxide thickness is 7.5nm. The junction depth of the source/drain region is $0.18\mu m$. The substrate doping density is $10^{17} cm^{-3}$. The sheet resistance of the source/drain region is $300\Omega\mu m$. Considering the junction depth, the sheet resistance is $1.6K\Omega/\square$. As shown in Fig. 2.42[41], as compared to the results using the conventional model, the analytical model results considering the electron temperature are more accurate. In addition, the source/drain resistance effect is not negligible. Without considering it, the drain current is overestimated.

For a deep-submicron NMOS device, the output conductance characteristics can reflect the completeness of an analytical model due to the mutual influence between the electron temperature and the post-saturation region. Specifically, the output conductance is determined by the width of the post-saturation region, which is deeply influenced by the electron temperature. As shown in Fig. 2.43[41], the analytical model considering the electron temperature can accurately predict the output conductance of a deep-submicron device. As shown in the figure, the influence of the source resistance in the output conductance is limited to the triode region. As shown in the figure, the transition region from the triode region to the saturation region, the output conductance curve is smooth. However, if the saturation model is not sufficient as in the conventional model, the transition region of the output conductance curve from the triode region to the saturation is not smooth.

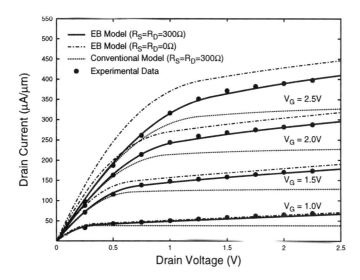

Fig. 2.42 IV characteristics of a $0.2\mu m$ NMOS device (Adapted from Ma and Kuo[41].)

2.16 VELOCITY OVERSHOOT

In a deep-submicron MOS device with a very small channel length, when carriers traveling through a lateral channel region with a rapidly changing electric field, velocity overshoot can occur—the traveling velocity exceeds the saturated velocity. Fig. 2.44 shows the electron drift velocity distribution in the lateral channel of the NMOS device biased at the onset of saturation $V_D = V_{DSAT}$ and $V_G = 1V$[42]. As shown in the figure, in the region near the drain in the lateral channel, the electron drift velocity exceeds its saturated velocity. Due to velocity overshoot, the transconductance behavior of the deep-submicron MOS device is different from that without velocity overshoot. Fig. 2.45 shows the transconductance versus the gate length of NMOS device with the velocity overshoot at 77K and 300K[44]. As shown in the figure, at a gate length of smaller than $0.1\mu m$, the transconductance exceeds its saturated value $v_{sat}C_{ox}$. Fig. 2.46 shows the saturated intrinsic transconductance versus the mobility to channel length ratio (μ_{eff}/L) of NMOS devices biased at $V_{GS} = V_{DS} = 1.5V$[45]. As shown in the figure, for a short channel device, when the inversion-layer mobility is high, velocity overshoot occurs.

2.17 MOS CAPACITANCES

For an NMOS device, in addition to the DC model, AC capacitance model is also important for transient analysis. The equivalent capacitance model of an NMOS device[47] is as shown in Fig. 2.47. The MOS capacitance model is composed of two intrinsic capacitances: the drain-gate capacitance (C_{DG}) and the source-gate

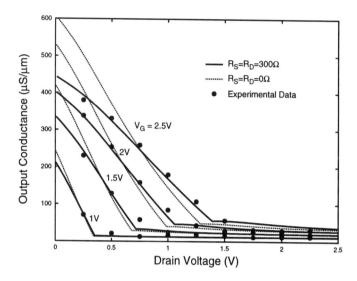

Fig. 2.43 Output conductance of the $0.2\mu m$ NMOS device (Adapted from Ma and Kuo[41].)

capacitance (C_{SG}) and the two extrinsic capacitances: the drain-body capacitance (C_{DB}) and the source-body capacitance (C_{SB}). The intrinsic capacitances affect the intrinsic performance of an MOS device. The extrinsic capacitances, which are similar to other parasitic capacitances, have secondary effects on the device performance.

(a) Intrinsic capacitances (C_{DG}, C_{SG})

The drain-gate capacitance (C_{DG}) and the source-gate capacitance (C_{SG}) are defined as[47][48]:

$$C_{SG} = \frac{\partial Q_S}{\partial V_G}, \qquad (2.59)$$

$$C_{DG} = \frac{\partial Q_D}{\partial V_G},$$

where Q_S/Q_D is the source/drain charge, which is the partitioned total charge under the gate. Defining Q_S and Q_D was a difficult task for developing a consistent MOS capacitance model in the 1970s. At the beginning, C_{SG} and C_{DG} were regarded as linear: $C_{SG} = C_{GS}$, $C_{DG} = C_{GD}$. Thus, C_{SG} and C_{DG} were obtained by $C_{SG} = \frac{\partial Q_G}{\partial V_S}$ and $C_{DG} = \frac{\partial Q_G}{\partial V_D}$. However, C_{DG} and C_{SG} are not linear, hence reciprocity does not hold. The above definitions of C_{DG} and C_{SG} assuming reciprocity are wrong. In this section, C_{DG} and C_{SG} are derived from the current continuity equation.

Consider an NMOS device as shown in Fig. 2.47. In order to simplify the analysis, the space charge is neglected. At location y in the channel of the device, the current

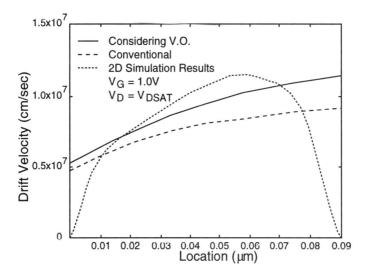

Fig. 2.44 Electron drift velocity distribution in the lateral channel of the NMOS device biased at the onset of saturation $V_D = V_{DSAT}$ and $V_G = 1V$. (Adapted from Kuo et al.[42].)

is:

$$I(y,t) = W\mu_n(y,t)Q_I(y,t)\frac{\partial V(y,t)}{\partial y}. \tag{2.60}$$

From the current continuity equation, in a small region dy at location y in the channel, in a unit time, the change of the total electrons in this region is equal to the gradient of the current profile:

$$\frac{\partial Q_I(y,t)}{\partial t} = -\frac{1}{qW}\frac{\partial I(y,t)}{\partial y}. \tag{2.61}$$

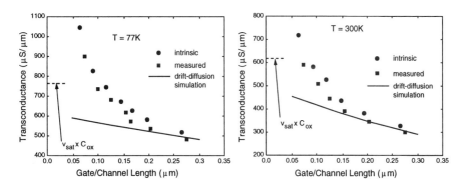

Fig. 2.45 Transconductance versus gate length of NMOS devices showing velocity overshoot at 77K and 300K. (From Sai-Halasz et al.[44], ©1988 IEEE.)

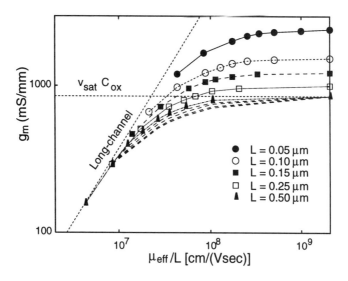

Fig. 2.46 Saturated intrinsic transconductance versus the mobility to channel length ratio (μ_{eff}/L) of NMOS devices biased at $V_{GS} = V_{DS} = 1.5V$. (From Pinto et al.[45], ©1993 IEEE.)

Integrating the above equation from source to y, an indefinite integral is obtained:

$$I(0,t) = qW\mu_n Q_I(y,t)\frac{\partial V}{\partial y} + qW \int_0^y \frac{\partial Q(y',t)}{\partial t}dy'. \tag{2.62}$$

Re-integrating this indefinite integral (Eq. (2.62)) from source to drain, one obtains:

$$I(0,t) = q\mu_n \frac{W}{L}\int_0^{V_D} Q(V(y),V_G)dV(y) + q\frac{W}{L}\int_0^L \int_0^y \frac{\partial Q_I(y',t)}{\partial t}dy'dy. \tag{2.63}$$

From the above equation, the transient current of the source terminal can be expressed in terms of the transport current and the derivative of the source charge with respect to time:

$$I_S(t) \equiv I(0,t) = I_T(t) - \frac{dQ_S(t)}{dt}, \tag{2.64}$$

where the transport current (I_T) is dependent on the terminal voltages, which is identical to the DC drain current:

$$I_T = q\mu_n \frac{W}{L}\int_0^{V_D} Q_I dV(y), \tag{2.65}$$

where $Q_I = C_{ox}W(V_G - V_T - V(y))$. From Eqs. (2.63) and (2.64), the source charge is composed of a double integral:

$$Q_S(t) = -q\frac{W}{L}\int_0^L \int_0^y Q_I(y',t)dy'dy. \tag{2.66}$$

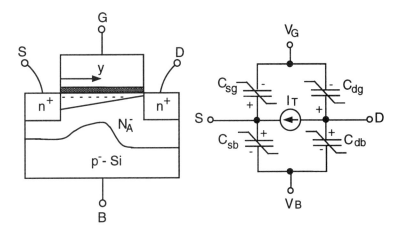

Fig. 2.47 Capacitance model of an NMOS device.

Using integration by parts, the double integral in the source charge can be simplified as:

$$Q_S(t) = -qW \int_0^L \left(1 - \frac{y}{L}\right) Q_I(y,t) dy, \qquad (2.67)$$

where y and dy in the above equation can be expressed as a function of $V(y)$ and $dV(y)$:

$$\begin{aligned} y &= \mu_n W C_{ox}\left((V_G - V_T)V(y) - \frac{1}{2}V(y)^2\right), \\ dy &= \frac{\mu_n W C_{ox}(V_G - V_T - V(y))}{I} dV(y). \end{aligned} \qquad (2.68)$$

From Eqs. (2.67) and (2.68), the source charge can be obtained. The source charge can be regarded as the weighted sum of the total electrons in the channel. From Eq. (2.67), at a location closer to the source, the weight of the electron is larger. At source, the weight is equal to 1. At drain, the weight is equal to 0. From Eq. (2.68), the source charge becomes:

$$\begin{aligned} Q_S &= -W \int_0^{V_{DS}} \frac{1 - \mu_n W C_{ox}}{IL}\left((V_{GS} - V_T)V - \frac{1}{2}V^2\right) \\ &\quad \cdot C_{ox}(V_{GS} - V - V_T) \cdot \frac{\mu_n W C_{ox}}{I}(V_{GS} - V_T - V) dV, \qquad (2.69) \\ &= \frac{-WLC_{ox}}{(V_{GS} - V_T)V_{DS} - \frac{1}{2}V_{DS}^2}\left\{(V_{GS} - V_T)^2 V_{DS}\right. \\ &\quad + \left[-2(V_{GS} - V_T) - \frac{(V_{GS} - V_T)^3}{(V_{GS} - V_T)V_{DS} - \frac{1}{2}V_{DS}^2}\right] \cdot \frac{1}{2}V_{DS}^2 \\ &\quad + \left[1 + \frac{5}{2}\frac{(V_{GS} - V_T)^2}{(V_{GS} - V_T)V_{DS} - \frac{1}{2}V_{DS}^2}\right] \cdot \frac{1}{3}V_{DS}^3 \end{aligned}$$

CMOS TECHNOLOGY AND DEVICES

$$-\frac{(V_{GS}-V_T)}{(V_{GS}-V_T)V_{DS}-\frac{1}{2}V_{DS}^2}\cdot\frac{1}{4}V_{DS}^4+\frac{1}{(V_{GS}-V_T)V_{DS}-\frac{1}{2}V_{DS}^2}\cdot\frac{1}{10}V_{DS}^5\bigg\}.$$

Using a similar method, the drain transient current is:

$$I_D(t) = I_T(t) + \frac{dQ_D(t)}{dt},$$
$$Q_D(t) = -qW\int_0^L \frac{y}{L}Q_I(y)dy, \quad (2.70)$$

where Q_D is the weighted total electrons in the channel. When computing Q_D, the electrons in the channel near the drain are weighted by a large factor and those far away from the drain are weighted by a small factor. From Eq. (2.70), the drain charge becomes:

$$Q_D = -W\int_0^{V_{DS}} \frac{\mu_n W C_{ox}}{LI}\left[(V_{GS}-V_T)V - \frac{1}{2}V^2\right]C_{ox}(V_{GS}-V-V_T)$$
$$\cdot\frac{\mu_n W C_{ox}}{I}(V_{GS}-V_T-V)dV,$$
$$= \frac{-WLC_{ox}}{[(V_{GS}-V_T)V_{DS}-\frac{1}{2}V_{DS}^2]^2}\bigg\{(V_{GS}-V_T)^3\frac{1}{2}V_{DS}^2$$
$$-\frac{5}{2}(V_{GS}-V_T)^2\frac{1}{3}V_{DS}^3+\frac{1}{2}(V_{GS}-V_T)V_{DS}^4-\frac{1}{10}V_{DS}^5\bigg\}. \quad (2.71)$$

The sum of Q_S and Q_D is equal to the quantity of the total electrons in the channel: $Q_S + Q_D = -W\int_0^L Q_I(y)dy$. From $Q_G = -(Q_S+Q_D) = -C_{ox}W\int_0^L (V_G - V_T - V(y))dy$, the gate charge is

$$Q_G = W\int_0^{V_{DS}} C_{ox}(V_{GS}-V_T-V)\frac{\mu_n W C_{ox}}{I}(V_{GS}-V_T-V)dV, \quad (2.72)$$
$$= \frac{WLC_{ox}}{(V_{GS}-V_T)V_{DS}-\frac{1}{2}V_{DS}^2}\left[(V_{GS}-V_T)^2 V_{DS} - (V_{GS}-V_T)V_{DS}^2 + \frac{1}{3}V_{DS}^3\right].$$

The above source charge, drain charge, and gate charge formulas are applicable for the device biased in the triode region. If the device biased in the saturation region, the above formulas are still applicable if the V_{DS} terms are replaced by $V_{GS} - V_T$. Then one obtains:

$$Q_S = \frac{-2}{5}WLC_{ox}(V_{GS}-V_T), \quad (2.73)$$
$$Q_D = \frac{-4}{15}WLC_{ox}(V_{GS}-V_T), \quad (2.74)$$
$$Q_G = \frac{2}{3}WLC_{ox}(V_{GS}-V_T). \quad (2.75)$$

In the subthreshold region, we have assumed that $Q_S = Q_D = Q_G = 0$ and $I = 0$. Based on the source charge, the drain charge, and the gate charge, the capacitances

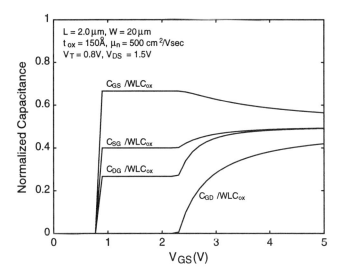

Fig. 2.48 C_{DG}, C_{GD}, C_{SG}, and C_{GS} versus V_{GS} of the NMOS device with a channel length of $2\mu m$, a channel width of $20\mu m$, a gate oxide of 15nm, and a threshold voltage of 0.8V, biased at $V_{DS} = 1.5V$.

of the MOS device are defined as:

$$C_{GS} = -\frac{dQ_G}{dV_S}, \qquad (2.76)$$

$$C_{GD} = -\frac{dQ_G}{dV_D}, \qquad (2.77)$$

$$C_{SG} = -\frac{dQ_S}{dV_G}, \qquad (2.78)$$

$$C_{DG} = -\frac{dQ_D}{dV_G}. \qquad (2.79)$$

Fig. 2.48 shows C_{DG}, C_{GD}, C_{SG}, and C_{GS} versus V_{GS} of the NMOS device with a channel length of $2\mu m$, a channel width of $20\mu m$, a gate oxide of 15nm, and a threshold voltage of 0.8V, biased at $V_{DS} = 1.5V$. As shown in the figure, the capacitances have been normalized by $C_O = C_{OX}WL$. Below the threshold voltage, all capacitances are zero. In the saturation region, C_{GS}, C_{SG}, and C_{DG} maintain their non-zero constant values—0.67, 0.4, and 0.27, respectively. In contrast, C_{GD} is zero in the saturation region. When it enters the triode region, C_{SG}, C_{DG} and C_{GD} increase as V_{GS} increases. At a large V_{GS}, C_{SG} and C_{DG} are close to 50% of the total gate oxide capacitance. As shown in the figure, when V_{GS} increases, C_{GS} decreases in the triode region. Generally speaking, for an MOS circuit, by imposing an input voltage pulse at the gate electrode, the transients of the output voltage and current are to be calculated. Compared to C_{GD} and C_{GS}, C_{DG} and C_{SG} are more frequently used in the VLSI circuits since the input voltage is usually imposed at the gate.

68   CMOS TECHNOLOGY AND DEVICES

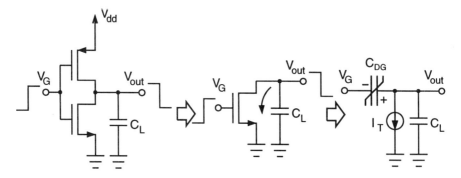

Fig. 2.49 Equivalent circuit of the NMOS device for transient analysis.

The gate-drain voltage C_{GD} is defined as the derivative of the gate charge with respect to the drain voltage:

$$C_{GD} = \frac{\partial Q_G}{\partial V_D}, \tag{2.80}$$

where Q_G is the gate charge:

$$Q_G = C_{ox} W \int_0^L (V_G - V_T - V(y)) dy. \tag{2.81}$$

C_{GD} is defined as the influence of the drain voltage in the gate charge. Due to the nonlinearity of the capacitances, C_{GD} is not equal to C_{DG}—the reciprocity does not hold. As shown in Fig. 2.48, in the saturation region, C_{GD} is zero, while C_{DG} is not. In analyzing the transient of a digital circuit, C_{DG} is more frequently used.

The capacitance model of an MOS device is used for transient analysis of an MOS circuit. As shown in Fig. 2.49, in an inverter circuit including an NMOS device and an output load capacitance, the NMOS device is used to drive the output load during the transient. The input signal is usually imposed at the gate electrode. The output voltage during the transient of the circuit is to be calculated. Based on the equivalent circuit as shown in Fig. 2.49, writing the node equation at the output node, a differential equation in terms of the output voltage is obtained:

$$\frac{dQ_D}{dt} + I_T(V_{out}, V_G) + C_L \frac{dV_{out}}{dt} = 0. \tag{2.82}$$

From the capacitance model, one obtains:

$$\frac{dQ_D}{dt} = \frac{\partial Q_D}{\partial V_G} \frac{dV_G}{dt} = C_{DG} \frac{dV_G}{dt}. \tag{2.83}$$

From Eqs. (2.82) and (2.83), one obtains a differential equation:

$$C_{DG}(V_{out}, V_G) \frac{dV_G}{dt} + I_T(V_{out}, V_G) + C_L \frac{dV_{out}}{dt} = 0. \tag{2.84}$$

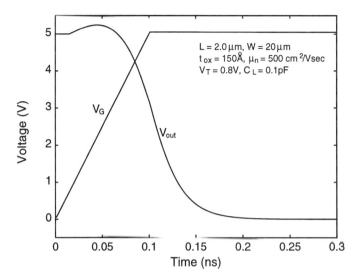

Fig. 2.50 Transient waveforms during the pulldown of a CMOS inverter using an NMOS device with a channel length of $2\mu m$, a channel width of $20\mu m$, a gate oxide of 15nm, and a threshold voltage of 0.8V. The output load is 0.1pf

Solving this differential equation, the transient waveform of the output voltage can be obtained. Fig. 2.50 shows the transient waveforms during the pulldown of a CMOS inverter using an NMOS device with a channel length of $2\mu m$, a channel width of $20\mu m$, a gate oxide of 15nm, and a threshold voltage of 0.8V. The output load is 0.1pf. The output transient waveform is obtained by solving the differential equation described before.

(b) Extrinsic capacitances (C_{DB}, C_{SB})

The extrinsic capacitances are passive. As shown in Fig. 2.51, the drain-body capacitance (C_{DB}) and the source-body capacitance (C_{SB}) are the parasitic capacitances between the drain/source and the substrate. C_{DB} and C_{SB} are due to the space charge in the depletion region of the reverse biased pn junctions. The drain/source-body capacitances C_{DB} and C_{SB} can be divided into two parts[55]: the junction capacitance and the sidewall capacitance:

$$C_{SB} = \frac{C_j A_S}{(1 - V_{SB}/\Phi_j)^{m_j}} + \frac{C_{jsw} P_S}{(1 - V_{SB}/\Phi_j)^{m_{jsw}}}, \quad (2.85)$$

$$C_{DB} = \frac{C_j A_D}{(1 - V_{DB}/\Phi_j)^{m_j}} + \frac{C_{jsw} P_D}{(1 - V_{DB}/\Phi_j)^{m_{jsw}}},$$

where C_j is the unit-area junction capacitance, A_S/A_D is the area of the source/drain region, and C_{jsw} is the unit-length junction capacitance at the sidewall. Under the field oxide surrounding the active region, there is the channel stop region. Therefore, the sidewall junction capacitance can be higher than the junction capacitance. P_S/P_D

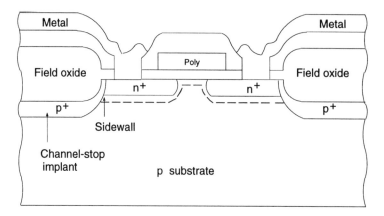

Fig. 2.51 Cross-section of an NMOS device.

is the length of the perimeter of the sidewall surrounding the source/drain region. ϕ_j is the built-in voltage. m_j and m_{jsw} are the coefficients used to describe the slope of the doping profile at the source/drain junction and the sidewall junction, respectively. Typically, $m_j = 0.5$ and $m_{jsw} = 0.33$.

2.18 HOT CARRIER EFFECTS

The channel length of a CMOS device made by mass production was shrunk from 2μm in 1980 to 0.25μm in 1995. Based on this rate, in 2010, the channel length will become 0.05μm. However, the operating voltage was not shrunk accordingly. In 1980, the supply voltage was 5V. In 1995, it was 1.5V for the 0.25μm device. In 2010, the supply voltage will be 1V. Therefore, the electric field in the lateral channel increases quickly. From 1980 to 1995, the average electric field in the lateral channel increased 2.4 times. From 1980 to 2010, it may increase 8 times. Since the electric field in the lateral channel increases rapidly as the device is scaled down, the high electric field effect is getting more and more important. As shown in Fig. 2.52, in a submicron NMOS device operating in the saturation region, the electric field in the region from the saturation point to the drain is especially high. Therefore, the electrons traveling through this region are accelerated. They may collide with the lattice. As a result, electron/hole pairs are generated. The generated electrons move toward the drain under the electric field. They may also acquire energy and collide with the lattice to generate other electron/hole pairs. The regeneration process will repeat itself. This is called impact ionization. Impact ionization brings in a sudden increase in the drain current. As a result, the device may break down—avalanche breakdown. The generated holes, which move toward the substrate contact due to the negative substrate bias, become substrate current. In the region with a high electric field, the generated electrons move toward the gate due to the positive gate bias. When the electrons reach the Si/SiO_2 interface, they may be trapped at the interface,

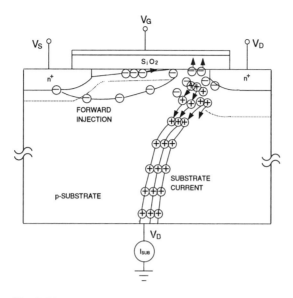

Fig. 2.52 MOS device biased under a high electric field.

and thus become the interface charge. Some electrons with higher energy directly cause interface-states (traps) generation. As a result, the threshold voltage of the local region may change. Consequently, the IV characteristics may change, which is serious when the device operates for a long time. This is the so-called aging, which causes reliability problems of the device. As shown in Fig. 2.52, due to the high electric field in the device, there are two effects. One is the aging effect due to the trapping of the electrons and the interface state generation at the Si/SiO_2 interface. The other is the impact ionization effect. These two effects are called the hot electron (carrier) effects. At a small drain voltage, the aging effect due to electrons trapped at the Si/SiO_2 interface and the interface state generation is more noticeable. When the stress time is long, the IV characteristics will change as shown in Fig. 2.53[50]. If the source and the drain are reversed after stress, IV characteristics in the triode region may change noticeably. This is due to the asymmetry in the device structure when trapping of the electrons at the oxide interface occurs. At a larger drain voltage, impact ionization is more important. Hot electron effects make the device and thus the related circuits unstable. As shown in Fig. 2.53, when the device is biased at a larger V_D, the difference in the IV characteristics before and after the stress (which is due to impact ionization) is smaller than that biased at a smaller V_D (which is due to trapping of electron and interface state generation at the interface). This implies that impact ionization exists at the beginning of the operation. In contrast, trapping of electrons and interface state generation at the oxide interface become more serious when the stress time is longer. The hot electron effects can be monitored by the substrate current and the gate current. As shown in Fig. 2.54[51], when the gate voltage is large, the electron current due to the hot electron effect and the gate electron current at the gate are almost identical. This implies that for a large gate voltage, the electrons,

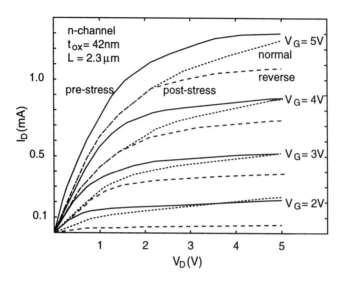

Fig. 2.53 IV characteristics of an NMOS device after stress. (From Schwerin et al.[50], ©1987 IEEE.)

which are generated by the hot electron effects, flow toward the Si/SiO$_2$ interface. Then they are trapped at the interface. Thus, at a large gate voltage, impact ionization is less important. Under this situation, the hot electron effect is dominated by the aging effect due to the trapping of electrons and the interface state generation at the Si/SiO$_2$ interface. On the other hand, for a small gate voltage, the electron current generated by the hot electron effect is close to the gate hole current measured at the gate.

In fact, the gate current and the substrate current are much smaller than the drain current. As shown in Fig. 2.55[52], the substrate current is less than 1% of the drain current. The substrate current is parabolically proportional to V_{GS}. The substrate current has a peak value at a certain V_{GS}. The substrate current is also proportional to V_{DS}. The hot electron effects can be monitored by the substrate current. The substrate current (I_{sub}) and the gate current (I_g) can be expressed by[52]:

$$I_{sub} = C_1 I_D e^{-\Psi_i/q\lambda E_m},$$
$$I_g = C_2 I_D e^{-\Psi_b/q\lambda E_m}, \quad (2.86)$$

where λ is the electron mean free path, Ψ_i is the minimum energy in eV to generate impact ionization ($\Psi_i \cong 1.3eV$), Ψ_b is the barrier energy at the Si/SiO$_2$ interface ($\Psi_b \cong 3.2eV$), and E_m is the average lateral electric field between drain and the saturation point ($E_m = \frac{V_D - V_{DSAT}}{l} = \frac{V_D - V_{DSAT}}{0.2 t_{ox}^{1/3} x_j^{2/3}}$, x_j is the junction depth, and t_{ox} is the gate oxide thickness). From the above equations, the average lateral electric field (E_m) determines the substrate current and the gate current. With a larger V_{DS},

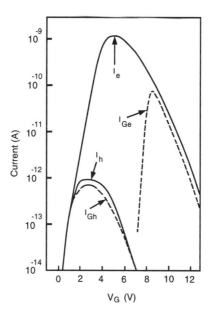

Fig. 2.54 Total electron and hole currents in an NMOS device due to hot electron effects (From Hofmann et al.[51], ©1984 IEEE.)

the substrate current and the gate current become larger.

In addition to the hot electron effects in the NMOS device, deep-submicron PMOS devices also have hot hole effects[53][54]. Under an identical electric field, the hot hole effects in the PMOS device are relatively smaller. For an NMOS device, when the channel length is smaller than $1\mu m$, the hot electron effects cannot be overlooked. However, for a PMOS device, only when the channel length is smaller than $0.5\mu m$, the hot hole effects will be noticeable. Consider the results for CMOS devices, which have an effective channel length of $1.8\mu m$, a gate oxide of 35nm, biased at $V_{DD} = 7V$ as shown in Fig. 2.56. The substrate current of the PMOS device is at least 1000 times smaller than that of the NMOS device, which implies that the hot carrier effect of the PMOS device is smaller than that of the NMOS device[3]. The smaller hot carrier effect in the PMOS device is because for generating hot holes, the valence band barrier between Si and SiO_2 is 3.6V. In contrast, for generating hot electrons, the conduction band barrier between Si and SiO_2 is 3.2V. Therefore, from hot carrier effect point of view, PMOS devices are more stable and reliable than NMOS devices.

With a channel length of larger than $1\mu m$, the performance of the PMOS device is worse than that of the NMOS device. However, in the deep-submicron regime, PMOS devices may not be worse. One of the reasons is that in the deep-submicron regime, the reliability of the PMOS device is better since it is less affected by the hot carrier effect. In the deep-submicron regime, due to the large lateral electric field

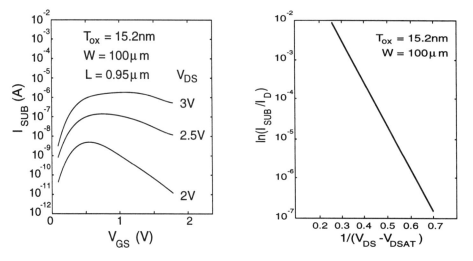

Fig. 2.55 Substrate current in an NMOS device (From Hu et al.[52], ©1985 IEEE.)

in the channel, for both electron and hole, it is beyond its respective critical field as shown in Fig. 2.57. As a result, both electrons in the NMOS device and holes in the PMOS device are traveling at the saturated velocity. At a low electric field, the hole mobility is smaller than the electron mobility. At an electric field greater than the critical electric field, the difference between the electron mobility and the hole mobility becomes smaller. Therefore, in the deep-submicron regime, the current driving capability of the PMOS device may not be much smaller than that of the NMOS device.

2.19 BSIM SPICE MODELS

BSIM SPICE models[55][56] are for small-geometry MOS devices. In the BSIM model, the vertical electric field dependent mobility, the carrier velocity saturation, the drain induced barrier lowering, the shared depletion charge, the effect of the nonuniformly doping profile, the channel length modulation, the subthreshold conduction, and the geometry dependence effects are included.

Table 2.6 shows the parameters used in the BSIM model. In the BSIM model, considering the sensitivity due to the process variations in gate length and gate width, an electric parameter is referred to the process parameters as follows:

$$P = P_0 + \frac{P_L}{L_i - DL} + \frac{P_W}{W_i - DW}, \qquad (2.87)$$

where P_0, P_L, and P_W are process parameters related to the electrical parameter P, L_i and W_i are the channel length and channel width designed, and DL and DW are the changes in the features accounting for process variations. Sensitivities of the

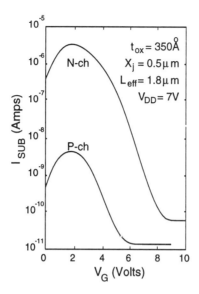

Fig. 2.56 Substrate currents in NMOS and PMOS devices (From Chatterjee et al.[3], ©1980 IEEE.)

electric parameters with respect to V_{BS} and V_{DS} are denoted by X2 and X3.

In the BSIM1 model, five parameters are needed in the threshold voltage model. Three parameters are used in the drain current model, and one parameters used in the subthreshold current model. The threshold voltage model is:

$$V_{TH} = V_{FB} + \phi_s + K_1\sqrt{\phi_s - V_{BS}} - K_2(\phi_s - V_{BS}) - \eta V_{DS}, \qquad (2.88)$$

where V_{FB} is the flat-band voltage, ϕ_s is the surface inversion potential ($\phi_s = 2\phi_f$), K_1 is the body effect coefficient ($K_1 = \frac{\sqrt{2q\epsilon_{si}N_A}}{C_{ox}}$), and K_2 is the source/drain depletion charge sharing coefficient. η refers to the drain induced barrier lowering (DIBL) effect:

$$\eta = \eta_0 + \eta_B V_{BS} + \eta_D(V_{DS} - V_{DD}), \qquad (2.89)$$

where η_0, η_B, and η_D are the zero-bias drain-induced barrier lowering coefficient, the sensitivities of the drain induced barrier lowering effect to the substrate bias, and the drain bias, respectively.

In the linear region ($V_{GS} > V_{TH}$, $V_{DS} < V_{DSAT}$), the drain current is:

$$I_D = \frac{\mu_0}{(1 + U_0(V_{GS} - V_{TH}))} \frac{C_{ox}\frac{W_{eff}}{L_{eff}}}{(1 + \frac{U_1}{L_{eff}}V_{DS})}\left((V_{GS} - V_{TH})V_{DS} - \frac{a}{2}V_{DS}^2\right), \qquad (2.90)$$

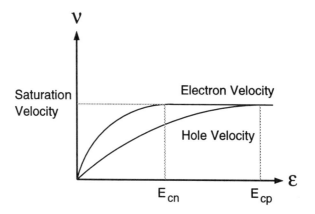

Fig. 2.57 Electron and hole traveling velocity versus electric field.

where $a = 1 + (gK_1/2\sqrt{\phi_s - V_{BS}})$, $g = 1 - 1/(1.744 + 0.8364(\phi_s - V_{BS}))$. In the saturation region ($V_{GS} > V_{TH}$, $V_{DS} \geq V_{DSAT}$), the drain current is:

$$I_D = \frac{\mu_0}{(1 + U_0(V_{GS} - V_{TH}))} \frac{C_{ox}\frac{W_{eff}}{L_{eff}}}{2aK}(V_{GS} - V_{TH})^2, \qquad (2.91)$$

where $K = (1 + v_c + \sqrt{1 + 2v_c})/2$, and $v_c = \frac{U_1}{L_{eff}}\frac{(V_{GS}-V_{TH})}{a}$. If $v_c \ll 1$, then $K \cong 1 + \frac{U_1}{L_{eff}}\frac{(V_{GS}-V_{TH})}{a}$. If $v_c \gg 1$, then $K \cong \frac{U_1}{L_{eff}}\frac{(V_{GS}-V_{TH})}{2a}$ and $I_D \cong C_{ox}v_{sat}W_{eff}(V_{GS} - V_{TH})$. In the above equation, the mobility model has been included. In the mobility model, U_0 and U_1 are used to account for the degradation due to the vertical and the horizontal electric fields, respectively. a is associated with the bulk doping effect and g is related to the average body effect on the drain current. The parameters U_0 and U_1, which are bias dependent, are:

$$U_0(V_{DS}, V_{BS}) = U_{0Z} + U_{0B}V_{BS},$$
$$U_1(V_{DS}, V_{BS}) = U_{1Z} + U_{1B}V_{BS} + U_{1D}(V_{DS} - V_{DD}). \qquad (2.92)$$

μ_0 is obtained by quadratic interpolation via three data points—μ_0 at $V_{DS} = 0V$, μ_0 at $V_{DS} = V_{DD}$, and the sensitivity of μ_0 to the drain bias at $V_{DS} = V_{DD}$, with $\mu_0(V_{DS} = 0V) = \mu_Z + \mu_{ZB}V_{BS}$ and $\mu_0(V_{DS} = V_{DD}) = \mu_s + \mu_{sB}(V_{BS})$. The saturation drain voltage V_{DSAT} is:

$$V_{DSAT} = \frac{(V_{GS} - V_{TH})}{a\sqrt{K}}. \qquad (2.93)$$

If $U_1/L_{eff}(V_{GS} - V_{TH}) \ll 1$, $V_{DSAT} \cong \frac{(V_{GS}-V_{TH})}{a}$. If $U_1/L_{eff}(V_{GS} - V_{TH}) \gg 1$, $V_{DSAT} \cong \sqrt{\frac{2(V_{GS}-V_{TH})}{a}\frac{L_{eff}}{U_1}}$.

Parameters	Units	Meaning
$V_{FB}(VFB)$, $V_{FBl}(LVFB)$, $V_{FBw}(WVFB)$	V	Flat-band voltage
$\phi_s(PHI)$, $\phi_{sl}(LPHI)$, $\phi_{sw}(WPHI)$	V	Surface inversion potential
$K_1(K1)$, $K_{1l}(LK1)$, $K_{1w}(WK1)$	$V^{1/2}$	Body effect coefficient
$K_2(K2)$, $K_{2l}(LK2)$, $K_{2w}(WK2)$	–	Drain/source depletion charge sharing coefficient
$\eta_0(ETA)$, $\eta_{0l}(LETA)$, $\eta_{0w}(WETA)$	–	Zero-bias drain-induced barrier lowering coefficient
$\mu_Z(MUZ)$, $\delta_l(DL)$, $\delta_w(DW)$	cm^2/Vs	Zero-bias mobility
$U_{0Z}(U0)$, $U_{0Zl}(LU0)$, $U_{0Zw}(WU0)$	V^{-1}	Zero-bias transverse-field mobility degradation coefficient
$U_{1Z}(U1)$, $U_{1Zl}(LU1)$, $U_{1Zw}(WU1)$	$\mu m V^{-1}$	Zero-bias velocity saturation coefficient
$\mu_{ZB}(X2MZ)$, $\mu_{ZBl}(LX2MZ)$, $\mu_{ZBw}(WX2MZ)$	cm^2/V^2s	Sensitivity of mobility to substrate bias at $V_{DS}=0$
$\eta_B(X2E)$, $\eta_{bl}(LX2E)$, $\eta_{Bw}(WX2E)$	V^{-1}	Sensitivity of DIBL to substrate bias
$\eta_D(X3E)$, $\eta_{Dl}(LX3E)$, $\eta_{Dw}(WX3E)$	V^{-1}	Sensitivity of DIBL at $V_{DS}=V_{DD}$
$U_{0B}(X2U0)$, $U_{0Bl}(LX2U0)$, $U_{0Bw}(WX2U0)$	V^{-2}	Sensitivity of transverse-field mobility degradation to substrate bias
$U_{1B}(X2U1)$, $U_{1Bl}(LX2U1)$, $U_{1Bw}(WX2U1)$	$\mu m V^{-2}$	Sensitivity of velocity saturation effect to substrate bias
$\mu_s(MUS)$, $\mu_{sl}(LMS)$, $\mu_{sw}(WMS)$	cm^2/Vs	Mobility at zero substrate bias and at $V_{DS}=V_D$
$\mu_{sB}(X2MS)$, $\mu_{sBl}(LX2MS)$, $\mu_{sBw}(WX2MS)$	cm^2/Vs	Sensitivity of mobility to substrate bias at $V_{DS}=V_{DD}$
$\mu_{sD}(X3MS)$, $\mu_{sDl}(LX3MS)$, $\mu_{sDw}(WX3MS)$	cm^2/Vs	Sensitivity of mobility to drain bias at $V_{DS}=V_{DD}$
$U_{1D}(X3U1)$, $U_{1Dl}(LX3U1)$, $U_{1Dw}(WX3U1)$	$\mu m V^{-2}$	Sensitivity of velocity saturation effect at $V_{DS}=V_{DD}$
$t_{ox}(TOX)$, $T_{temp}(TEMP)$, $V_{DD}(VDD)$	$\mu m (CV)$	Gate-oxide thickness
$TEMP$		Temperature at which the process parameters are measured
V_{DD}		Measurement bias range
$n_0(N0)$, $LN0$, $WN0$	–	Zero-bias subthreshold slope coefficient
$n_B(NB)$, LNB, WNB	–	Sensitivity of subthreshold slope to substrate bias
$n_D(ND)$, LND, WND	–	Sensitivity of subthreshold slope to drain bias
$CGD0$, $CGS0$, $CGB0$	F/m	Gate-drain,-source,-body overlap capacitance
$XPART$, $DUM1$, $DUM2$	–	Gate-oxide capacitance model flag.
$R_{sh}(RSH)$	$\Omega/square$	Sheet resistance/square
$C_j(CJ)$	F/m^2	Zero-bias bulk junction bottom capacitance/unit area
$C_{jw}(CJW)$	F/m	Zero-bias bulk junction bottom capacitance/unit length
$I_{js}(IJS)$	A/m^2	Bulk junction saturation current/unit area
$P_j(PJ)$	V	Bulk junction bottom potential
$P_{jw}(PJW)$	V	Bulk junction sidewall potential
$M_j(MJ)$	–	Bulk junction bottom grading coefficient
$M_{jw}(MJW)$	–	Bulk junction sidewall grading coefficient
$W_{df}(WDF)$	m	Default width of the layer
$\delta_l(DL)$	m	Average variation of size

Table 2.6 BSIM SPICE models. (Adapted from Massobrio and Antognetti[56].)

In the BSIM-1 model, the total drain current is composed of the strong-inversion drain current described above and the weak-inversion drain current. The weak-inversion drain current is expressed as:

$$I_D = \left(\frac{1}{I_{exp}} + \frac{1}{I_{limit}}\right)^{-1},$$

$$I_{exp} = \mu_0 C_{ox} \frac{W_{eff}}{L_{eff}} \left(\frac{kT}{q}\right)^2 e^{1.8} e^{(V_{GS}-V_{TH})/\frac{nkT}{q}}\left(1 - e^{-V_{DS}/\frac{kT}{q}}\right),$$

$$I_{limit} = \frac{\mu_0 C_{ox}}{2} \frac{W_{eff}}{L_{eff}} \left(\frac{3kT}{q}\right)^2, \qquad (2.94)$$

where n is dependent on V_{DS} and V_{BS}:

$$n(V_{DS}, V_{BS}) = n_0 + n_B V_{BS} + n_D V_{DS}, \qquad (2.95)$$

where n_0, n_B, n_D are the zero-bias subthreshold slope coefficient, the sensitivities of the subthreshold slope to the substrate bias, and the drain bias, respectively.

In the BSIM-1 model, in the accumulation region, the gate charge (Q_G) in the accumulation region is $Q_G = W_{eff}L_{eff}C_{ox}(V_{GS} - V_{FB} - V_{BS})$. The body charge is equal to $Q_B = -Q_G$. In the subthreshold region, the gate charge is $Q_G = W_{eff} L_{eff} C_{ox} \frac{K_1^2}{2}(-1 + (1 + \frac{4(V_{GS}-V_{FB}-V_{BS})}{K_1^2})^{\frac{1}{2}})$. The body charge is equal to $Q_B = -Q_G$. In the linear region, the gate charge (Q_G), the body charge (Q_B), the source charge (Q_S) and the drain charge (Q_D) are

$$Q_G = W_{eff} L_{eff} C_{ox} \left(V_{GS} - V_{FB} - \phi_s - \frac{V_{DS}}{2} + \frac{V_{DS}}{12} \frac{\alpha_x V_{DS}}{(V_{GS} - V_{TH} - \frac{\alpha_x}{2} V_{DS})} \right),$$

$$Q_B = W_{eff} L_{eff} C_{ox} \left(-V_{TH} + V_{FB} + \phi_s + \frac{(1-\alpha_x)}{2} V_{DS} \right.$$
$$\left. - \frac{(1-\alpha_x) V_{DS}}{12} \frac{\alpha_x V_{DS}}{(V_{GS} - V_{TH} - \frac{\alpha_x}{2} V_{DS})} \right),$$

$$Q_S = -W_{eff} L_{eff} C_{ox} \left[\frac{V_{GS} - V_{TH}}{2} + \frac{\alpha_x V_{DS}}{12} \frac{\alpha_x V_{DS}}{(V_{GS} - V_{TH} - \frac{\alpha_x}{2} V_{DS})} \right.$$
$$- \frac{\alpha_x V_{DS}}{(V_{GS} - V_{TH} - \frac{\alpha_x}{2} V_{DS})^2} \cdot \left(\frac{(V_{GS} - V_{TH})^2}{6} - \frac{\alpha_x V_{DS}(V_{GS} - V_{TH})}{8} \right.$$
$$\left. \left. + \frac{(\alpha_x V_{DS})^2}{40} \right) \right],$$

$$Q_D = -W_{eff} L_{eff} C_{ox} \left[\frac{V_{GS} - V_{TH}}{2} - \frac{\alpha_x V_{DS}}{2} + \frac{\alpha_x V_{DS}}{(V_{GS} - V_{TH} - \frac{\alpha_x}{2} V_{DS})^2} \right.$$
$$\left. \cdot \left(\frac{(V_{GS} - V_{TH})^2}{6} - \frac{\alpha_x V_{DS}(V_{GS} - V_{TH})}{8} + \frac{(\alpha_x V_{DS})^2}{40} \right) \right] \quad (2.96)$$

where $\alpha_x = a(1 + \frac{U_1}{L_{eff}}(V_{GS} - V_{TH}))$. In the saturation region, they are

$$Q_G = W_{eff} L_{eff} C_{ox} \left(V_{GS} - V_{FB} - \phi_s - \frac{V_{GS} - V_{TH}}{3\alpha_x} \right),$$

$$Q_B = W_{eff} L_{eff} C_{ox} \left(V_{FB} + \phi_s - V_{TH} + \frac{(1-\alpha_x)(V_{GS} - V_{TH})}{3\alpha_x} \right),$$

$$Q_S = -\frac{2}{5} W_{eff} L_{eff} C_{ox} (V_{GS} - V_{TH}),$$

$$Q_D = -\frac{4}{15} W_{eff} L_{eff} C_{ox} (V_{GS} - V_{TH}). \quad (2.97)$$

For transient analysis, at time t_0 the voltage at all nodes of the equivalent circuit is known: V_{x1}^0. If the input applied signal is known, the transient waveform of each node is to be computed. During the transient analysis, it's based on time increment method—from a time with all node voltages already known to compute the node voltages at the next time step. From the formula: $\int_{t_0}^{t_1} i dt = Q(t_1) - Q(t_0)$, all related equations can be built in an array:

$$\frac{h}{2}(i_{y1}^1 + i_{y0}) = (Q_{y1} - Q_{y0}) + \Sigma_x \left(\frac{dQ_y}{dV_x} \right)|_{x1}^0 (V_{x1}^1 - V_{x1}^0) \quad (2.98)$$

where $dQ_y/dV_x = C_{yx}$ is the capacitance of the related SOI MOS device, V_{x1}^0 is the estimated voltage value of the previous iteration at node x_1, V_{x1}^1 is the estimated voltage value of the current iteration at node x_1, h is the time step ($h = t_1 - t_0$), i_{y0} is the estimated current of the previous iteration, and i_{y1}^1 is the current to be estimated in this current iteration. After building all node equations, a node voltage vector can be obtained. Then based on the numerical implicit integration algorithm, iterations have been used to compute the node voltage vector. If the result of the current iteration has a change much smaller than the result of the previous iteration, the convergence of the computation has been reached—the voltage at each node has been successfully obtained at a time step. Then the iteration process for computing the node voltage for the next time step can be initiated. If the time step is too large, as compared to the change in the input waveform, the iterations of node equations cannot reach a convergence. Then a cutback in the next time step is required.

2.20 SUMMARY

In this chapter, CMOS technology and devices have been described. Starting from the evolution of the CMOS technology, a 0.25μm CMOS technology with shallow trench isolation (STI) and lightly doped drain (LDD) techniques have been explained. Then, BiCMOS technology is analyzed, followed by the 0.1μm CMOS technology and SOI CMOS technology. CMOS device behaviors in terms of threshold voltage, body effect, short and narrow channel effects, mobility, velocity overshoot, drain current including electron temperature effects, and capacitances have been depicted. In the final portion of this chapter, SPICE models for CMOS devices were described. In the next chapter, CMOS static logic circuits are described.

REFERENCES

1. *International Solid-State Circuits Conference (ISSCC)* Digest.

2. 1997 report of National Technology Roadmap for Semicondcutors, *http://notes.sematch.org/97pelec.htm*.

3. P. K. Chatterjee, W. R. Hunter, T. C. Holloway, and Y. T. Lin, "The Impact of Scaling Laws on the Choice of n-Channel or p-Channel for MOS VLSI," *IEEE Elec. Dev. Let.*, **1**(10), 220–223 (1980).

4. R. H. Dennard, F. H. Gaensslen, H.-N. Yu, V. L. Rideout, E. Bassous, and A. R. Leblanc, "Design of Ion-Implanted MOSFET's with Very Small Physical Dimensions," *IEEE J. Sol. St. Ckts.*, **9**(5), 256–268 (1974).

5. M. Ono, M. Saito, T. Yoshitomi, C. Fiegna, T. Ohguro, and H. Iwai, "A 40nm Gate Length n-MOSFET," *IEEE Trans. Elec. Dev.*, **42**(10), 1822–1830 (1995).

6. M. Lenzlinger and E. H. Snow, "Fowler-Nordheim Tunneling into Thermally Grown SiO_2," *J. Appl. Phys.*, **40**(1), 278–283 (1969).

7. T. Ohzone, H. Shimura, K. Tsuji, and T. Hirao, "Silicon-Gate n-Well CMOS Process by Full Ion-Implantation Technology," *IEEE Trans. Elec. Dev.*, **27**(9), 1789–1795 (1980).

8. M. Bohr, S. S. Ahmed, S. U. Ahmed, M. Bost, T. Ghani, J. Greason, R. Hainsey, C. Jan, P. Packan, S. Sivakumar, S. Thompson, J. Tsai, and S. Yang, "A High Performance $0.25\mu m$ Logic Technology Optimized for 1.8V Operation," *IEDM Dig.*, 847–850 (1996).

9. H. Sayama, T. Kuroi, S. Shimizu, M. Shirahata, Y. Okumura, M. Inuishi, and H. Miyoshi, "Low Voltage Operation of Sub-Quarter Micron W-Polycide Dual Gate CMOS with Non-Uniformly Doped Channel," *IEDM Dig.*, 583–586 (1996).

10. M. Rodder, Q. Z. Hong, M. Nandakumar, S. Aur, J. C. Hu, and I.-C. Chen, "A Sub-$0.18\mu m$ Gate Length CMOS Technology for High Performance (1.5V) and Low Power (1.0V)," *IEDM Dig.*, 563–566 (1996).

11. W.-H. Chang, B. Davari, M. R. Wordeman, Y. Taur, C. C.-H. Hsu, and M. D. Rodriguez, "A High-Performance $0.25\text{-}\mu m$ CMOS Technology: I-Design and Characterization," *IEEE Trans. Elec. Dev.*, **39**(4), 959–966 (1992).

12. B. Davari, W.-H. Chang, K. E. Petrillo, C. Y. Wong, D. Moy, Y. Taur, M. R. Wordeman, J. Y.-C. Sun, C. C.-H. Hsu, and M. R. Polcari, "A High-Performance $0.25\text{-}\mu m$ CMOS Technology: II–Technology," *IEEE Trans. Elec. Dev.*, **39**(4), 967–975 (1992).

13. M. S. C. Luo, P. V. G. Tsui, W.-M. Chen, P. V. Gilbert, B. Maiti, A. R. Sitaram, and S.-W. Sun, "A $0.25\mu m$ CMOS Technology with 45Å NO-nitrided Oxide," *IEDM Dig.*, 691–694 (1995).

14. C. Chen, J. W. Chou, W. Lur, and S. W. Sun, "A Novel $0.25\mu m$ Shallow Trench Isolation Technology," *IEDM Dig.*, 837–840 (1996).

15. P. J. Tsang, S. Ogura, W. W. Walker, J. F. Shepard, and D. L. Critchlow, "Fabrication of High-Performance LDDFET's with Oxide Sidewall-Spacer Technology," *IEEE Trans. Elec. Dev.*, **29**(4), 590–596 (1982).

16. S. Ogura, P. J. Tsang, W. W. Walker, D. L. Critchlow, and J. F. Shepard, "Design and Characteristics of the Lightly Doped Drain-Source (LDD) Insulated Gate Field-Effect Transistor," *IEEE Trans. Elec. Dev.*, **27**(8), 1359–1367 (1980).

17. J. H. Sim and J. B. Kuo, "An Analytical Delayed-Turn-Off Model for Buried-Channel PMOS Devices Operating at 77K," *IEEE Trans. Elec. Dev.*, **39**(4), 939–947 (1992).

18. G. J. Hu, C. Y. Ting, Y. Taur, R. H. Dennard, "Design and Fabrication of p-Channel FET for $1\mu m$ CMOS Technology," *IEDM Dig.*, 710–713 (1982).

19. H. Iwai, G. Sasaki, Y. Unno, Y. Niitsu, M. Norishima, Y. Sugimoto, and K. Kanzaki, "0.8μm BiCMOS Technology with High f_T Ion-Implanted Emitter Bipolar Transistor," *IEDM Dig.*, 28–31 (1987).

20. A. R. Alvarez, *BiCMOS Technology and Applications*, Kluwer, Boston, 1989.

21. M. Kubo, I. Masuda, K. Miyata, and K. Ogiue, "Perspective on BiCMOS VLSI's," *IEEE J. Sol. St. Ckts.*, **23**(1), 5–11 (1988).

22. T. M. Liu, G. M. Chin, D. Y. Jeon, M. D. Morris, V. D. Archer, H. H. Kim, M. Cerullo, K. F. Lee, J. M. Sung, K. Lau, T. Y. Chiu, A. M. Voshchenkov, and R. G. Swartz, "A Half-micron Super Self-aligned BiCMOS Technology for High Speed Applications," *IEDM Dig.*, 23–26 (1992).

23. D. L. Harame, E. F. Crabbe, J. D. Cressler, J. H. Comfort, J. Y.-C. Sun, S. R. Stiffler, E. Kobeda, J. N. Burghartz, M. M. Gilbert, J. C. Malinowski, A. J. Dally, S. Ratanaphanyarat, M. J. Saccamango, W. Rausch, J. Cotte, C. Chu, and J. M. C. Stork, "A High Performance Epitaxial SiGe-Base ECL BiCMOS Technology," *IEDM Dig.*, 19–22 (1992).

24. R. H. Yan, K. F. Lee, D. Y. Jeon, Y. O. Kim, B. G. Park, M. R. Pinto, C. S. Rafferty, D. M. Tennant, E. H. Westerwick, G. M. Chin, M. D. Morris, K. Early, P. Mulgrew, W. M. Mansfield, R. K. Watts, A. M. Voshchenkov, J. Bokor, R. G. Swartz, and A. Ourmazd, "High-Performance 0.1-μm Room Temperature Si MOSFETs," *Sym. VLSI Tech. Dig.*, 86–87 (1992).

25. S. Asai and Y. Wada, "Technology Challenges for Integration Near and Below 0.1μm," *IEEE Proc.*, **85**(4), 505–520 (1997).

26. P. K. Vasudev and P. M. Zeitzoff, "Si-ULSI with a Scaled-Down Future: Trends and Challenges for ULSI Semiconductor Technology in the Coming Decade," *IEEE Circuits & Devices Magazine*, **14**(3), 19–29 (1998).

27. J.-P. Colinge, *Silicon-on-Insulator Technology: Materials for VLSI*, Kluwer, Boston, 1991.

28. K. Tokunaga and J. C. Sturm, "Substrate Bias Dependence of Subthreshold Slopes in Fully Depleted Silicon-on-Insulator MOSFET's," *IEEE Trans. Elec. Dev.*, **38**(8), 1803–1807 (1991).

29. T. Ohno, Y. Kado, M. Harada, and T. Tsuchiya, "Experimental 0.25-μm-Gate Fully Depleted CMOS/SIMOX Process Using a New Two-Step LOCOS Isolation Technique," *IEEE Trans. Elec. Dev.*, **42**(8), 1481–1486 (1995).

30. H. H. Hosack, T. W. Houston, and G. P. Pollack, "SIMOX Silicon-on-Insulator: Materials and Devices," *Sol. St. Tech.*, **33**(12), 61–66 (1990).

31. J.-P. Colinge, "Reduction of Kink Effect in Thin-Film SOI MOSFET's," *IEEE Elec. Dev. Let.*, **9**(2), 97–99 (1988).

32. L. D. Yau, "A Simple Theory to Predict the Threshold Voltage of Short-Channel IGFET's," *Sol. St. Elec.*, **17**(10), 1059–1063 (1974).

33. G. Merckel, "A Simple Model of the Threshold Voltage of Short and Narrow Channel MOSFETs," *Sol. St. Elec.*, **23**(12), 1207–1213 (1980).

34. L. A. Akers and J. J. Sanchez, "Threshold Voltage Models of Short, Narrow and Small Geometry MOSFET's: A Review," *Sol. St. Elec.*, **25**(7), 621–641 (1982).

35. B. Davari, C. Koburger, T. Furukawa, Y. Taur, W. Nobel, A. Megdanis, J. Warnock, and J. Mauer, "A Variable-Size Shallow Trench Isolation (STI) Technology with Diffused Sidewall Doping for Submicron CMOS," *IEDM Dig.*, 92–95 (1988).

36. S. C. Sun and J. D. Plummer, "Electron Mobility in Inversion and Accumulation Layers on Thermally Oxidized Silicon Surfaces," *IEEE Trans. Elec. Dev.*, **27**(8), 1497–1508 (1980).

37. G. Merckel, J. Borel, and N. Z. Cupcea, "An Accurate Large-Signal MOS Transistor Model for Use in Computer-Aided Design," *IEEE Trans. Elec. Dev.*, **19**(5), 681–690 (1972).

38. N. Shigyo and R. Dang, "Analysis of an Anomalous Subthreshold Current in a Fully Recessed Oxide MOSFET Using a Three-Dimensional Device Simulator," *IEEE Trans. Elec. Dev.*, **32**(2), 441–445 (1985).

39. K. Sonoda, K. Taniguchi, and C. Hamaguchi, "Analytical Device Model for Submicrometer MOSFET's," *IEEE Trans. Elec. Dev.*, **38**(12), 2662–2668 (1991).

40. S.-Y. Ma and J. B. Kuo, "Concise Analytical Model for Deep Submicron N-Channel Metal-Oxide-Semiconductor Devices with Consideration of Energy Transport," *Jpn. J. Appl. Phys.*, **33**(1B), 550–553 (1994).

41. S.-Y. Ma and J. B. Kuo, "A Closed-Form Physical Drain Current Model Considering Energy Balance Equation and Source Resistance for Deep Submicron N-Channel Metal-Oxide-Semiconductor Devices," *Jpn. J. Appl. Phys.*, **33**(11A), 5647–5653 (1994).

42. J. B. Kuo, Y.-W. Chang, and Y. G. Chen, "An Analytical Velocity Overshoot Model for $0.1\mu m$ N-Channel Metal-Oxide-Silicon Devices Considering Energy Transport," *Jpn. J. Appl. Phys.*, **35**(5A), 2573–2577 (1996).

43. P. H. Woerlee, C. A. H. Juffermans, H. Lifka, W. Manders, H. Pomp, G. Paulzen, A. J. Walker, and R. Woltjer "A Low Power $0.25\mu m$ CMOS Technology," *IEDM Dig.*, 31–34 (1992).

44. G. A. Sai-Halasz, M. R. Wordeman, D. P. Kern, S. Rishton, and E. Ganin, "High Transconductance and Velocity Overshoot in NMOS Devices at the $0.1\text{-}\mu m$ Gate-Length Level," *IEEE Elec. Dev. Let.*, **9**(9), 464–466 (1988).

45. M. R. Pinto, E. Sangiorgi, and J. Bude, "Silicon MOS Transconductance Scaling into the Overshoot Regime," *IEEE Elec. Dev. Let.*, **14**(8), 375–378 (1993).

46. J. R. Brews, "Subthreshold Behavior of Uniformly and Nonuniformly Doped Long-Channel MOSFET," *IEEE Trans. Elec. Dev.*, **26**(9), 1282–1291 (1979).

47. D. E. Ward and R. W. Dutton, "A Charge-Oriented Model for MOS Transistor Capacitances," *IEEE J. Sol. St. Ckts.*, **13**(5), 703–708 (1978).

48. Y. P. Tsividis, *Operation and Modeling of the MOS Transistor*, McGraw-Hill, New York, 1988.

49. B. J. Sheu, "MOS Transistor Modeling and Characterization for Circuit Simulation," *UC Berkeley Tech. Rep.* (1985).

50. A. Schwerin, W. Hansch, and W. Weber, "The Relationship Between Oxide Charge and Device Degradation: A Comparative Study of n- and p- Channel MOSFET's," *IEEE Trans. Elec. Dev.*, **34**(12), 2493–2500 (1987).

51. K. R. Hofmann, W. Weber, C. Werner, and G. Dorda, "Hot Carrier Degradation Mechanism in N-MOSFETs," *IEDM Dig.*, 104–107 (1984).

52. C. Hu, S. C. Tam, F.-C. Hsu, P.-K. Ko, T.-Y. Chan, and K. W. Terrill, " Hot-Electron-Induced MOSFET Degradation–Model, Monitor, and Improvement," *IEEE Trans. Elec. Dev.*, **32**(2), 375–385 (1985).

53. M. Koyanagi, A. G. Lewis, J. Zhu, R. A. Martin, T.-Y. Huang, and J. Y. Chen, "Investigation and Reduction of Hot Electron Induced Punchthrough (HEIP) Effect in Submicron PMOSFETs," *IEDM Dig.*, 722–725 (1986).

54. M. Koyanagi, A. G. Lewis, R. A. Martin, T.-Y. Huang, and J. Y. Chen, "Hot-Electron-Induced Punchthrough (HEIP) Effect in Submicrometer PMOSFET's," *IEEE Trans. Elec. Dev.*, **34**(4), 839–844 (1987).

55. B. J. Sheu, D. L. Scharfetter, P.-K. Ko, and M.-C. Jeng, "BSIM: Berkeley Short-Channel IGFET Model for MOS Transistors," *IEEE J. Sol. St. Ckts.*, **22**(4), 558–566 (1987).

56. G. Massobrio and P. Antognetti, *Semiconductor Device Modeling with SPICE*, McGraw-Hill, New York, 1993.

Problems

1. Compare the advantages between CMOS and BiCMOS technologies. Why is CMOS technology the number one technology for VLSI?

2. Compare the tradeoffs between bulk and SOI CMOS technologies.

3. What are the differences between partially-depleted and fully-depleted SOI CMOS devices?

4. Calculate the threshold voltage of an NMOS device with an N^+ polysilicon gate, a channel length of $0.25\mu m$, a gate oxide of 55Å, a junction depth of $0.15\mu m$, and a doping density of $10^{18} cm^{-3}$.

5. What are the tradeoffs of using shallow trench isolation (STI) and LOCOS for CMOS devices?

6. What are the challenges in realizing a CMOS VLSI technology with a channel length smaller than $0.1\mu m$?

7. Why should the subthreshold slope in terms of mV/dec of an MOS device be as small as possible? How is the subthreshold slope for an MOS device reduced?

8. What is the influence of the velocity saturation on the performance of an MOS device? What is the effect of velocity overshoot?

3
CMOS Static Logic Circuits

Until now in this book, processing technology and device behavior of CMOS VLSI have been described. In this chapter, fundamental CMOS static logic circuits are presented. Starting from the basic CMOS circuits, CMOS differential logic circuits are described. Then CMOS pass-transistor logic circuits are introduced, followed by the BiCMOS static logic circuits. Recently, SOI (silicon-on-insulator) technology has been intensively used to integrate CMOS VLSI circuits. In the following section, low-voltage SOI CMOS logic circuits, which use dynamic threshold voltage techniques, are explained. Low-voltage and low-power have become the trend for CMOS VLSI. The next section is dedicated to low-voltage and low-power CMOS static circuits. Several low-voltage CMOS circuit techniques are introduced, followed by the low-power CMOS circuit techniques.

3.1 BASIC CMOS CIRCUITS

CMOS technology has been used to integrate VLSI circuits for decades. Initially, CMOS technology with metal-gate devices in n/p wells was used. In the 1980s, CMOS technology with submicron polysilicon-gate devices in twin-tubs had been used for mass production of VLSI circuits. In the 1990s, CMOS technology entered the deep-submicron era. Deep-submicron CMOS devices had been used to integrate multi-million-transistor microprocessor chips and 64M-bit DRAMs. CMOS technology has become the mainstream technology for integrating VLSI digital circuits owing to its advantages in noise margin, conciseness, and low power consumption. In this section, fundamental CMOS static logic circuits are reviewed. Starting from the basic CMOS inverter, DC and transient analysis are presented, followed by the fundamental CMOS logic circuits. As CMOS technology goes toward the $0.1\mu m$ regime, low-voltage has been the trend. In the final portion of this section the difficulties of the CMOS logic circuits for low-voltage operations are reviewed.

3.2 CMOS INVERTERS

In the early days, MOS logic circuits were based on using resistors as the load. Since resistors could not be economically integrated on a chip, transistors had been designed to replace the resistor load. At that time, CMOS was not available. In addition, enhancement-type and inversion-type devices were not available in the chip simultaneously. Only single-channel, enhancement-type devices were available for designing digital circuits. The transistor biased in the linear region had been used for the load initially. Owing to the progress in the processing technology, in the 1970s, NMOS technology could offer enhancement-type and depletion-type devices at the same time. Using the depletion-type device as the load, the performance of the logic circuits had been improved. However, the asymmetry of the enhancement/depletion NMOS logic circuit with respect to the power supply voltages was a major drawback. In addition, the non-zero DC power dissipation further limited its capability to be used for VLSI. In the 1980s, owing to the availability of CMOS technology, PMOS was used as the load. The advantages of the complementary circuit technique is in the symmetry of the device structure with respect to the supply voltages. In addition, its full swing also leads to zero DC power dissipation.

The advantages of CMOS technology for the logic circuits were well recognized before the rise of CMOS technology. At the beginning, in CMOS technology, metal gate was used. The well in metal-gate CMOS technology was deep. For latchup consideration, separation between NMOS and PMOS devices is large. Due to the low device density, the early-stage CMOS technology could not compete with depletion-load NMOS technology. In the late 1970s, CMOS was still regarded as an expensive technology. The strong point of a CMOS logic circuit is its symmetry with respect to the power supply voltage. Both active pull-down and pull-up can be realized—the passive pull-up in NMOS technology was replaced by the active pull-up PMOS device. During the pull-down process, the PMOS device is turned off. As a result, V_{OL} can reach 0V at zero DC power dissipation. During pull-up, the output voltage can reach V_{DD}. Therefore, full swing and no DC power consumption become the specialties of CMOS technology.

For a CMOS inverter as shown in Fig. 3.1 based on a $0.8\mu m$ CMOS technology, $k'(=\mu C_{ox})$ is $30\mu A/V^2$ for NMOS and $15\mu A/V^2$ for PMOS. The threshold voltage is $V_{TN} = 0.8V$ for NMOS and $V_{TP} = -0.8V$ for PMOS. The supply voltage is $V_{DD} = 5V$. Using aspect ratios of $(W/L)_n = (2\mu m/0.8\mu m)$ for NMOS and $(W/L)_p = (4\mu m/0.8\mu m)$ for PMOS, β becomes: $\beta_n = k'_n(W/L)_n = 75\mu A/V^2$ for NMOS and $\beta_p = k'_p(W/L)_p = 75\mu A/V^2$ for PMOS. In a CMOS inverter circuit, the aspect ratio (W/L) of the PMOS device is usually designed to be two times that

CMOS INVERTERS   87

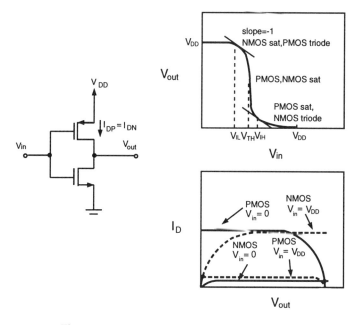

Fig. 3.1 I/O transfer curve of the CMOS inverter.

of the NMOS such that the difference in the mobilities can be compensated. As a result, the I/O transfer curve can be symmetric with respect to center of the voltage swing, hence the noise margin can be maximized. In addition, the high-to-low and the low-to-high propagation delay times can be compatible. In the following paragraphs, DC and transient behaviors of a CMOS inverter are analyzed.

(a) V_{OH}, V_{OL}

Due to the active pull-up and pull-down structure, when the input is high ($V_{in} = V_{DD}$), the output voltage is low (0V) and the PMOS device is off. For a CMOS circuit, the output load is usually capacitive. Without a DC path, the lowest output voltage can reach $V_{OL} = 0V$ at no power dissipation. Similarly, $V_{OH} = V_{DD}$.

(b) V_{IL}, V_{IH}

When the input is $V_{in} = V_{IL}$, the NMOS device is in the saturation region and the PMOS device is in the triode region. By $I_{Dn(sat)} = I_{Dp(triode)}$, one obtains:

$$\frac{\beta_n}{2}(V_{IL} - V_{TN})^2 = \frac{\beta_p}{2}\left[2(V_{DD} - V_{IL} - |V_{TP}|)(V_{DD} - V_{out}) - (V_{DD} - V_{out})^2\right]. \tag{3.1}$$

Differentiating Eq. (3.1) with respect to V_{IL}, one obtains:

$$\beta_n(V_{IL} - V_{TN}) = \beta_p\left[-(V_{DD} - V_{out}) - (V_{DD} - V_{IL} - |V_{TP}|)\frac{dV_{out}}{dV_{IL}}\right.$$
$$\left. + (V_{DD} - V_{out})\frac{dV_{out}}{dV_{IL}}\right]. \qquad (3.2)$$

Using $\frac{dV_{out}}{dV_{IL}} = -1$, from Eq. (3.2), one obtains:

$$V_{out} = \frac{(\frac{\beta_n}{\beta_p} + 1)V_{IL} - \frac{\beta_n}{\beta_p}V_{TN} + |V_{TP}| + V_{DD}}{2}. \qquad (3.3)$$

From Eqs. (3.3) and (3.1), using the device parameters, $V_{IL} = 2.075V$.

At V_{IH}, the NMOS device is in the triode region and the PMOS device is in the saturation region. By $I_{Dn(triode)} = I_{Dp(sat)}$, one obtains:

$$\frac{\beta_n}{2}\left[2(V_{IH} - V_{TN})V_{out} - V_{out}^2\right] = \frac{\beta_p}{2}(V_{DD} - V_{IH} - |V_{TP}|)^2. \qquad (3.4)$$

Differentiating Eq. (3.4) with respect to V_{IH}, one obtains:

$$\beta_n\left[(V_{IH} - V_{TN})\frac{dV_{out}}{dV_{IH}} + V_{out} - V_{out}\frac{dV_{out}}{dV_{IH}}\right] = -\beta_p(V_{DD} - V_{IH} - |V_{TP}|). \qquad (3.5)$$

From $\frac{dV_{out}}{dV_{IH}} = -1$, from Eq. (3.5), one obtains:

$$V_{out} = \frac{(\frac{\beta_p}{\beta_n} + 1)V_{IH} + \frac{\beta_p}{\beta_n}|V_{TP}| - V_{TN} - \frac{k_P}{k_N}V_{DD}}{2}. \qquad (3.6)$$

From Eqs. (3.6) and (3.4), using the device parameters, $V_{IH} = 2.925V$.

Note that in the above DC analysis, the lateral diffusion (LD) has been neglected. In fact, considering the lateral diffusion, the effective channel length of a CMOS device is $L_{eff} = L - 2L_D$.

From the equations for V_{IH} and V_{IL}, by having $\beta_n = \beta_p$, the I/O transfer curve can be symmetric with respect to the center of the signal swing. Therefore, $NM_H = V_{OH} - V_{IH} = NM_L = V_{IL} - V_{OL}$, which maximizes the noise margin performance. (Note that in a logic circuit, if one of NM_H and NM_L is particularly bad, the noise margin is degraded by the worse one.) The logic threshold voltage of the CMOS inverter circuit is defined as the input voltage when both NMOS and PMOS devices are in the center of the transition region—NMOS and PMOS are in saturation. Therefore, by equating their drain currents:

$$\frac{\beta_n}{2}(V_{TH} - V_{TN})^2 = \frac{\beta_p}{2}(V_{DD} - V_{TH} - |V_{TP}|)^2, \qquad (3.7)$$

the logic threshold voltage of the inverter is:

$$V_{TH} = \frac{V_{DD} - |V_{TP}| + V_{TN}\sqrt{\frac{\beta_n}{\beta_p}}}{1 + \sqrt{\frac{\beta_n}{\beta_p}}}. \tag{3.8}$$

From the above equation, the logic threshold voltage of the CMOS inverter is dependent on the $\beta = \frac{\beta_n}{\beta_p}$ ratio. By increasing the β ratio, the logic threshold voltage can be decreased. When the β ratio is $\frac{\beta_n}{\beta_p} = 1$, $V_{TH} \cong \frac{V_{DD}}{2}$.

Fig.3.2(a) shows the I/O transfer curves of the CMOS inverter and the NMOS inverters implemented by the resistive-load, the linear-load, the saturation-load, and the depletion-load techniques. Among four NMOS circuits, the depletion-load inverter has the steepest transition region, followed by the resistive load, and the linear-load. The worst case is the saturation-load inverter. Among four cases of the NMOS inverters, the lowest output voltage (V_{OL}) cannot reach 0V. Except the saturation-load, the highest output voltage (V_{OH}) is V_{DD}. The noise margin of the depletion-load inverter is the best among four. The resistive-load is next, followed by the linear-load. The saturation-load has the worst noise margin performance. From mid-1970s to the early 1980s, the depletion-load NMOS technology was the mainstream technology for digital MOS ICs. At an identical load current level, Fig.3.2(b) shows the comparison of the load lines implemented by the resistor, the linear device, the saturation device, and the depletion device. Among four load lines, the depletion-load is the best, followed the resistive-load, and the linear-load. The saturation-load is the worst.

From the above analysis, the use of the single-channel MOS device cannot result in ideal logic circuit performance. The weak point of the logic circuits implemented by single-channel MOS devices is the asymmetry in the pull-up and pull-down structure. At the output node, pull-down is usually driven by the input transistor—the active pull-down. However, pull-up is not active. Instead, it is passive pull-up. Due to the passive pull-up structure, even during the pull-down process, the passive pull-up device is still on. Therefore, V_{OL} cannot reach 0V. In addition, DC power consumption cannot be zero. This was the weakest point of the NMOS logic circuits until the rise of CMOS technology. As shown in Fig.3.2, in the CMOS inverter, the PMOS device functions like an active pull-up device. Therefore, from the output node, the circuit structure of the CMOS inverter is symmetric. Hence, its load line and the I/O transfer curve are close to the ideal ones.

Fig. 3.3 shows the pull-down transient analysis of the CMOS inverter. As shown in the figure, when a voltage step from low to high is imposed at the input to the CMOS inverter, the output will change from high to low. The propagation delay between the output and the input is the high-to-low propagation delay time (t_{PHL}). As shown in the figure, t_{PHL} is computed as the time from the middle of the input swing to the middle of the output swing. When the input changes from low to high, PMOS turns off and NMOS turns on. The output voltage will move along

90 CMOS STATIC LOGIC CIRCUITS

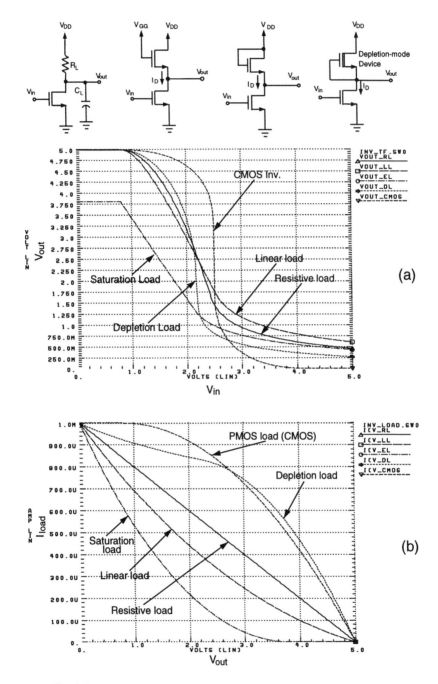

Fig. 3.2 Characteristics of inverters. (a) I/O transfer curves; (b) load lines.

CMOS INVERTERS

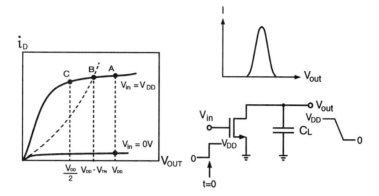

Fig. 3.3 Pulldown transient analysis of the CMOS inverter.

the NMOS IV curve from point A toward the origin. t_{PHL} is the time for the operating point to move from point A to point C. Between point A and point C, the condition of the NMOS device changes. The analysis of t_{PHL} is divided into two segments: (1) Between point A and point B, $V_{DD} > V_{out} > V_{DD} - V_{TN}$. The NMOS device is in the saturation region. During this segment, the propagation delay time is called t_{PHL1}. (2) Between point B and point C, the NMOS device is in the triode region. During this segment, the propagation delay time is called t_{PHL2}.

(1) Between point A and point B, NMOS is in saturation. Using the drain current formula, one obtains the propagation delay time in this segment:

$$t_{PHL1} = \frac{C_L(V_{DD} - (V_{DD} - V_{Tn}))}{\frac{\beta_n}{2}(V_{DD} - V_{Tn})^2}, \quad (3.9)$$

where C_L is the effective load capacitance at the output. From Eq. (3.9), the drain current of the NMOS device in this segment is a constant, which is independent of the output voltage.

(2) Between point B and point C, NMOS is in the triode region. The drain current of the NMOS device is dependent on the output voltage. The propagation delay time in this segment is:

$$t_{PHL2} = \int_{V_{DD}-V_{Tn}}^{\frac{V_{DD}}{2}} \left(-\frac{C_L}{I_{Dn}}\right) dV_{out}. \quad (3.10)$$

Using the triode region drain current formula, Eq. (3.10) becomes:

$$t_{PHL2} = -\frac{C_L}{\beta_n(V_{DD} - V_{Tn})} \int_{V_{DD}-V_{Tn}}^{\frac{V_{DD}}{2}} \frac{dV_{out}}{\frac{V_{out}^2}{2(V_{DD}-V_{Tn})} - V_{out}}. \quad (3.11)$$

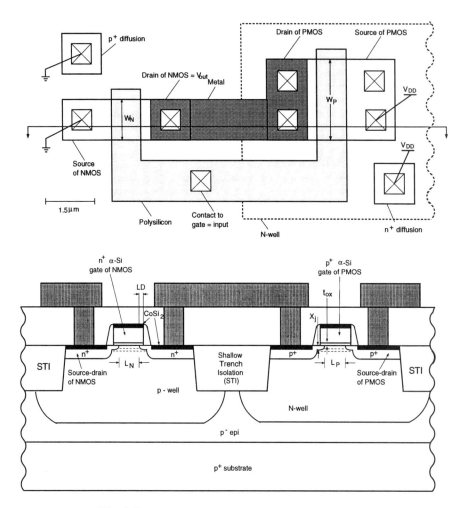

Fig. 3.4 Layout and cross-section of the CMOS inverter.

From $\int \frac{dx}{ax^2-x} = ln(1 - \frac{1}{ax})$, Eq. (3.11) becomes:

$$t_{PHL2} = \frac{C_L}{\beta_n(V_{DD} - V_{Tn})} ln \frac{3V_{DD} - 4V_{Tn}}{V_{DD}}. \quad (3.12)$$

From Eqs. (3.9) and (3.12), one obtains:

$$t_{PHL} = \frac{2C_L}{\beta_n(V_{DD} - V_{Tn})} \left(\frac{V_{Tn}}{V_{DD} - V_{Tn}} + 0.5 ln \frac{3V_{DD} - 4V_{Tn}}{V_{DD}} \right), \quad (3.13)$$

where C_L can be computed by considering the layout of the circuit as shown in Fig. 3.4. C_L is the sum of the parasitic capacitances in the output terminal. C_L includes :

$$C_L = C_{dgn} + C_{dgp} + C_{dbn} + C_{dbp} + C_p + C_{gn} + C_{gp}. \quad (3.14)$$

C_L includes the NMOS drain-gate capacitance (C_{dgn}), the PMOS drain-gate capacitance (C_{dgp}), the NMOS drain-body junction capacitance (C_{dbn}), the PMOS drain-body junction capacitance (C_{dbp}), the interconnect parasitic capacitance (C_p), and the load capacitance of the next stage (C_{gn}, C_{gp}). As shown in Fig. 3.4, the drain-body junction capacitance (C_{db}) includes the junction capacitance (C_j) and the sidewall capacitance (C_{jsw}). In this example, the interconnect capacitance is $C_p = 10fF$. Based on a $0.8\mu m$ CMOS technology with the SPICE parameters as shown in Fig. 3.5, the load capacitance is computed. The drain-body junction capacitance is

$$C_{db} = \frac{CJ \cdot AD}{(1 - V_{BD}/\phi_J)^{MJ}} + \frac{CJSW \cdot PD}{(1 - V_{BD}/\phi_J)^{MJSW}}, \quad (3.15)$$

where $\phi_J = \frac{kT}{q} ln(\frac{N_{sub}}{n_i}) + 0.56V$. $CJ, CJSW, MJ$, and $MJSW$ are the parameters as specified in SPICE device models. AS and AD are the area of the source and the drain regions, respectively. PS and PD are the perimeter of the sidewall regions of the source and the drain, respectively. AS, AD, PS, and PD can be found from the layout as shown in Fig. 3.4. For an MOS device biased in the triode region, the gate-drain capacitance is

$$C_{dg} = C_{ox} \left[1 - \left(\frac{V_{GS} - V_T}{2(V_{GS} - V_T) - V_{DS}} \right)^2 \right] + C_{GDO} W. \quad (3.16)$$

In the saturation region, the gate-drain capacitance is $C_{dg} = C_{GDO} W$, where C_{GDO} is the overlapped capacitance ($C_{GDO} = C_{ox} L_D$). C_{ox} is unit-area oxide capacitance ($C_{ox} = \frac{\epsilon_{ox}}{t_{ox}}$). Both C_{db} and C_{dg} are nonlinear functions of V_{out}. In order to simplify the calculation, the effective C_{db} and C_{dg} are computed based on the average values at $V_{out} = V_{DD}$ and $V_{out} = V_{DD}/2$. From Eq. (3.15), C_{dbn} is:

$$\begin{aligned} C_{dbn} &= \frac{1}{2}(C_{dbn}|_{V_{out}=V_{DD}} + C_{dbn}|_{V_{out}=V_{DD}/2}), \quad (3.17) \\ &= \frac{CJ_n \cdot AD_n}{2(1 + V_{DD}/\phi_{Jn})^{MJ_n}} + \frac{CJSW_n \cdot PD_n}{2(1 + V_{DD}/\phi_{Jn})^{MJSW_n}} \end{aligned}$$

$$+ \frac{CJ_n \cdot AD_n}{2(1 + 0.5V_{DD}/\phi_{Jn})^{MJ_n}} + \frac{CJSW_n \cdot PD_n}{2(1 + 0.5V_{DD}/\phi_{Jn})^{MJSW_n}}$$
$$= 2.34fF,$$

where AD_n is the area of the NMOS drain region ($AD_n = 4.8\mu m^2$), and PD_n is the length of the perimeter surrounding the NMOS drain region ($PD_n = 8.8\mu m$).

$$C_{dbp} = \frac{CJ_p \cdot AD_p}{2} + \frac{CJSW_p \cdot PD_p}{2} + \frac{CJ_p \cdot AD_p}{2(1 + 0.5V_{DD}/\phi_{Jp})^{MJ_p}}$$
$$+ \frac{CJSW_p \cdot PD_p}{2(1 + 0.5V_{DD}/\phi_{Jp})^{MJSW_p}}$$
$$= 8.57fF, \tag{3.18}$$

where AD_p is the area of the PMOS drain region ($AD_p = 9.6\mu m^2$), and PD_p is the length of the surrounding perimeter of the PMOS drain region ($PD_p = 12\mu m$). Therefore, C_{dgn} and C_{dgp} are:

$$C_{dgn} = \frac{C_{oxn}}{2}\left[1 - \left(\frac{V_{DD} - V_{Tn}}{2(V_{DD} - V_{Tn}) - \frac{1}{2}V_{DD}}\right)^2\right] + C_{GDOn}W_n = 1.743fF,$$
$$C_{dgp} = C_{GDOp}W_p = 0.46fF. \tag{3.19}$$

From Eq. (3.14), $C_L = 23.11fF$. From Eq. (3.13), with an effective channel length of $L_{eff} = L - 2LD = 0.68\mu m$, $t_{PHL} = 0.068ns$. Considering the rise time of the input signal ($t_r = 0.1ns$), the high-to-low propagation delay time is:

$$t_{PHL} = \sqrt{t_{PHL}^2 + \left(\frac{t_r}{2}\right)^2} = 0.084ns. \tag{3.20}$$

From the SPICE simulation (0.078ns) as shown in Fig. 3.5, similarly, the low-to-high propagation delay time (t_{PLH}) can also be computed. The overall propagation delay time is the average of t_{PHL} and t_{PLH}:

$$t_P = \frac{t_{PHL} + t_{PLH}}{2}. \tag{3.21}$$

As shown in Fig. 3.5, SPICE simulation of the CMOS inverter circuit has been carried out. The MOS device models used in the SPICE simulation are as described in Chapter 2. In the circuit cards, the connections of the CMOS devices are defined. In each circuit card, the node numbers for the drain, the gate, the source, and the body are defined. AS/AD defines the area of the source/drain in the device. PS/PD defines the perimeter of the sidewall in the source/drain area. The length and the area used in the SPICE simulation are based on the MKS system in terms of meters. As shown in Fig. 3.5, the simulated result has been obtained.

As shown in Fig. 3.3, the CMOS inverter consumes power only during the transient. At steady state, no power is consumed in the CMOS inverter. By designing

```
.model n NMOS ( VTO=0.8 KP=30U GAMMA=0.3 LD=0.06U TOX=18N UO=156.45
+              NSUB=5E16 PHI=0.779 XJ=0.5U LAMBDA=0.1385 RS=10
+              RD=10 JS=1E-9 CGBO=1.2E-10 CGSO=3.1E-10
+              CGDO=3.1E-10 CJ=345U CJSW=300P MJ=0.5 MJSW=0.33 )

.model p PMOS ( VTO=-0.8 KP=15U GAMMA=0.673 LD=0.06U TOX=18N
+              UO=78.23 NSUB=3E16 PHI=0.752 XJ=0.6U LAMBDA=0.0552
+              RS=30 RD=30 JS=1E-9 CGBO=1.28E-10 CGSO=3.3E-10
+              CGDO=3.3E-10 CJ=610U CJSW=390P MJ=0.5 MJSW=0.33 )

V_Vdd vdd 0 5V
V_Vin vin 0 pulse ( 0 5 1n 0.1n 0.1n 1.9n 4n )

mp vout vin vdd vdd p ( w=4u l=0.8u as=9.6p ad=9.6p ps=12.8u pd=12.8u)
mn vout vin 0   0  n ( w=2u l=0.8u as=4.8p ad=4.8p ps=8.8u pd=8.8u)
cl vout 0 10F

.OPTION POST
.tran .001ns 5ns
.END
```

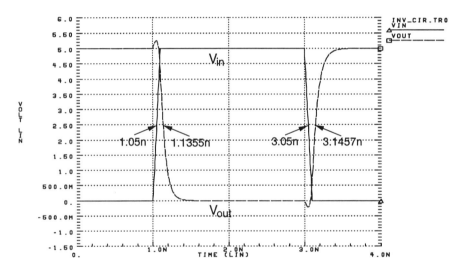

Fig. 3.5 Transient analysis of the CMOS inverter using SPICE simulation.

$(W/L)_p = 2(W/L)_n$ to compensate for the difference between electron and hole mobilities, t_{PHL} and t_{PLH} can be compatible. Symmetric rise/fall times t_{PHL}/t_{PLH} facilitate an easy circuit design. In addition, the I/O transfer curve can be symmetric with respect to the center of the signal swing—noise margin can be optimized.

From the t_{PHL} formula, by reducing the load capacitance (C_L) and increasing β_n and β_p, the propagation delay can be reduced. The progress in CMOS technology brings in shrinkage of the layout of the CMOS circuits, hence C_L is reduced. k_n and k_p can be increased by shrinking the thickness of the gate oxide and increasing the aspect ratio (W/L). For a more advanced CMOS technology, the gate oxide is thinner. However, for a thinner gate oxide, mobility may be degraded, which offsets the advantage. By increasing the aspect ratio (W/L), the parasitic capacitance at the

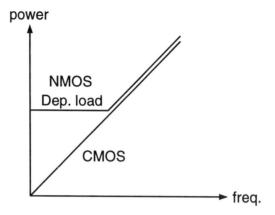

Fig. 3.6 Power dissipation versus operating frequency of CMOS and NMOS depletion-load inverters.

output node may rise, which is not helpful for reducing the propagation delay time. In addition, power consumption may increase.

The advantages of the CMOS logic circuits are full swing and no DC power consumption. In addition, since the body contacts of both NMOS and PMOS devices are connected to their respective source contacts, no body effect exists in the CMOS inverter. As shown in Fig. 3.2, compared with other NMOS inverters, the CMOS inverter has the steepest transition region in the I/O transfer curve since there is no body effect in the PMOS load. In other NMOS inverters, due to the body effect of the passive load, the slope in the transition region of the I/O transfer curve is less steep. Therefore, the noise margin is degraded.

The power dissipation of a CMOS inverter is linearly proportional to its operating frequency. As shown in Fig. 3.3, only during the switching transient, there is power consumption. At steady state, no DC power is consumed except for the leakage. In contrast, the NMOS inverter consumes DC power. Therefore, the CMOS inverter conserves power. When the circuit is not switching at standby, no power consumes except the leakage current. As shown in Fig. 3.6, when operation frequency rises, there is power consumption. This is one of the advantages of CMOS logic circuits. This is the reason why CMOS becomes the mainstream technology for VLSI.

3.2.1 CMOS Static Logic Circuits

In addition to the inverter, in the CMOS logic circuits, NOR and NAND are also important. Fig.3.7 shows a CMOS NOR and NAND circuit. The logic function of the NOR gate is $\overline{A + B}$ and the logic function of the NAND gate is $\overline{A \cdot B}$. As shown in the figure, in the CMOS NOR/NAND circuit, with respect to the output node, the circuit structure is not symmetric. Therefore, t_{PHL} and t_{PLH} are different. The

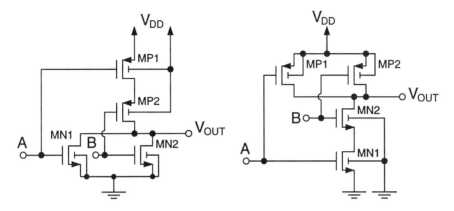

Fig. 3.7 CMOS NOR and NAND circuits.

t_{PHL} of the CMOS NOR gate varies depending on the input conditions. When both inputs A and B are high, two NMOS devices are on, hence the output voltage is pulled low by the two pull-down NMOS devices. When only one of A and B is high, only one NMOS device is on. Hence, the output voltage is pulled down at a speed slower than the previous case. When both A and B are low, both PMOS devices are on. In addition to the serial connection, the smaller hole mobility further degrades t_{PLH}. In order to make t_{PLH} and t_{PHL} compatible, the aspect ratio of the PMOS devices is four times that of the NMOS devices. Therefore, the number of inputs to the NOR gate cannot be too large, otherwise the difference between t_{PLH} and t_{PHL} can be too large. In addition, an increase in the parasitic capacitance at the output and the input capacitance of the PMOS device degrades the speed performance. Generally speaking, the number of the inputs to NAND/NOR had better not exceed three.

As described before, the CMOS inverter does not suffer from body effect. However, in the NOR/NAND gate circuit, body effect still exists. In a NOR circuit, the PMOS device in the middle (MP2) still has body effect. In the NAND circuit, the NMOS device in the middle (MN2) still has body effect. When designing CMOS circuits, whether NOR or NAND gates should be emphasized depending on the CMOS technology. If n-well CMOS technology is used, due to a more heavily doped substrate for the n-well technology, the body effect of the PMOS device is stronger than that of the NMOS device. From reducing body effect point of view, NAND gates should be used as often as possible since the PMOS devices in the NOR gates suffer from the body effect more than the NMOS devices in the NAND gates. Similarly, using the p-well CMOS technology, NOR gate should be used since the NMOS devices in the NAND gates may suffer from body effect more.

In addition to NAND/NOR, exclusive OR and exclusive NOR (XOR/XNOR) are frequently used in the logic circuits. As shown in Fig. 3.8, XOR is $AB' + A'B$ and XNOR is $AB + A'B'$. If using NAND and NOR gates, three gates are required to implement XOR/XNOR. As shown in Fig. 3.8, a single-stage CMOS gate can also

98 CMOS STATIC LOGIC CIRCUITS

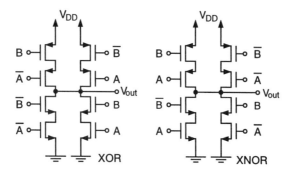

Fig. 3.8 CMOS XOR and XNOR circuits.

be used to implement XOR/XNOR. Any logic function can be realized using the single-stage CMOS gates. However, if the complexity of the single-stage CMOS gate is too high, its delay may be too high. Under this situation, it may be better to implement it with several stages.

3.2.2 Difficulties for Low-Power and Low-Voltage Operations

During the past 20 years, MOS devices have been continuously scaled down. For research-level CMOS devices, in 1972, the channel length was 6μm. In 1992, it was 0.1μm. Gate oxide was scaled from 100nm in 1972 to 3nm in 1992. Threshold voltage was scaled from 0.8V to 0.5V. For the past 20 years, channel length has been scaled down 60 times and gate oxide 30 times. However, power supply voltage was scaled down only three times and threshold voltage two times. In 1995, MOS devices with a channel length of 0.05μm were developed. As described in Chapter 2, the down-scaling of MOS devices may have been approaching the physical limit. The influence of the scaling on the device performance was discussed. Here, the influence of the scaling in the circuit performance is analyzed.

As described in Chapter 2, scaling of the future deep-submicron CMOS devices, the constant voltage (CV) scaling law will still be the trend with some modifications— the gate oxide thickness and the junction depth stay fixed. Based on CV scaling law, the lateral and the vertical dimensions of an MOS device are scaled down by s times ($s > 1$). Threshold voltage and power supply voltage are maintained. As a result, current increases s times and power dissipation increases s times. Propagation delay decreases s^2 times and power delay product decreases s times. This analysis does not consider high electric field effects. Therefore, down-scaling of current and propagation delay may have been overestimated.

Since supply voltage is not scaled down, the lateral electric field in the device increases s times. Therefore, carriers in the channel tend to move at the saturated velocity. Considering velocity saturation, the drain current is:

$$I_D = WC_{ox}(V_{GS} - V_T)v_{sat}. \tag{3.22}$$

Based on the above equation, under CV scaling law, considering velocity saturation, down-scaling does not increase the drain current at the above-mentioned rate. Since the propagation delay is proportional to $C_L \Delta V/I_D$, if the current stays unchanged, the propagation delay only shrinks by s times, instead of s^2 as mentioned above. The power dissipation does not increase instead of an increase of s times as mentioned above. The power delay product still shrinks s times as predicted under CV scaling law. In the same die area, the number of transistors increases s^2 times. Therefore, the power dissipation per unit area increases s^2 times. Therefore, power dissipation is a serious problem in VLSI circuits using deep-submicron CMOS devices.

Without using a special cooling system, generally speaking, power dissipation of a VLSI chip should not exceed 1–10W no matter how many transistors are in the chip. For a VLSI DRAM/SRAM chip this power dissipation limit may not cause a difficulty since the read/write operation of the DRAM/SRAM does not involve many transistors. Most transistors in DRAM/SRAMs are used in the memory cells array, which does not consume a substantial amount of power when not accessed. Although in a VLSI DRAM/SRAM chip, the number of transistors is huge, most of the power is consumed in the input buffers, output buffers, sense amps, and word line drivers. Therefore, when the integration size of DRAM/SRAM is increased, the power dissipation limit may not be the biggest difficulty. On the other hand, the increased junction leakage due to the increased junction temperature, which is the result of power consumption, may degrade the data retention of a capacitor in a DRAM memory cell.

On the other hand, power dissipation limit may become a big problem for a VLSI microprocessor chip. Unlike DRAM/SRAM, power dissipation of a microprocessor chip is more uniformly distributed throughout the chip. The scaling law described before is more applicable for microprocessors. Due to the yield consideration, there is a limit on the die area of a VLSI chip. When the device is scaled down, if the die area does not change, the number of transistors increases s^2 times. As the number of the total transistors in a chip grows, the limit in the power dissipation will be reached soon. Therefore, how to reduce the power dissipation from system design point of view is a major challenge for designing the next-generation microprocessor.

Lowering the power supply voltage is a straightforward, efficient way to reduce power consumption. When the power supply voltage is lowered, $V_{GS} - V_{TH}$ also drops. The drain current, which is a function of $I_{Dsat} \propto (V_{GS} - V_{TH})^\gamma$, where γ is $1 \leq \gamma \leq 2$, is lowered. The propagation delay of the circuit is a function of $t_{pd} \propto C_L \Delta V/I_{Dsat} \propto (V_{GS} - V_{TH})^{-\gamma} \propto (V_{DD} - V_{TH})^{-\gamma}$. When the power supply voltage is lowered, the propagation delay time increases. Although the threshold voltage can be scaled down to shrink the propagation delay, an increase in the

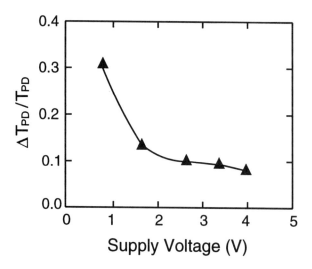

Fig. 3.9 $\Delta t_{pd}/t_{pd}$ versus supply voltage of a CMOS inverter integrated using a 0.8μm CMOS technology. (From Sun and Tsui[1], ©1995 IEEE.)

subthreshold leakage also causes the rise in power dissipation.

The voltage bounce ($\Delta V \cong L\frac{dI}{dt}$) due to the change in the supply current ($\frac{dI}{dt}$) on the supply line may further limit stability and performance of low-voltage operation. Since metal lines tend to be finer and finer in the future, the voltage bounce problem will become more and more serious. The other factor limiting the downscale of the supply voltage is due to the increased delay time variation from the change in the threshold voltage after the supply voltage is shrunk. As described before, the propagation delay time is a function of the difference between supply voltage and the threshold voltage: $t_{pd} \propto 1/(V_{DD} - V_T)^\gamma$, $1 \leq \gamma \leq 2$. Differentiating the propagation delay time with respect to the threshold voltage, one obtains: $\Delta t_{pd} \propto \Delta V_T/(V_{DD} - V_T)^{\gamma+1}$. Therefore, when the supply voltage (V_{DD}) is scaled down, the propagation delay time variation Δt_{pd} due to the variation in the threshold voltage ΔV_T may increase. As a result, circuit design becomes more difficult. Fig. 3.9 shows the $\Delta t_{pd}/t_{pd}$ versus supply voltage of a CMOS inverter integrated using a 0.8μm CMOS technology[1]. As shown in the figure, when the supply voltage is greater than 2V ($V_{DD} > 2V$), $\Delta t_{pd}/t_{pd}$ is approximately at its minimum value. When the supply voltage is smaller than 2V ($V_{DD} < 2V$), due to the threshold voltage variation, the propagation delay time variation due to the change in the supply voltage increases substantially. For a circuit operating at a low supply voltage, reliable and stable performance is required. When the supply voltage is scaled down, the threshold voltage also needs to be shrunk accordingly. In order to maintain a stable performance, a well control of the variation in the threshold voltage is one of the biggest challenges for future low-voltage VLSI technology.

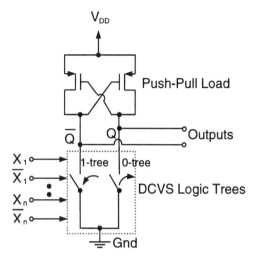

Fig. 3.10 Structure of the differential cascode voltage switch (DCVS) logic circuit.

When scaling down next-generation deep-submicron CMOS technology, power supply voltage V_{DD} may need to be further scaled down—from 1.5V to 1V—in order to obtained good device reliability. Considering subthreshold slope, threshold voltage cannot be scaled down easily. When power supply voltage is scaled down to 1V, use of current CMOS circuits may encounter difficulties since the gate voltage overdrive is too small. For future deep submicron CMOS technology using 1V supply voltage, current CMOS logic circuits may need to be modified.

3.3 CMOS DIFFERENTIAL STATIC LOGIC

In the previous section, conventional CMOS static logic circuits have been described. In this section, CMOS differential static logic circuits are analyzed. Specifically, three types of the differential static logic circuits are described. First, the differential cascode voltage switch (DCVS) logic circuits are analyzed, followed by the differential split-level (DSL) logic circuits and the differential cascode voltage switch with pass-gate (DCVSPG) logic circuits. In the final portion of this section, push-pull cascode logic (PPCL) circuits are described.

3.3.1 Differential Cascode Voltage Switch (DCVS) Logic

Differential cascode voltage switch (DCVS) logic is also called cascode voltage switch logic (CVSL)[2]. Fig. 3.10 shows the structure of the differential cascode voltage switch (DCVS) logic circuit[3]. The DCVS logic circuit is composed of two parts—the DCVS logic tree and the PMOS push-pull load. The DCVS logic tree, which is implemented in NMOS devices, is in the bottom portion of the DCVS logic

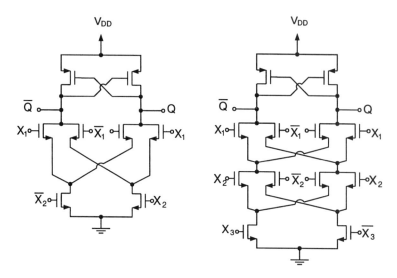

Fig. 3.11 DCVS XOR circuits. (Adapted from Chu and Pulfrey[3].)

circuit. Since the complementary results are available at the output nodes, both 0-tree and 1-tree are required in the DCVS logic tree. The outputs of the DCVS logic trees are connected to the PMOS push-pull load, which is at the top of the DCVS logic circuit. The operation of the DCVS logic circuit is described below. When input signals x_1, $\overline{x_1}$, x_2, $\overline{x_2}$,... x_n, $\overline{x_n}$ are imposed, either 0-tree or 1-tree will provide a conducting path between the outputs and ground. If 1-tree is on, the output $\overline{Q}$ is pulled to low. If 0-tree is on, the output Q is pulled to low. The push-pull load, which has a cross-coupled structure of two PMOS devices, is used to provide a pull-up load for the DCVS logic circuit.

Fig. 3.11 shows an XOR gate implemented by the DCVS logic circuit[3]. For the two-input XOR circuit, its logic function ($x_1 \oplus x_2 = x_1\overline{x_2} + \overline{x_1}x_2$) can be implemented by cascoding two levels of the two input transistors, which have the AND function. At each of the two output nodes, two of the cascoded AND logic paths are connected together, which provides the OR function. As shown in the figure, each of the lower-level results can be used as the inputs to the two upper-level paths. For the three-input exclusive OR circuit, its logic function ($x_1 \oplus (x_2 \oplus x_3)$) can be implemented by cascoding three levels of the three input transistors. Due to the recursive nature, the XOR circuit implemented by the DCVS logic circuit technique is concise.

Fig. 3.12 shows a logic function: $x_1x_2 + x_2x_3 + x_3x_1$ realized by the DCVS logic circuit[3]. The logic function $x_1x_2 + x_2x_3 + x_3x_1$ can be rewritten according to the K-map as shown in Fig. 3.12(a) as $x_2x_3 + x_1(\overline{x_2}x_3 + x_2\overline{x_3})$. $x_1(\overline{x_2}x_3 + x_2\overline{x_3})$. It is shown in Fig. 3.12(b) using a wired OR function at the output of the bottom level. From the K-map, the 0-tree portion—$x_1x_2 + x_2x_3 + x_3x_1$ can be rewritten

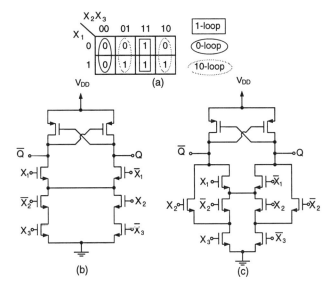

Fig. 3.12 A logic function: $x_1x_2 + x_2x_3 + x_3x_1$ realized by the DCVS logic circuit. (Adapted from Chu and Pulfrey[3].)

as $\overline{x_2x_3} + \overline{x_1}(\overline{x_2}x_3 + x_2\overline{x_3})$, where $\overline{x_1}(\overline{x_2}x_3 + x_2\overline{x_3})$ is the 0-tree in Fig. 3.12(b). Thus, $x_1(\overline{x_2}x_3 + x_2\overline{x_3})$ from the 1-tree and $\overline{x_1}(\overline{x_2}x_3 + x_2\overline{x_3})$ from the 0-tree can share $(\overline{x_2}x_3 + x_2\overline{x_3})$. Assume that a 1-cell represents a cube x_1p and another 0-cell represents cube $\overline{x_1}p$. Thus, p can be assumed to represent a 10-cell. For example, as shown in the figure, p is assumed to be $\overline{x_2}x_3$ and $x_2\overline{x_3}$— a 10-cell. From 10-cell and 01-cell, they can be simplified to become 10-tree and 01-tree. Two adjacent 10-cells(01-cells) become a loop—10-loop(01-loop). Another simplification method is similar to the conventional K-map approach. First, 01-loop is simplified to be as shown in Fig. 3.12(b) followed by adding 1-loop and 0-loop. Also shown in Fig. 3.12 is the complete circuit after adding 1-loop (x_2x_3) and 0-loop ($\overline{x_2} \cdot \overline{x_3}$). The 1-logic portion is connected to $\overline{Q}$ and the 0-logic portion is connected to Q. Thus, it becomes the situation as shown in Fig. 3.12(c). Based on the above approach, from the modified K-map method, a concise DCVS logic arrangement can be obtained.

3.3.2 Differential Split-Level (DSL) Logic

After creation of the DCVS logic circuit, improvements have been done. For the DCVS logic circuit, its speed performance is determined by the voltage swing at the output nodes. A smaller voltage swing at the output nodes can speed up the circuit. Following this concept, the differential split-level (DSL) logic circuit as shown in Fig. 3.13 has been created[4]. Different from the DCVS logic circuit, two extra NMOS devices with their gates biased at $\frac{V_{DD}}{2} + V_{TN}(\frac{V_{DD}}{2})$ are connected to the top push-pull load, which is composed of two PMOS devices as described in the DCVS

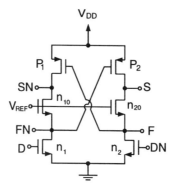

Fig. 3.13 Differential split-level (DSL) logic circuit. (Adapted from Pfennings et al.[4].)

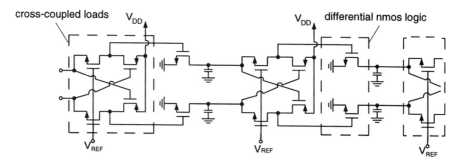

Fig. 3.14 Cascading of the DSL logic circuits. (Adapted from Pfennings et al.[4].)

logic circuit. By adding these NMOS devices to the push-pull load, the maximum voltages of F and FN are limited to about $V_{DD}/2$. As a result, the speed performance of the DSL logic circuit is better than that of the DCVS logic circuit if F and FN are used as the outputs instead of S and SN.

Fig. 3.14 shows the cascading of the DSL logic circuits[4]. As shown in the figure, following the concept of the basic DSL logic circuit describe in Fig. 3.13, in the case of DSL logic circuits, some modifications have been done. The inputs to the cascaded DSL logic circuit are located at the nodes between the push-pull load and the logic tree, instead of the inputs to the gates of the transistors in the logic tree as shown in Fig. 3.13. In addition, the outputs are from the drain ends of the logic trees (F and FN), instead of from the nodes between the push-pull load and the logic tree (S and SN) as shown in Fig. 3.13. By examining the circuit diagram of the cascaded DSL logic circuits as shown in Fig. 3.14 carefully, we can understand that both Figs. 3.13 and 3.14 are related. In the cascaded DSL logic circuit, the output node of a DSL logic circuit is connected as the input to the next DSL logic circuit. In Fig. 3.13, the swing of F and FN is $\frac{V_{DD}}{2}$. Thus, in Fig. 3.14, the signal swing at the output of the DSL is $\frac{V_{DD}}{2}$, instead of V_{DD} for the non-standard DCVS. Thus, a

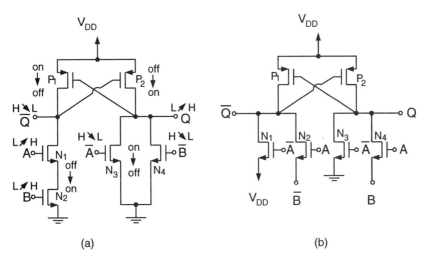

Fig. 3.15 (a) Conventional DCVS logic circuit; and (b) the differential cascode voltage switch with pass-gate (DCVSPG) logic circuit. (Adapted from Lai and Hwang[5].)

better speed performance can be expected. However, its power dissipation is higher since the pull-up PMOS devices cannot be turned off completely.

3.3.3 Differential Cascode Voltage Switch with Pass-Gate (DCVSPG) Logic

The DCVS logic circuit has some problems. As shown in Fig. 3.15[5], the DCVS logic circuit is asymmetric with the respect to the output nodes between the push-pull load and the logic tree. As a result, pull-up and pull-down transients at the output nodes are not symmetric. Consider the following transient situation. Initially, both inputs A and B are low, therefore, the path between the output node $\overline{Q}$ and ground is broken. At this time, the output node $\overline{Q}$ is pulled up to V_{DD} by the PMOS device P_1. In addition, PMOS device P_2 is off. Both NMOS devices N_3 and N_4 are on, hence, the output Q is pulled down to ground. When both inputs A and B switch to high, NMOS devices N_3 and N_4 turn off. Therefore, the output node Q is floating—its previous state (low) is maintained. Hence, PMOS device P_1 is still turned on. At this time, both N_1 and N_2 are on. Therefore, the voltage at node $\overline{Q}$ is determined by the aspect ratio between N_1/N_2 and P_1. If the aspect ratio is appropriate, such that $\overline{Q}$ drops to below $V_{DD} - |V_{tp}|$, P_2 slowly turns on. Therefore, the voltage at the output node Q rises slowly—rise time of Q is large. If the width of P_2 is small, the pull-up capability of the push-pull load at the node Q is not sufficient. In addition, simultaneous turn-on of P_1, and N_1/N_2 also increases short-circuit power consumption of the circuit. When the output $\overline{Q}$ discharges from high to low, it is a ratioed logic. Thus, the strength of the pull-up PMOS devices should be individually adjusted in each n-logic tree. Therefore, the temporary floating of Q causes problems for the DCVS logic circuit. Fig. 3.15(b) shows the DCVS with pass-gate (DCVSPG) logic

106 CMOS STATIC LOGIC CIRCUITS

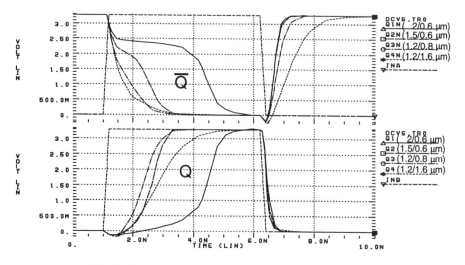

Fig. 3.16 Transients of a 2-input DCVS AND/NAND gate.

circuit, which is derived from the DCVS logic circuit with its problems overcome. As shown in the figure, pass transistors have been used in the logic trees. In 0-tree, N_4 is not connected to ground any time. Instead, N_4 is connected to the input B. As a result, when input B is low, Q is grounded as in the DCVS logic circuit. When input B is high, if input A is high, the output Q is pulled up to high by both P_2 and the pass transistor N_4. In contrast, in the DCVS logic circuit, the output Q is pulled up to high by P_2 only. Unlike DCVS, there is no temporary floating problem in the DCVSPG. Therefore, using the pass transistor structure in the logic tree, the pull-up capability of the DCVSPG logic circuit has been enhanced.

Fig. 3.16 shows the transients of a 2-input DCVS AND/NAND gate as shown in Fig. 3.15(a). As shown in the figure, when the inputs (A,B)=(1,1), N_1 and N_2 are on; and N_3 and N_4 are off. Thus, node Q is floating. When $\overline{Q}$ discharges to turn on PMOS P_2, node Q is pulled up and the whole regenerative feedback path is activated. In this study, four aspect ratios of the pull-up PMOS device—$W/L = 2\mu m/0.6\mu m(Q_1)$, $1.5\mu m/0.6\mu m(Q_2)$, $1.2\mu m/0.8\mu m(Q_3)$, and $1.2\mu m/1.6\mu m(Q_4)$ have been designed. As shown in the figure, when the aspect ratio of the PMOS device is smaller, it is easier to discharge, but the pull-up is slower. If the aspect ratio of the pull-up PMOS P_1 becomes too large, there is a possibility that $\overline{Q}$ cannot discharge to turn on PMOS P_2. For example, in Case 1 with an aspect ratio of $2\mu m/0.6\mu m$ for the PMOS device, when (A,B)=(1,1), P_1, N_1, and N_2 form a ratioed logic. After discharge, the voltage at node $\overline{Q}$ is at about 2.4V. Thus, PMOS P_2 is weakly turned on. As shown in the figure, when the voltage at node Q rises until the conductance ratio of P_1 to N_1 and N_2 decreases, node $\overline{Q}$ continues discharging and node Q is pulled up at a faster speed. After the feedback regenerative path is built, the logic function is completed. Thus, the aspect ratio of the PMOS device

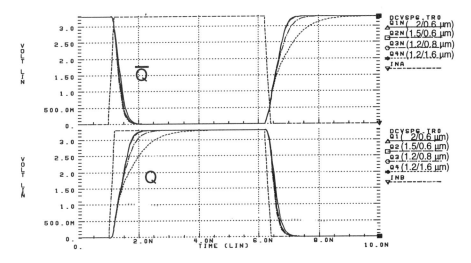

Fig. 3.17 Transient of the 2-input DCVSPG AND/NAND gate.

cannot be too large.

Fig. 3.17 shows the transients of the 2-input DCVSPG AND/NAND gate as shown in Fig. 3.15(b). As shown in Fig. 3.17, since N_4 turns on, node Q is pulled up. Thus, the conductance of P_1 is reduced substantially and the regenerative feedback path can be constructed earlier. As a result, the size of the pull-up PMOS device does not affect the pull-down substantially. Instead, the size of the pull-up PMOS device only affects pull-up. Therefore, DCVSPG is closer to ratioless logic as compared to DCVS—easier to design.

Fig. 3.18 shows another differential logic circuit—push-pull cascode logic (PPCL) [6]. The operation of PPCL is similar to DCVSPG. In DCVSPG, pass transistor logic circuits are used. In PPCL, the PMOS devices in the CMOS static logic circuits are replaced by the NMOS devices such that the CMOS static logic circuits are transformed into NMOS 1-tree and 0-tree. Consider the logic tree for $A \cdot B$. Replacing the two cascoded NMOS devices by the PMOS devices and the inputs A and B replaced by $\overline{A}$ and $\overline{B}$, it is a CMOS $A \cdot B$ gate. The PMOS cross-coupled load is used to compensate the logic-1 output, which cannot be pulled up to V_{DD} by the NMOS, such that a regenerative path can be constructed to enhance push/pull at the output. Thus, PPCL has the advantages as for DCVSPG. The weakness of PPCL is that the device count of 1-tree and 0-tree is identical to that of the conventional CMOS static logic circuits. Therefore, the layout area of PPCL is larger. Thus, the idea of using NMOS logic tree to reduce the PMOS load capacitance in DCVS has been offset.

108 CMOS STATIC LOGIC CIRCUITS

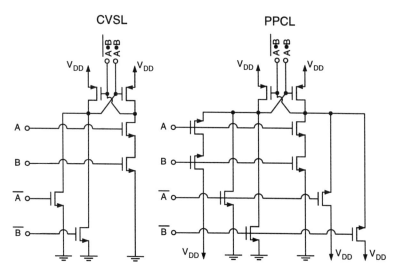

Fig. 3.18 Push-pull cascode logic (PPCL). (Adapted from Partovi and Draper[6].)

3.4 CMOS PASS-TRANSISTOR LOGIC

In the previous section, CMOS differential logic circuits have been described. In this section, CMOS pass-transistor logic circuits are described.

3.4.1 Pass-Transistor Logic Fundamentals

Pass-transistor logic circuits are based on switches, which are implemented by MOS devices. In a pass-transistor logic circuit, input signals can be used as the control signals and the pass signal referred to pass transistors as shown in Fig. 3.19(a)[7]. As in standard CMOS static logic circuits, at steady state, no power dissipation is needed for the pass-transistor logic circuits. The design procedure of pass-transistor logic circuits is described here. The carry signal in an adder circuit ($f = A \cdot B + B \cdot C + C \cdot A$) is used as an example. If two input signals A and B are used as the control signals, then the input signal C is the pass signal. With two control signals, there are four rows of the signal paths connected to the output. Each of the four rows of the signal paths is controlled by the two input control signals: A and B. In the first row, the control signals are $\overline{A}$ and $\overline{B}$, therefore, the input pass signal is $f(A = 0, B = 0, C) = 0$. In the second row, the control signals are $\overline{A}$ and B, therefore, the input pass signal is $f(A = 0, B = 1, C) = C$. In the third row, the control signals are A and $\overline{B}$, and the input pass signal is $f(A = 1, B = 0, C) = C$. In the fourth row, the control signals are A and B and the input pass signal is $f(A = 1, B = 1, C) = 1$. The results are as shown in Fig. 3.19(b). Fig. 3.19(c) shows a logic function $f(A, B, C) = AB + BC + CA$ realized by CMOS pass-transistor logic circuits. As shown in the figure, since the NMOS pass transistors cannot effectively pass logic 1, by adding PMOS pass transistors both logic 0 and 1

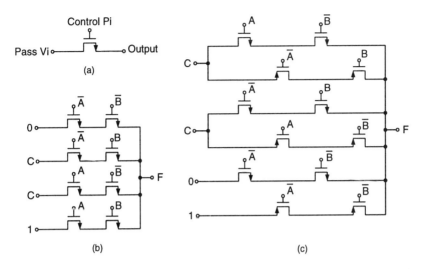

Fig. 3.19 Pass-transistor logic circuits for realizing the carry signal in an adder: $f = A \cdot B + B \cdot C + C \cdot A$.

can be effectively passed. When pass variable '0' is to be passed, NMOS pass transistors are used. When pass variable '1' is to be passed, PMOS should be used. When the value of the pass variable is not certain, both NMOS and PMOS pass transistors should be used. In addition, the control variable for the PMOS pass transistors should be complementary to that for the NMOS. The design procedure described above can be used for realizing pass-transistor logic circuits. For n inputs, if m inputs ($m < n$) are used as the control signals, then, there are 2^m rows of the input paths. The input signal of each row needs to be individually determined using the method described above. If the number of the inputs is too large, the signal path may be too long. Then the RC delay associated with the signal path may degrade the speed performance. Under this situation, it had better be designed with several pass-transistor logic blocks relayed by buffers. Each block may contain up to four input signals. At the end of each block, it is followed by a buffer, which is used to amplify the signal propagated via the signal path in each block. Therefore, after the buffer, the propagated signal is reconstructed to be used as another input pass signal for the next block.

For the NMOS pass-transistor logic circuit as shown in Fig. 3.20(a), the maximum level in the signal path cannot go beyond $V_{DD} - V_T$, if in the circuit the body contacts of all NMOS devices are connected to 0V and the level of the input signals is from 0V to V_{DD}. Considering the body effect, the maximum level in the signal path cannot go beyond $V_{DD} - V_T(V_{SB})$, where V_{SB} is the source-body voltage. If the body bias is negative, the body effect of the NMOS devices may further degrade the maximum level. Therefore, the maximum level of the input signal to the buffer is also $V_{DD} - V_T$. Consequently, the noise margin and the speed performance of the buffer may be limited when the power supply voltage is scaled down. This is a major drawback of the NMOS pass-transistor logic circuit. In order to overcome this

110   CMOS STATIC LOGIC CIRCUITS

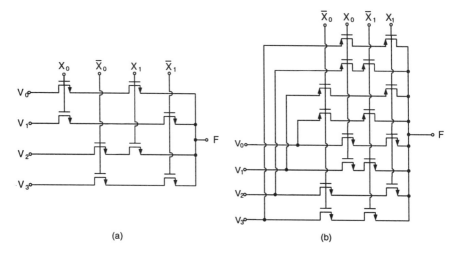

Fig. 3.20 (a) NMOS pass-transistor logic circuit; (b) CMOS pass-transistor logic circuit.

difficulty, CMOS pass-transistor logic circuit as shown in Fig. 3.20(b) has been used. As shown in the figure, instead of NMOS pass transistors, CMOS pass transistors are used. Therefore, a full swing of V_{DD} exists in the signal path. Hence, the body effect has no effect on the noise margin of the circuit. Fig. 3.20(b) shows the CMOS 4-to-1 pass-transistor logic circuit. As shown in the figure, there are two sets of the signal paths in the circuit. One set is realized by the NMOS pass transistors. The other is realized by the PMOS transistors with their control signals reversed as compared to the NMOS ones. Using this approach, a full-swing output with the static signal property is available. Note that the property of a static signal is referred to that the signal is always tied to 0 or 1 instead of floating as in the dynamic logic circuits described in the next chapter.

3.4.2 Other Pass-Transistor Logics

Fig. 3.21 shows the differential pass-transistor logic (DPTL) 4-to-1 selector circuit[8]. As shown in the circuit, the DPTL circuit is composed of two parts. One is the NMOS pass-transistor logic circuit. The other is a differential buffer. The NMOS pass-transistor logic circuit as the DCVS logic circuit described before has 1-tree and 0-tree, which are made of pass transistors. Different from DCVS logic circuit, where 1-tree and 0-tree are two mutually disjoint logic trees, 1-tree and 0-tree in DPTL are identical but the pass variables for 1-tree and 0-tree are complementary. The outputs of the 1-tree and 0-tree are the two complementary output signals to be amplified by the differential buffer. As shown in the figure, the differential amplifier can be either static or clocked. In the static differential amplifier, a DCVS buffer load as described in the DCVS section has been used. In the clocked differential amplifier, the dynamic latch techniques, which will be described in the next chapter, are used.

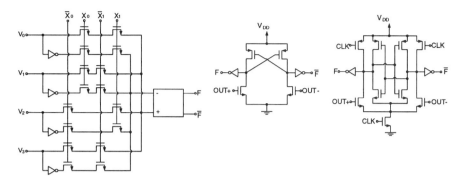

Fig. 3.21 Differential pass-transistor logic (DPTL) 4-to-1 selector circuit. (Adapted from Pasternak et al.[8].)

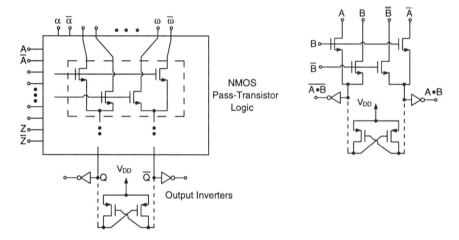

Fig. 3.22 The complementary pass-transistor logic (CPL) circuit. (Adapted from Yano et al.[9].)

Similar to the DPTL circuit, Fig. 3.22 shows the complementary pass-transistor logic (CPL) circuit[9]. CPL circuits are also derived from the DCVS logic circuit. As shown in the figure, the CPL circuit is similar to the DPTL circuit except that the outputs of the 1-tree and 0-tree are individually connected to their respective buffers, instead of being connected to the differential amplifier in the DPTL circuit. If the cross-coupled PMOS load is added to CPL, it is almost identical to DCVSPG described before. Fig. 3.23 shows the basic logic modules realized by the CPL circuits[9]. Following a similar procedure as for the standard pass-transistor logic circuits, 1-tree and 0-tree logic paths can be realized.

Fig. 3.24 shows the swing restored pass-transistor logic (SRPL) AND/NAND gate[10]. As shown in the figure, a cross-coupled inverter latch is added to the output of the NMOS pass-transistor logic circuit. Thus a full swing is available at the output.

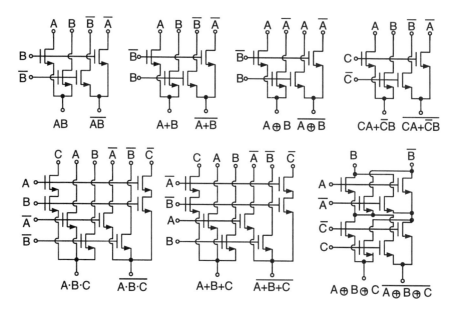

Fig. 3.23 Basic logic modules realized by the CPL circuits. (Adapted from Yano et al.[9].)

Due to the regenerative feedback in the inverter latch, the pull-down of the output can be expedited. Due to the regenerative feedback, during the switching of the output node, the NMOS pass-transistor tree is competing with the inverter latch—still a ratioed logic. Therefore, it is not suitable for operation with a low supply voltage. Due to the body effect, the NMOS pass transistors have difficulties in producing a high output. When the output switches from low to high, the pull-down NMOS device in the latch competes with the pull-up path in the NMOS pass-transistor tree. When the supply voltage is lowered, the pull-up capability of the NMOS tree is degraded. Therefore, its propagation delay lengthens.

CPL circuits described above are based on NMOS devices, which may have body effects. As a result, no full-swing exists in the signal paths. To overcome this difficulty, the double pass-transistor (DPL) circuits as shown in Fig. 3.25 have been created[11]. In addition to NMOS pass transistors, PMOS pass transistors are also available in the DPL circuit. Compared to the standard CMOS pass-transistor logic circuits, in the DPL circuit, the pass signal and the control signal are reversed. Therefore, two variables with an identical value can be passed at the same time at the output as shown in Fig. 3.25. A smaller equivalent resistance leads to a faster speed as compared to the CMOS pass-transistor logic circuit. As shown in the figure, using the DPL circuit, various basic logic functions have been realized. The drawback in the DPL circuits is that sometimes it is difficult to realize complicated logic functions with more than two inputs.

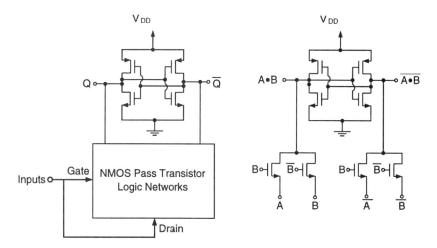

Fig. 3.24 Swing restored pass-transistor logic (SRPL) AND/NAND gate. (Adapted from Parameswar et al. [10].)

3.4.3 CAD for Pass-Transistor Logic

Due to lack of design methodology to generate general logic functions and lack of synthesis tools and cell libraries for logic designs, pass-transistor logic had not been widely used in the past. CAD tools have not been used to realize pass-transistor logic circuits until recently. Pass-transistor based cell library and synthesis tools have been constructed. The basic function of a pass-transistor based cell is its multiplexer function and the open-drain structure. The basic pass transistor cell has the flexibility of transistor-level circuit designs and the compatibility with conventional cell-based designs. A simple cell library with only seven cells is needed to synthesize any pass-transistor logic circuits. Among these seven basic cells, four of them are inverting buffers with different strengths, which can be classified into two categories as shown in Fig. 3.26[12]. The inverting buffers function as relays for those pass-transistor logic trees. As shown in the figure, both inverting buffers have a feedback transistor with its gate signal from the output to help enhance the pull-up capability associated with the internal pass-transistor logic paths. In addition to the inverting buffers, there are three other types of basic cells as shown in the figure. The first basic cell (Y1) functions as a multiplexer, which is controlled by the input signal C to determine which of the two input signals A and B is selected. The second basic cell (Y2) is an evolved version of Y1 with two control signals C_1 and C_2 and three input signals A, B, D. There are two selections in this circuit. The first selection is to determine which of A and B is selected by C_1. At the second selection, which is controlled by the control signal C_2, the input signal D or the output of the first selection is selected as the output. The third basic cell (Y3) has four input signals and two control signals. At the first selection level, two input signals from two pairs of the input signals are selected, respectively. At the second selection level, one of the two outputs from the first selection is determined as the final output. Using

114 CMOS STATIC LOGIC CIRCUITS

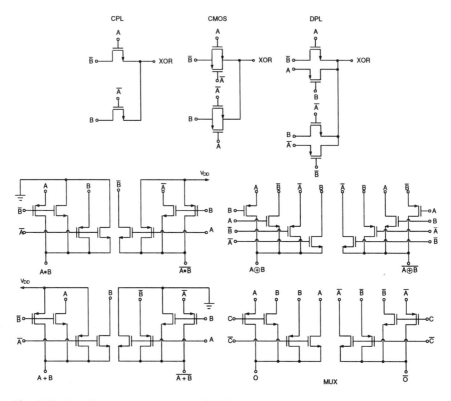

Fig. 3.25 Double pass-transistor logic (DPL) circuits. (Adapted from Suzuki et al. [11].)

these basic cells, all pass-transistor logic circuits can be realized using binary decision algorithm. Fig. 3.27 shows the design procedure in realizing a logic function: $f = w + x + (y \cdot z + \bar{y} \cdot \bar{z})$[12]. This design procedure is based on the binary decision diagram (BDD). As shown in the flow chart, there are three binary decisions involved. At the location with the first binary decision, if w is 1, then the output of the function is 1— the procedure is done. If w is not 1 ($\bar{w}$ is true), the location with the second binary decision is tested. At the location with the second binary decision, if $\bar{w}x$ is 1, then the output is 1—the procedure is done. Otherwise, $\bar{x}$ is true and the third decision is initiated. At the location with the third binary decision, if $\bar{w} \cdot \bar{x} \cdot y$ is 1, then the output of the function is z. Otherwise, $\bar{y}$ is true and the output of the overall function is $\bar{z}$. Inverting buffers are inserted at a designated interval, and appropriate basic cells (Y1 and Y2) are inserted to replace the appropriate segments in the flow chart. Due to the inserted inverting buffers, the pass variables should be reversed such that the whole logic function does not change.

The binary decision diagram (BDD) for designing LEAP pass-transistor logic can be used for all pass-transistor logic designs. Fig. 3.28 shows the logic binary trees of a function $f = a \cdot b \cdot c + a \cdot \bar{b} \cdot \bar{c} + \bar{a} \cdot \bar{b} \cdot \bar{c}$ based on binary decision diagram

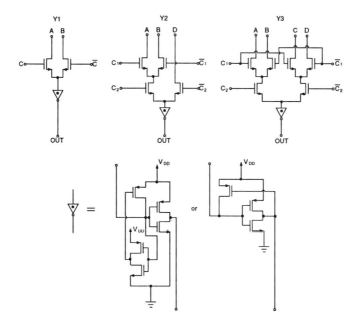

Fig. 3.26 Basic cells used in a top-down pass-transistor-based CAD—LEAP (Lean Integration with Pass-Transistors). (Adapted from Yano et al.[12].)

(BDD)[13]. As shown in the figure, its binary decision tree can be simplified. Operation 1 is referred to the situation with the binary decision that A_1 and A_2 are equivalent ($A_1 = A_2 = x \cdot B + \overline{x} \cdot C$). Thus, A_1 and A_2 can be merged. Operation 2 means that $A = x \cdot B + \overline{x} \cdot B$. Therefore, A and B can be merged—the control variable x can be discarded. Fig. 3.29 shows the procedure to realize a logic function: $f = a \cdot b \cdot c + a \cdot \overline{b} \cdot \overline{c} + \overline{a} \cdot \overline{b} \cdot \overline{c}$[13]. As shown in the figure, via operation 1 to merge all possible nodes, the binary tree has been simplified to the most concise binary decision tree. Then, the binary decision tree is transformed into an NMOS logic tree, where the control variable is mapped to an NMOS device with the control variable controlling its gate. 0 is mapped to ground and 1 is to V_{DD}. Adding the pass variables finalizes the whole logic circuit design.

3.5 BiCMOS STATIC LOGIC CIRCUITS

BiCMOS is an important technology derived from the mainstream CMOS technology. In this section, fundamentals of the BiCMOS static logic circuits are described.

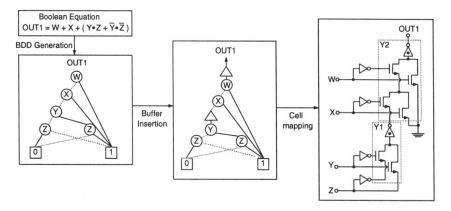

Fig. 3.27 Design procedure in realizing a logic function: $f = w + x + (y \cdot z + \bar{y} \cdot \bar{z})$. (Adapted from Yano et al. [12].)

3.5.1 Standard BiCMOS

Fig. 3.30 shows the circuit diagram of a standard 5V BiCMOS inverter, which is designed to drive a large capacitive load. As shown in the figure, NMOS device N_1 and bipolar device Q_2 constitute the BiNMOS device to enhance the driving capability during the pull-down transient. PMOS device P_1 and bipolar device Q_1 form a BiPMOS device, which is used to strengthen the driving capability during the pull-up transient. The npn bipolar devices Q_1 and Q_2 form a totem-pole circuit configuration. NMOS device N_0 is used to help evacuate the mobile minority carriers from the base of Q_1 during the turn-off process of Q_1. NMOS device N_2 is used to help removal of minority carriers from the base of Q_2 during the turn-off process of Q_2.

As shown in Fig. 3.30, this BiCMOS inverter is a design based on using a supply voltage of 5V. Fig. 3.31 shows the speed performance of the BiCMOS inverter. Also shown in the figure is the speed performance of the CMOS circuit plotted for comparison. From the figure, for a load capacitance of greater than 0.2pF, the speed of the BiCMOS circuit is higher. At a load capacitance of 1pF, the speed advantage of the BiCMOS circuit over the CMOS one can be two times due to the enhancement in the current driving capability from the BiNMOS and BiPMOS devices. However, as the load capacitance is smaller than 0.1pF, BiCMOS may not be faster.

Fig. 3.32 shows the transient waveforms of the BiCMOS inverter as shown in Fig. 3.30, where a set of SPICE Level-3 device models for a 0.8μm BiCMOS technology are used. As shown in the figure, during the pull-up transient, when the bipolar device Q_1 turns on, its V_{be} stays at a fixed value, which implies that a stable collector current is supplied to charge the capacitive load. As a result, both the output voltage and the base voltage go up. One behavior worth pointing out is that the base voltage may exceed the supply voltage (5V) during the pull-up transient. This is the

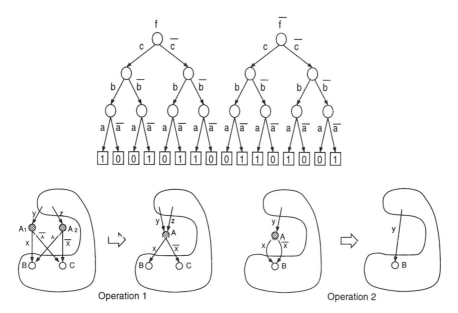

Fig. 3.28 Logic binary trees of a function $f = a \cdot b \cdot c + a \cdot \bar{b} \cdot \bar{c} + \bar{a} \cdot \bar{b} \cdot \bar{c}$ based on binary decision diagram (BDD). (Adapted from Kuroda and Sakurai [13].)

so-called 'internal voltage overshoot' [16], which can be explained as follows: During the pull-up transient, when the bipolar device Q_1 turns on, its base is full of electrons. Its output node is charged by the emitter current of Q_1. When the output voltage goes up, the base voltage rises accordingly. When the output voltage is close to 5V, due to the base-emitter capacitance coupling effect, the base voltage may exceed 5V— the internal voltage overshoot. At the internal voltage overshoot, the drain voltage of the PMOS device may be higher than the potential of the n-well. As a result, the junction between the drain and the n-well may be forward biased. Current may flow into the n-well, which increases the possibility of latchup. When the internal voltage overshoot occurs at the base, the bipolar device Q_1 enters the saturation region. From Fig. 3.32, when the forward transit time (τ_f) becomes smaller, the internal voltage overshoot shrinks and the output swing also becomes smaller. After internal voltage overshoot, the output voltage is approaching 5V. However it can't reach 5V due to $V_{CE(sat)}$ of the bipolar device at saturation. As shown in Fig. 3.32, during the pull-down transient, the output voltage cannot reach 0V due to $V_{CE(sat)}$ of the bipolar device at saturation. Therefore, the output swing of the BiCMOS circuit cannot be full-swing as the power supply. This is a major drawback of the BiCMOS circuit, which is especially serious for low-voltage operation. In addition, due to the bottleneck in the turn-off process of the bipolar device, the switching speed of the overall BiCMOS circuit is degraded. The turn-off of the bipolar device is restricted by the evacuation of electrons from the base, which is via the base terminal. The attachment of the NMOS devices N_2 and N_0 to the base in the circuit is to provide an evacuation path for the electrons during the turn-off process such that the rate of

118 CMOS STATIC LOGIC CIRCUITS

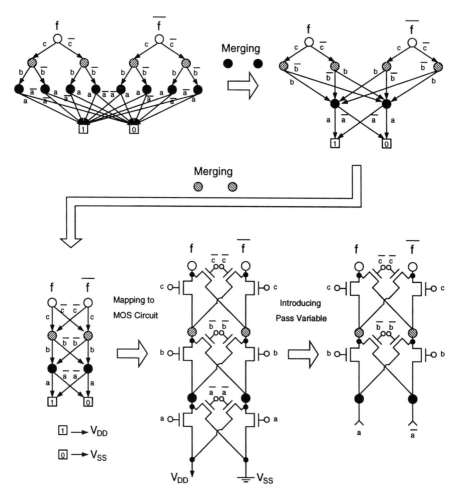

Fig. 3.29 Procedure to realize a logic function: $f = a \cdot b \cdot c + a \cdot \bar{b} \cdot \bar{c} + \bar{a} \cdot \bar{b} \cdot \bar{c}$ using DPL. (From Kuroda and Sakurai[13], ©1995 IEICE.)

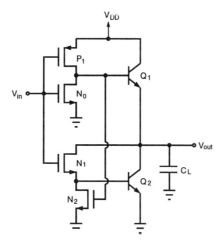

Fig. 3.30 5V BiCMOS inverter.

evacuation can be speeded up.

For the BiCMOS inverter circuit, one drawback is that due to the totem-pole configuration at the output formed by the bipolar devices, the circuit structure is not symmetric with respect to the power supply voltages. Since stable pnp bipolar devices are difficult to realize, the structures of BiNMOS and the BiPMOS devices are not symmetric to each other, which affects the circuit operation. In Fig. 3.30, the BiPMOS device is made of Q_1 and P_1, which are used in the pull-up transient. Therefore, Q_1 and P_1 are called the BiPMOS pull-up device. The BiNMOS device, which is made of Q_2 and N_1, is used in the pull-down transient. Therefore, Q_2 and N_1 are called the BiNMOS pull-down device.

As shown in Fig. 3.33, both the BiPMOS pull-up and the BiNMOS pull-down devices have an inverting capability. The basic concept of BiPMOS and BiNMOS, which came from bipolar Darlington pair, is targeted to increase the current driving capability of the MOS device. Indeed, compared to the MOS device, the transconductance of the BiPMOS and the BiNMOS devices is enhanced. However, the BiCMOS digital circuit operates at large signals. As a result, the enhanced small-signal transconductance cannot be fully utilized for the large-signal operation. In addition, the side effect from the buildup and the removal processes of the bipolar devices further limits the advantages of the BiCMOS circuits. Therefore, the speed performance is enhanced only 2–3 times. When the power supply voltage is lowered, the speed performance becomes lower.

In the BiPMOS device, the input is at the gate of the PMOS device. The source of the PMOS device and the collector of the bipolar device are connected to V_{DD}. The drain of the PMOS device is connected to the base of the bipolar device. The

120 CMOS STATIC LOGIC CIRCUITS

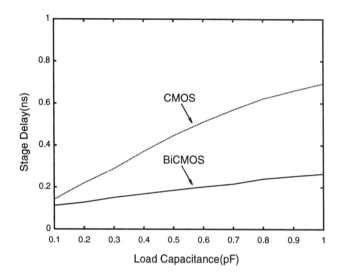

Fig. 3.31 Propagation delay versus the load capacitance of the BiCMOS inverter.

output is at the emitter of the bipolar device. Due to $V_{CE(sat)}$ of the bipolar device at saturation, the output voltage cannot reach V_{DD}. In the BiNMOS device, the input is at the gate of the NMOS device. The emitter of the bipolar device is connected to ground. The source of the NMOS device is connected to the base of the bipolar device. Both the drain of the NMOS device and the collector of the bipolar device are connected to the output. Also due to $V_{CE(sat)}$, the output voltage cannot reach ground. The non-full-swing of the output voltage of the BiCMOS inverter prevents itself from the low-voltage applications.

As described in the previous paragraphs, internal voltage overshoot may exist at the base node of the BiPMOS device during the pull-up transient. However, there is no internal voltage 'undershoot' at the base node of BiNMOS device during the pull-down transient. The main reason is that the BiNMOS pull-down device is not symmetric to the BiPMOS pull-up device. If the pnp bipolar device can be used in the BiNMOS pull-down structure, the 'internal voltage undershoot' at the base node of BiNMOS device during the pull-down transient may also occur.

Pull-up and pull-down speeds of the BiCMOS inverter are influenced by the change in the internal electron distribution in the bipolar devices. Fig. 3.34 shows the total charge in the bipolar devices during the pull-down and the pull-up transients of the BiCMOS inverter circuit based on 2D device-level simulation[14][15]. As shown in the figure, during the pull-down transient, the bipolar device turns on quickly. After turn-on of the bipolar device, the quantity of the total electrons in the base increases rapidly. After the quantity of the electrons reaches the peak, it decreases quickly, although the bipolar device still remains on. Therefore, during the pull-down transient, when the bipolar device is on, the quantity of the total electrons in the base does not

BiCMOS STATIC LOGIC CIRCUITS   121

```
.MODEL nch NMOS (LEVEL=3 NFS=2E11 UO=556 VTO=0.8 TOX=15N NSUB=5.3E16
+ THETA=0.078 XJ=0.530U WD=0.05U LD=0.06U VMAX=150K ETA=0.025 KAPPA=0.085
+ ACM=2 RSH=55 LDIF=0.25U TPG=1.0 PB=0.85 J=350U CJSW=245P MJ=0.39
+ MJSW=0.25 CGSO=380P CGDO=380P HDIF=1.9U DELTA=0.8932)

.MODEL pch PMOS (LEVEL=3 NFS=1E11 UO=180 VTO=-0.8 TOX=15N NSUB=3.3E16
+ THETA=0.12 XJ=0.06U WD=0.025U LD=0.06U VMAX=158K ETA=0.037 KAPPA=1.2
+ DELTA=1.9 ACM=2 RSH=90 LDIF=0.25U TPG=-1.0 PB=0.81 CJ=595U CJSW=370P
+ MJ=0.46 MJSW=0.30 CGSO=400P CGDO=400P HDIF=1.9U)

.MODEL bjt NPN (IS=4E-17 NF=1 BF=100 IKF=1.5E-3 IKR=1E-3 NE=1.5
+ BR=1 NR=1 NC=1.5 RB=120 RE=20 RC=75 VJE=0.7 MJE=0.44 VJC=0.75
+ MJC=0.5 VJS=0.7 MJS=0.5 EG=1.11 CJE=20F CJC=22F CCS=47F TF=10P TR=100P)

Vdd 100 0 dc 5V
Vin 1 0 PULSE(5 0 1N 0.1N 0.1N 2N 4N)
M1 2 1 100 100 PCH W=16U L=1U AD=32P AS=32P PD=20U PS=20U
M2 2 1 0 0 NCH W=4U L=1U AD=8P AS=8P PD=8U PS=8U
M3 3 1 4 0 NCH W=8U L=1U AD=16P AS=16P PD=12U PS=12U
M3 4 2 0 0 NCH W=4U L=1U AD=8P AS=8P PD=8U PS=8U
Q1 100 2 3 BJT
Q2 3 4 0 BJT
CL 3 0 1P
.IC V(3)=0.5
.TRAN 10P 5P
.PROBE
.END
```

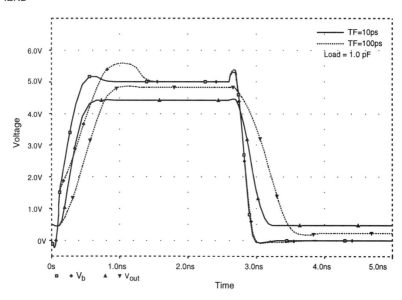

Fig. 3.32 Transients of the BiCMOS inverter circuit.

122 CMOS STATIC LOGIC CIRCUITS

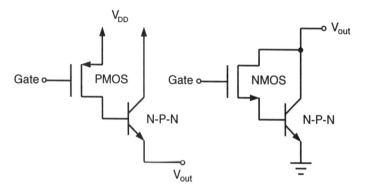

Fig. 3.33 BiPMOS pullup device and BiNMOS pulldown device.

stay at a fixed value. Instead, even during the on state, charge buildup and removal can be seen. During the pull-up transient, a similar situation can be identified. During the transients, the buildup and the removal of the electrons in the base of the bipolar devices strongly affect the speed performance of an overall BiCMOS inverter circuit.

Until now, the basic concepts of BiCMOS circuits have been described. In order to obtain insights into the internal voltage overshoot phenomenon and its implication on the speed performance of a BiCMOS circuit, the pull-up transient analysis of the BiCMOS inverter circuit is described. For a BiCMOS inverter circuit as shown in Fig. 3.30, Fig. 3.35 shows the waveforms during the pull-up transient [16]. During the pull-up transient, the most important devices are the PMOS device (P_1) and the bipolar device (Q_1) as shown in Fig. 3.30. During the pull-up transient, the input voltage drops from 5V to 0V. When the input reaches the final state, the bipolar device is in the initial base charge build-up period. During this period, the base voltage goes up quickly. When the bipolar device turns on, the quantity of the electrons in the base increases steadily. From this figure, the slope of the output curve is always positive, which implies that the load capacitance at the output is continuously charged by the emitter current of Q_1. Due to the base-emitter capacitance, the base voltage may exceeds 5V—the internal voltage overshoot. At the peak of the internal voltage overshoot, the slope of the base curve is zero:

$$\frac{dV_b}{dt} = 0. \qquad (3.23)$$

Since the base voltage is the sum of the output voltage and the base-emitter voltage ($V_b = V_{out} + V_{be}$), therefore, at the peak, the slope of the base-emitter voltage curve is opposite to the slope of the output voltage curve:

$$\frac{dV_{be}}{dt} = -\frac{dV_{out}}{dt}. \qquad (3.24)$$

Since the output voltage is always increasing ($\frac{dV_{out}}{dt} > 0$), at the peak of the internal voltage overshoot, the bipolar device is removing its minority carriers from the base

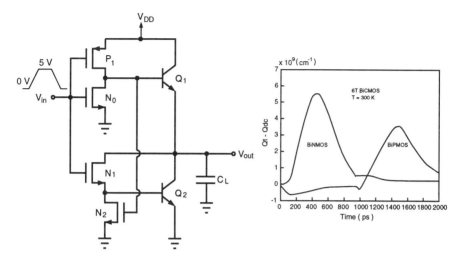

Fig. 3.34 Quantity of the total electrons in the bipolar device of the BiCMOS inverter during the transients.

($\frac{dV_{be}}{dt} < 0$) via the PMOS device P1 since the drain voltage of P1 is larger than V_{DD}—the drain current flows from the base of Q_1 to V_{DD}. This implies that the time when the base has the most electrons is earlier than the time when the internal voltage overshoot reaches its peak. When the base voltage exceeds 5V, the bipolar device is saturated. After the peak of the internal voltage overshoot, the output voltage gradually reaches its final value. During the pull-up transient, the internal voltage overshoot is an important phenomenon. In addition, the base charge buildup and removal process determines the switching speed.

Fig. 3.36 shows a BiCMOS NAND circuit and a general BiCMOS logic circuit. As shown in the figure, the BiPMOS and the BiNMOS devices are still the basic components. Only two extra PMOS devices connected in parallel and two extra NMOS devices connected in series are used to accommodate the logic function at the inputs. In addition, two NMOS devices, instead of one, are connected to the base of Q_1 to help the charge removal process during the turn-off transient. Since the electron mobility is larger than the hole mobility, a proper design of the aspect ratios for the PMOS and the NMOS devices is needed to facilitate a symmetric switching time during pull-up and pull-down transients. In the general BiCMOS logic circuit as shown in the figure, N-logic#1 is used to keep the bipolar device Q_1 off when necessary and N-logic#2 is used to turn on Q_2. Thus the device aspect ratio of the N-logic#2 should be larger than that of N-logic#1.

124 CMOS STATIC LOGIC CIRCUITS

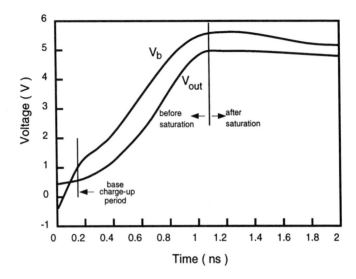

Fig. 3.35 Pull-up transient of the BiCMOS inverter. (Adapted from Lu and Kuo[16])

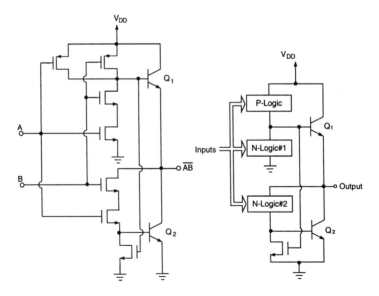

Fig. 3.36 BiCMOS NAND circuit and a general BiCMOS logic circuit.

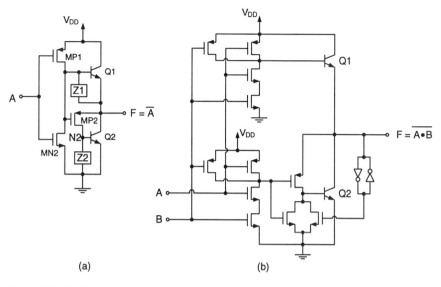

Fig. 3.37 (a) Quasi-complementary BiCMOS inverter; and (b) its NAND gate. (Adapted from Yano et al. [17].)

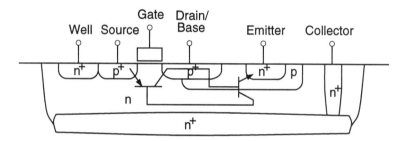

Fig. 3.38 Cross-section of the merged BiPMOS device. (Adapted from Ritts et al. [18].)

3.5.2 Sub-3V BiCMOS

Fig. 3.37 shows the quasi-complementary BiCMOS inverter and its NAND gate[17]. As shown in the figure, instead of the BiNMOS device in the pull-down path as shown in the 5V BiCMOS structure, the BiPMOS device has been used. In the BiNMOS device, due to the V_{BE} drop of the bipolar device and the threshold voltage of the NMOS device in addition to the non-full-swing of the input, the gate overdrive voltage of the MOS device is small if the supply voltage is shrunk. Using this BiPMOS device to replace the BiNMOS device, the small gate overdrive problem can be reduced. Since the V_{SG} of the BiPMOS device (MP2) is referred to the output node instead of the input node, the BiPMOS pull-down device can work well at a reduced supply voltage. In the quasi-complementary BiCMOS inverter, both pull-up and pull-down paths use the BiPMOS devices. As shown in Fig. 3.38, the bipolar device and the PMOS device can be merged in the same n-well[18].

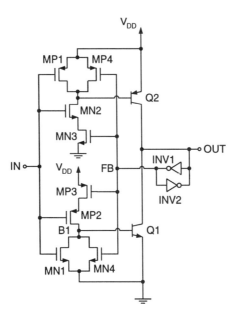

Fig. 3.39 1.5V BiCMOS buffer circuit with transient feedback. (Adapted from Hiraki et al. [19].)

3.5.3   1.5V BiCMOS Logic with Transient Feedback

Fig. 3.39 shows the 1.5V BiCMOS buffer circuit with transient feedback[19]. As shown in the figure, the BiPMOS pull-down device, which is composed of the PMOS device $MP2$ and the npn bipolar device Q_1, has been used in the pull-down structure. In the pull-up structure, a complementary BiNMOS pull-up device, which is composed of the NMOS device $MN2$ and the pnp bipolar device Q_2, has been used. In addition, a feedback circuit, which is composed of two CMOS inverters connected back to back, functioning as a latch, has been added to provide the feedback signal FB and to enhance the stability of OUT. As described before, the BiPMOS pull-down device has the full-swing capability since the npn bipolar device can be saturated. Due to the saturation of the npn bipolar device Q_1, the pull-up of OUT via the pnp device Q_2 is affected. Therefore, the speed performance is degraded. By adding a transistor $MN4$ controlled by the feedback signal FB to help evacuate the minority carriers from the base of the npn bipolar device Q_1 after OUT has been pulled down, the pull-up transient of Q_2 is not affected by the influence from the saturation of Q_1. The BiPMOS pull-down device and the BiNMOS pull-up device are fully complementary. Before a step from high to low is imposed at the input, the output of the buffer is maintained at high by the latch. At this time, the feedback signal FB is low, $MP3$ is on and $MN4$ is off. Both the pnp and the npn bipolar devices are turned off. After the input becomes low, $MN1$ is off and $MP2$ is on. Therefore, the drain current of $MP2$ and $MP3$ is charging the base of the npn bipolar device Q_1. As a result, the npn bipolar device Q_1 turns on. Consequently,

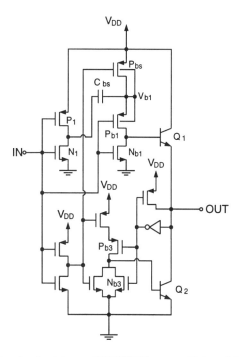

Fig. 3.40 1.5V full-swing bootstrapped BiCMOS inverter. (Adapted from Chik and Salama [20].)

the output is being pulled to low. At the same time, $MP1$ becomes on and $MN2$ is off. Therefore, the base voltage of the pnp bipolar device is V_{DD}. Therefore, the pnp bipolar bipolar device Q_2 is off. When the output is low, via the latch feedback, the feedback signal FB switches from low to high. Hence, $MN4$ turns on to remove excess minority carriers from the base of the npn bipolar device Q_1—this is the so-called 'transient saturation technique'. Therefore, the npn bipolar device is turned off. Without the feedback circuit, the bipolar device cannot turn off when the pull-down transient is over. Consequently, during the following pull-up transient, the removal of the minority carriers stored in the base of the pull-down npn bipolar device may not degrade the speed of the pull-up transient. For the pull-down npn bipolar device, it is turned on only during the pull-down transient. When the output settles to a stable low value, the pull-down npn bipolar device is turned off. Pull-up transient has a complementary behavior. For simplicity, the pull-up transient is not analyzed.

3.5.4 1.5V Bootstrapped BiCMOS Logic

Fig. 3.40 shows the 1.5V full-swing bootstrapped BiCMOS inverter[20]. As shown in the circuit, it is based on the BiPMOS devices for both the pull-down and the pull-up structures as described in the quasi-complementary BiCMOS inverter. In addition,

128 CMOS STATIC LOGIC CIRCUITS

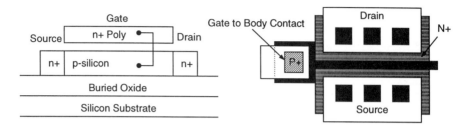

Fig. 3.41 SOI CMOS device structure with its body tied to gate. (Adapted from Assaderaghi et al. [24].)

a feedback circuit and a bootstrap technique have been used. During the pull-down transient, the pull-down BiPMOS device, which is composed of P_{b3} and the bipolar device Q_2, turns on. When the output settles down to a stable state, via the feedback inverter, N_{b3} turns off the bipolar device Q_2—transient saturation technique. The bootstrap technique, which requires the bootstrap capacitor C_{bs}, has been used to raise the output high level during the pull-up transient. When the input is high, N1 is on, hence, the left side of the bootstrap capacitor C_{bs} is discharged to ground. At the same time, since P_{bs} is on, the right side of the bootstrap capacitor C_{bs} is pulled to V_{DD}. Since input is high, N_{b1} is on and V_{b1} is low. In addition, P_{b1} is off. Therefore, the bootstrap capacitor C_{bs} contains an amount charge, which is equal to $C_{bs}V_{DD}$. When the input becomes low, P_1 turns on and N_1 turns off. At the same time, P_{b1} is on and N_{b1} is off. In addition, P_1 forces the left side of the bootstrap capacitor C_{bs} to V_{DD} suddenly. Due to the conservation of charge, the right side of the C_{bs} will rise to surpass V_{DD}. Since P_{b1} is on, V_{b1} will also surpass V_{DD}. As a result, the bipolar device Q_1 can provide its emitter current to charge the output node to near V_{DD}. The value of the bootstrap capacitor determines the amount of the voltage overshoot at the base node of Q_1. In order to avoid forward bias of the source/body junction in P_{bs} and P_{b1}, the body of them is not connected to V_{DD}. Instead, they are connected to the right side of C_{bs}. Different from the pull-down transient, the bipolar device Q_1 does not turn off completely after settle-down of the pull-up transient.

3.6 SOI CMOS STATIC LOGIC

SOI CMOS technology has been used to integrate VLSI circuits. Compared to bulk CMOS devices, SOI CMOS devices have smaller short-channel effects, smaller parasitic capacitances, and better subthreshold slopes owing to the buried oxide isolation structure. SOI CMOS devices have potentials in realizing high-speed, low-voltage, and low-power VLSI circuits. In this section, SOI CMOS static logic circuits are described.

Fig. 3.41 shows the SOI NMOS device structure with it body tied to gate using the dynamic threshold voltage technique[24]. As shown in the figure, the body is

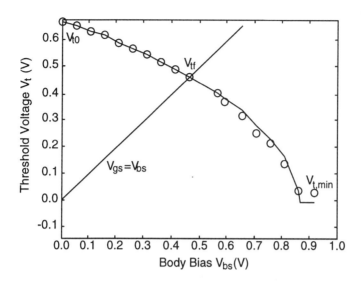

Fig. 3.42 Threshold voltage versus body-source voltage of an SOI NMOS device. (From Assaderaghi et al.[24], ©1997 IEEE.)

connected to the gate— $V_{bs} = V_{gs}$. Therefore, by controlling its gate voltage, the body potential of an SOI CMOS device can be controlled. As shown in Fig. 3.42[24], the threshold voltage of the SOI NMOS device is lowered when a positive body bias is applied. Specifically, if the body is tied to the gate, when the gate voltage is high, its threshold voltage can be lowered by about 0.18V, which can provide a larger drain current. As shown in Fig. 3.41, by tying the body to the gate, due to a drop in the threshold voltage, the gate overdrive voltage becomes larger. As a result, the speed performance of the inverter can be enhanced. In addition, owing to lowering of the threshold voltage during the transient, this SOI CMOS inverter is suitable for low-voltage operation. On the other hand, this SOI circuit also has weaknesses. Since the body is connected to the gate, the previous stage, which drives the gate, needs to absorb the extra body leakage current. In addition, the supply voltage is limited to below 0.6V such that the forward bias of the body-source junction can be avoided. Otherwise, when the body-source junction is forward biased, a large drain current may flow through the source contact, thus power dissipation is increased. Due to the increase in the leakage current, driving of the previous stage seems to be more difficult.

In order to overcome the weakness of the SOI CMOS inverter with its body tied to gate as shown in Fig. 3.41, Fig. 3.43 shows an SOI CMOS inverter with the dynamic threshold technique[25]. As shown in the figure, two auxiliary transistors driven by the input voltage are added to connect the bodies to the drain. With this circuit configuration, the previous stage circuit, which drives the gate, does not need to absorb the body leakage current any more. In addition, the forward bias of the body-source/drain junction can be controlled to below 0.6V since the auxiliary transistors may turn on. Therefore, this circuit can work at a higher supply voltage ($> 0.6V$).

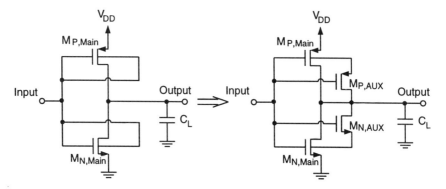

Fig. 3.43 SOI CMOS inverter with the dynamic threshold technique. (Adapted from Chung et al. [25].)

The operation of this SOI CMOS inverter is described here. During the pull-down transient, the input switches from low to high. Under this situation, both main and auxiliary NMOS devices turn on. If the output load capacitance is much larger than the body capacitance of the main transistor, the body potential of the main transistor is pulled to high earlier by the auxiliary transistor. (Note that at this time the output node is not pulled low yet.) Therefore, the threshold voltage of the main transistor drops due to the pull-high of the body potential. As a result, the larger drain current of the main transistor pulls the output to low faster. Owing to the turn-on of the auxiliary transistor, the body potential of the main transistor also drops in accordance with the fall of the output voltage. Therefore, the threshold voltage of the main transistor rises. When the output voltage reaches the ground level, the body potential is also 0V. At this time, the threshold voltage is restored to its zero back-gate-bias value. When the input switches to low, both main and auxiliary NMOS transistors turn off. Under this situation, the body potential of the main NMOS transistor remains at the ground level and floating and its threshold voltage maintains its maximum value—minimum leakage current. The internal potential distribution in the body of the main transistor depends on the distance from the body contact. Due to the RC effect, at a location farther away from the body contact in the silicon thin-film of the main transistor, its body potential is less influenced from the auxiliary transistor. Therefore, a main transistor with a smaller equivalent RC time constant of the silicon thin-film region can benefit more from the dynamic threshold effect. In addition, the body potential is also influenced by the coupling via the gate-body capacitance and the clock feedthrough via the auxiliary transistor. This SOI CMOS inverter circuit with the dynamic threshold technique is especially suitable for driving a large capacitive load.

The previous SOI CMOS inverter can be further improved. Fig. 3.44 shows an SOI CMOS buffer[26]. For the SOI CMOS inverter used as a buffer to drive a large load in a large VLSI system, there may be a chain of cascading inverters involved. The propagation delay between inverters may be substantial. As shown in Fig. 3.43,

SOI CMOS STATIC LOGIC   **131**

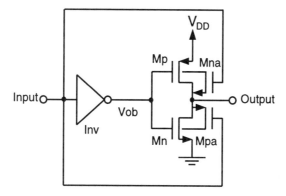

Fig. 3.44 SOI CMOS buffer. (Adapted from Houston [26].)

the auxiliary transistors, which are used to lower the threshold voltage of the main transistors, are driven by the input voltage. Due to the propagation delay, the function of the auxiliary transistor may be deferred. As a result, the speed enhancement of the SOI CMOS inverter as shown in Fig. 3.43 is not fully utilized. As shown in Fig. 3.44, if the gates of the auxiliary transistors are controlled by the input signal to the previous stage instead of the present stage, the auxiliary transistor can function in time to shrink the threshold voltage of the main transistor during the transient of the present stage[26]. As a result, the dynamic threshold technique can be fully utilized. In order to accommodate this new technique, the SOI circuit also needs to be modified accordingly. First, the auxiliary NMOS and the PMOS devices should be reversed such that the inverting function of the previous stage can be taken into account. Therefore, including the function of the inverter of the previous stage, the whole circuit becomes an SOI CMOS buffer using the dynamic threshold MOS devices.

Fig. 3.45 shows a multi-threshold SOI CMOS logic circuit suitable for operation at a low supply voltage such as 0.5V[27]. As shown in the figure, this SOI CMOS circuit is composed of two portions—the power switch transistor and the logic block. In the logic block, all transistors are fully-depleted SOI CMOS devices with a smaller magnitude in the threshold voltages. The supply line to the logic block is called virtual V_{DD}, which is connected to V_{DD} of 0.5V via the power switch transistor, which is a partially-depleted SOI MOS device with a larger threshold voltage. Its gate is controlled by the SL (sleep) signal. When SL is high, which indicates the circuit enters the sleep mode, the circuit does not have any active operation. Only when SL is low does the circuit enter the active mode. During the sleep mode, the power switch transistor is turned off to save power. The larger threshold voltage guarantees a smaller leakage current in the turned-off power switch transistor. As shown in the figure, a diode-connected auxiliary fully-depleted SOI MOS device based on the dynamic threshold technique has been used to tie the body of the partially-depleted power switch transistor to gate. As a result, during the turn-on period the threshold

132 CMOS STATIC LOGIC CIRCUITS

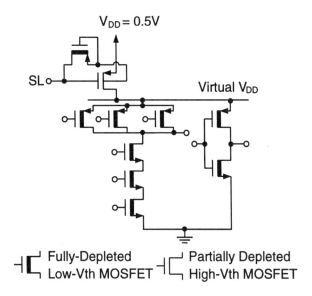

Fig. 3.45 Multi-threshold CMOS logic circuit suitable for operation at a low supply voltage such as 0.5V. (Adapted from Douseki et al. [27].)

voltage of the high-threshold power switch transistor can be lowered. Hence, when the virtual V_{DD} is connected to V_{DD}, its voltage difference between V_{DD} and virtual V_{DD} is small. Based on the multi-threshold CMOS/SIMOX circuit scheme, a 16-bit arithmetic logic unit (ALU) based on a $0.25\mu m$ CMOS technology has been integrated. At a supply voltage of 0.5V, a speed of 50MHz has been reached[27]. The power consumption is only 0.35mW. During the sleep mode, the power consumption is less than 5nW.

3.7 LOW-VOLTAGE CMOS STATIC LOGIC CIRCUIT TECHNIQUES

For next-generation deep-submicron CMOS VLSI circuits, low supply voltage and low power are the trend. In the past, for $1\mu m$ CMOS technology, a 5V supply voltage is used. The power supply voltage has been scaled down to 3.3V for $0.5\mu m$ CMOS technology. For sub-$0.1\mu m$ CMOS technology, 1.5V or below is necessary. A down-scaled supply voltage reduces power consumption. However, speed performance of a logic gate using deep-submicron CMOS devices decreases as the supply voltage is scaled down since the threshold voltage of a deep-submicron MOS device cannot be scaled down accordingly with the supply voltage. In this section, various low-voltage CMOS static logic circuit techniques are described.

3.7.1 Bootstrapped CMOS Driver

Bootstrap technique has been used in low-voltage CMOS VLSI circuits[28]. Fig. 3.46

LOW-VOLTAGE CMOS STATIC LOGIC CIRCUIT TECHNIQUES   133

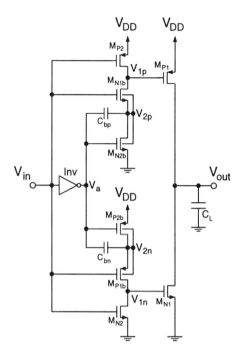

Fig. 3.46 1.5V full-swing bootstrapped CMOS driver circuit. (Adapted from Lou and Kuo [28].)

shows the 1.5V full-swing bootstrapped CMOS driver, which is composed of the fundamental segment and the bootstrap segment[28]. In the fundamental segment, there are two PMOS devices—M_{P1}, M_{P2} and two NMOS devices—M_{N1}, M_{N2}. In the bootstrap segment, there are two NMOS devices—M_{N1b}, M_{N2b} and two PMOS devices—M_{P1b}, M_{P2b}. In addition, a CMOS inverter and two bootstrap capacitances C_{bp} and C_{bn} are also included in the bootstrap segment, where M_{N1b}, M_{N2b}, and C_{bp} are used for the pull-up transient; M_{P1b}, M_{P2b}, and C_{bn} are used for the pull-down transient.

Fig. 3.47 shows the transient waveforms of the 1.5V full-swing bootstrapped CMOS driver circuit driving an output load of 10pF during the pull-up transient[28]. As shown in the figure, the right side of the bootstrap capacitor C_{bp} (V_{2p}) can go below 0V during the transient. During the pull-up transient, the operation of the full-swing bootstrapped CMOS driver circuit is divided into two periods regarding the bootstrap capacitor C_{bp}: (1) the charge build-up period, and (2) the bootstrap period. Fig. 3.48 shows the equivalent circuit of the 1.5V full-swing bootstrapped CMOS driver circuit at the time prior to the pull-up transient and after the input ramp-up period[28]. As shown in the figure, prior to the pull-up transient at t_1, the input is at 0V and at the output of the inverter V_a is at 1.5V. Therefore, M_{N1b} and M_{N2} are off; M_{N2b} is on. Consequently, at the output of the driver, V_{out}, which is indirectly

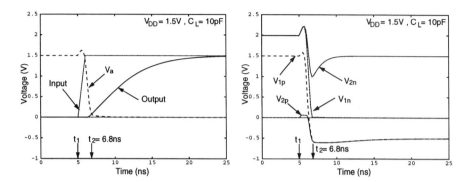

Fig. 3.47 Transient waveforms of the 1.5V full-swing bootstrapped CMOS driver circuit during the pull-up transient. (Adapted from Lou and Kuo[28].)

driven by M_{P1b} in the bootstrap segment, is at 0V. On the other hand, M_{N2b} and C_{bp} of the bootstrap segment are separated from the M_{P2} and M_{P1} of the fundamental segment. As a result, the bootstrap capacitor C_{bp} has a charge of $1.5C_{bp}$ since the left side is pulled to 1.5V and the right side V_{2p} to 0V (M_{N2b} is on). As shown in Fig. 3.48(b), after the input ramp-up period, the output of the inverter V_a changes to 0.075V. Therefore, the right side of the bootstrap capacitor C_{bp} is disconnected from ground since M_{N2b} is off. Instead, it's connected to the gate of M_{P1} since M_{N1b} is on. Due to the voltage change at the left side of the bootstrap capacitor C_{bp} from 1.5V to 0.075V ($V_a = 1.5V \rightarrow 0.075V$) from t_1 to t_2, the right side of the bootstrap capacitor C_{bp} changes from 0V to $-0.58V$ ($V_{2p} = 0V \rightarrow -0.58V$)—"the internal voltage undershoot". (Note that the extent of the internal voltage undershoot is determined by the ratio of C_{bp} to the parasitic capacitance at the right side of C_{bp}.) As a result, the output voltage can switch at a faster pace since the gate of M_{P1} is driven at $-0.56V$. As shown in Fig. 3.47, the output voltage can go up to the full-swing value—1.5V. Pull-down transient has a complementary configuration.

Fig. 3.49 shows the pull-up and the pull-down delay times versus load capacitance of the 1.5V full-swing bootstrapped CMOS driver during the pull-up and the pull-down transients[28]. All MOS transistors in Fig. 3.46 have a channel length of $0.8\mu m$. The channel widths of M_{P1} and M_{N1} are $80\mu m$ and $40\mu m$, respectively. The channel width of M_{N1b}, M_{P2} and M_{P2b} is $20\mu m$. The channel width of M_{P1b} is $40\mu m$. The channel width of M_{N2b} and M_{N2} is $10\mu m$. The capacitances of the two bootstrap capacitances C_{bp} and C_{bn} are 0.35pF and 0.25pF, respectively. The threshold voltage of the CMOS devices is $0.75V$ for NMOS and $-0.9V$ for PMOS. Also shown in the figure are the delay times for the conventional CMOS driver circuit using a PMOS device of $80\mu m/0.8\mu m$ and an NMOS device of $40\mu m/0.8\mu m$. As shown in the figure, the full-swing bootstrapped CMOS driver provides a switching speed improvement of 2.6× and 1.7× for the pull-up and pull-down transients, respectively. In the circuit, the body of M_{N1b} and M_{N2b} is connected to V_{2p}, instead of 0V. Therefore, the internal voltage undershoot at V_{2p} does not cause forward bias of

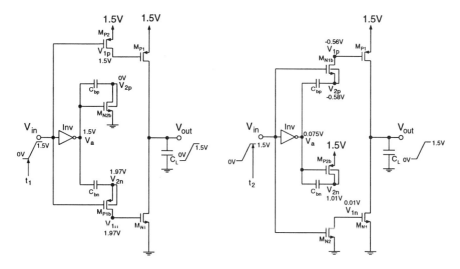

Fig. 3.48 Equivalent circuit of the 1.5V full-swing bootstrapped CMOS driver circuit at the time (a) prior to the pull-up transient; and (b) after the input ramp-up period. (Adapted from Lou and Kuo [28].)

the source/drain-substrate junction. (Note that an individual p-well is used for M_{N1b} and M_{N2b}.) Similarly, the body of M_{P1b} and M_{P2b} is connected to V_{2n} instead of 1.5V. Therefore, the internal voltage overshoot at V_{2n} does not cause forward bias of the source/drain-substrate junction.

3.7.2 Multi-Threshold Standby/Active Techniques

Lowering power supply voltage is an effective way to reduce power dissipation—low-power. However, propagation delay time increases when the supply voltage is lowered. At a low supply voltage, in order to maintain a high-speed performance, utilization of the CMOS devices with a smaller threshold voltage for the VLSI application is the trend. At a smaller threshold voltage, DC power dissipation due to the subthreshold leakage current becomes more important. Recently, several power-down techniques have been used to meet the low-power requirement. When the system is not actively used, its clock is stopped or slowed down. However, due to the standby leakage current, there is still power consumed even when circuit operation is halted. For example, for a mobile telephone set, most of the time it is in the standby (sleep) mode. Thus, the DC leakage power cannot be neglected in a mobile telephone set. Multi-threshold technique can be used to reduce this problem. Fig. 3.50 shows the circuit diagram of a multi-threshold CMOS circuit[30]. As show in the figure, two kinds of devices have been used. One kind is with a low threshold voltage. The other is with a high threshold voltage. The low-threshold devices are used to realize the logic block such that switching speed can be enhanced. The logic block is not connected to V_{DD} and ground directly. Instead, it is connected to the virtual power

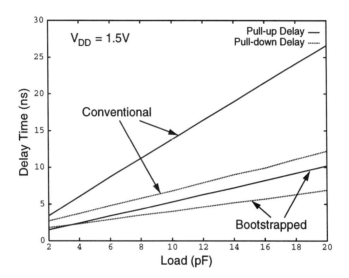

Fig. 3.49 Pull-up and pull-down delay times versus load capacitance of the 1.5V full-swing bootstrapped CMOS driver. (Adapted from Lou and Kuo [28].)

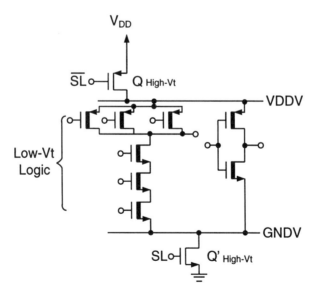

Fig. 3.50 Circuit diagram of a multi-threshold CMOS circuit. (Adapted from Shigematsu et al. [30].)

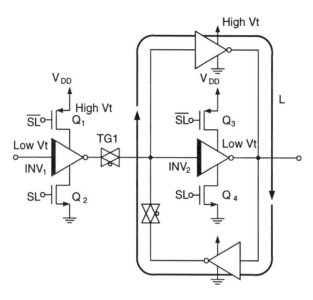

Fig. 3.51 Circuit diagram of a multi-threshold CMOS latch. (Adapted from Shigematsu et al.[30].)

lines V_{DDV} and $GNDV$, which are connected to V_{DD} and ground via the power switch transistors with a high threshold voltage, respectively. Via the sleep mode control (SL) signal, the circuit can be operating in active or standby (sleep) mode. When SL is high (active mode), the power transistors are on, and V_{DDV}/$GNDV$ functions as V_{DD}/ground of the logic block. Under this situation, the low-threshold devices in the logic block work at an enhanced speed. When SL is low (standby mode), the logic block is separated from the power lines V_{DD} and GND, controlled by the power switch transistors. Owing to the high threshold voltage of the power switch transistors, the standby leakage currents are small. During the standby operation, since the logic gates in the logic block are floating, data may be lost. Therefore, when the circuit enters the standby mode, some important data should be stored. The multi-threshold CMOS latch[30] as shown in Fig. 3.51 is created for this purpose. As shown in the figure, latch L, which is used to store the data during the standby mode, is connected to V_{DD} and GND directly via the high-threshold devices such that standby leakage can be minimized. CMOS transmission gate TG1, which is located in the critical path from the input to the output, is used to keep two logic blocks from shorting. Therefore, the high-threshold device should be used for TG1. Q_1–Q_4 are used to turn off two inverters in the standby mode. Therefore, high-threshold devices should also be used for Q_1–Q_4.

As described before, high-threshold devices can be used to reduce leakage current when they are off. However, when the circuit enters the active mode from the standby mode, speed performance may be degraded due to the influence from the high-threshold device. As a result, usually they are designed as the critical compo-

138 CMOS STATIC LOGIC CIRCUITS

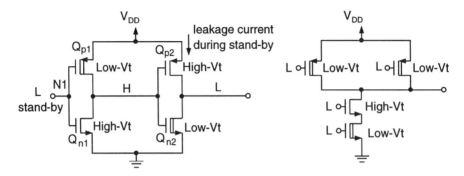

Fig. 3.52 Inverter and NAND gate using the multi-threshold logic circuit and the standby/active mode logic. (Adapted from Takashima et al. [31].)

nents during the standby mode operation. Due to their inferior driving capability, the use of the high-threshold devices should be minimized such that the speed performance of the active mode operation is not degraded. Fig. 3.52 shows the inverter and NAND gate using the multi-threshold logic circuit and the standby/active mode logic[31]. As shown in Fig. 3.52(a), during the standby mode, the input to the inverter is low, both Q_{n1} and Q_{p2} are off. Thus, both Q_{n1} and Q_{p2} need to have a high-threshold voltage. On the other hand, during the standby mode, Q_{p1} and Q_{n2} are on. Thus, low-threshold-voltage devices can be used.

Fig. 3.52(b) shows the case for the NAND gate. If during the standby mode operation, all inputs are tied to low. In order to minimize the standby leakage, it doesn't need all four devices to be high-threshold type. Since all NMOS devices are off, only one NMOS device, which is farthest from ground, needs to be high-threshold type. Under this circumstance, the switching performance will not be sacrificed too much during the active mode operation.

The multi-threshold approach is not the only way to implement the standby/active mode logic. The standby/active logic can also be realized using all low-threshold devices. Some gate inputs during the standby cycle are fixed (e.g., in some memory circuits). For the logic blocks, which are turned off during the standby cycle, they are connected to the virtual power lines. Using this arrangement, when entering from the standby mode to the active mode, the degradation of the speed performance is minimized. Fig. 3.53 shows the circuit diagram of the standby/active mode logic using all low-threshold devices[31]. As shown in the figure, the virtual power line concept has been adopted. Its concept is from the switched-source-impedance (SSI) CMOS circuit technique [32] as shown in Fig. 3.54. As shown in the figure, a resistance R_s is added. When switch S_s turns off, the leakage current is via R_s to ground. Thus, the source voltage of the NMOS device Mn increases—its body-source voltage (V_{sb}) becomes V_{SL} as shown. From the increase in the source voltage, its threshold voltage increases due to the body effect ($\Delta V_T = \gamma(\sqrt{V_{SL} + 2\phi_B} - \sqrt{2\phi_B})$). Therefore, the leakage current is decreased from I_{L0} to I_{L1}. In addition, V_{GS} changes from 0V

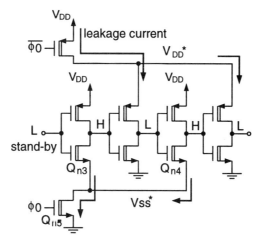

Fig. 3.53 Circuit diagram of the standby/active mode logic using all low-threshold devices. (Adapted from Takashima et al. [31].)

to $-V_{SL}$, thus its leakage current changes from I_{L1} to I_{L2}—a further decrease in the leakage current. As shown in Fig. 3.53, Q_{n5} has the function of SSI described in Fig. 3.54. During the standby mode, those devices, which are turned off, are connected to the virtual power lines, which are separated from the real power lines by the power switch transistors. When ϕ_0 is low in the standby mode, the leakage current from Q_{n3} and Q_{n4}, which are off, flows through the power transistor Q_{n5} to ground. Since Q_{n5} is also off, the leakage current from Q_{n3} and Q_{n4} raises the potential of the virtual ground line. Consequently, the gate-source voltage V_{GS} of Q_{n3} and Q_{n4} decreases and the threshold voltage rises. Therefore, the leakage current in Q_{n3} and Q_{n4} can be reduced substantially. Owing to its low threshold, during the active mode, the switching speed is high. One problem associated with the virtual power lines is the voltage bounce problem, which may lead to degradation of speed and noise margin. In addition, malfunction of the circuits may occur if the voltage bounce is too large.

The dynamic threshold MOS (DTMOS) described in the previous section can be used for the bulk CMOS technology. Fig. 3.55 shows the cross-section of a dynamic-threshold MOS (DTMOS) device[33] in bulk CMOS technology. This DTMOS device in bulk CMOS technology is derived from the DTMOS technique based on the SOI technology, which has been described in this chapter. As shown in the figure, in this bulk CMOS technology, a triple-well process is needed. Under the p-well, a deep n-well with its well connected to V_{DD} is placed. With the deep n-well, the p-well of the NMOS device—its body is connected to the gate voltage for the dynamic-threshold operation. However, with this deep n-well, the parasitic capacitance between the p-well and the deep n-well may degrade the speed performance. In contrast, in the SOI device, due to the buried oxide isolation structure, no such consideration should be included. Fig. 3.56 shows the cross-section of a bulk

140   CMOS STATIC LOGIC CIRCUITS

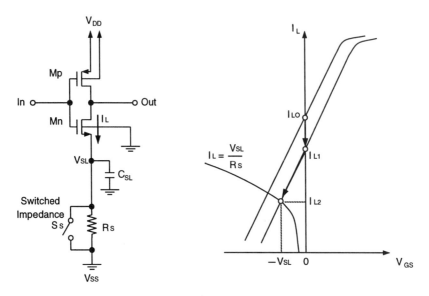

Fig. 3.54 Concept of the switched-source-impedance (SSI) CMOS circuit. (Adapted from Horiguchi et al. [32].)

dynamic-threshold MOS (DTMOS) device using a shallow/deep well technology[34]. As shown in the figure, shallow trench isolation has been used. In order to reduce the parasitic RC effect in the body contact path, a p^--p-p^- sandwiched structure has been adopted. In the p^--p-p^- structure, the center high-doping p-area is used to reduce the parasitic resistance in the body contact path. The upper p^- region is used to realize the low threshold voltage and to reduce the short channel effect. The lower p^- is used to reduce the parasitic capacitance between the shallow well and the deep well. By this p^--p-p^- sandwiched structure, performance of the bulk DTMOS device can be improved. Fig. 3.57 shows the subthreshold characteristics of the bulk DTMOS devices[34]. Also shown in the figure is the subthreshold behavior of the bulk NMOS device without the body connected to the gate. As shown in the figure, with the body tied to the gate, at the gate voltage of 0V, it has a smaller leakage current at the zero back-gate-biased threshold voltage. When the gate voltage rises, its body is forward biased, thus its threshold voltage becomes smaller. Thus, compared to the conventional bulk CMOS device, DTMOS has a steeper subthreshold curve, which is suitable for low-voltage operation. As for the circuit technique, it is similar to the one described in the previous section.

3.8 LOW-POWER CMOS CIRCUIT TECHNIQUES

In the previous section, low-voltage CMOS static logic circuit techniques have been described. Low-voltage and low-power are often related to each other. For a deep-submicron CMOS static logic circuit, the use of a low supply voltage is often also

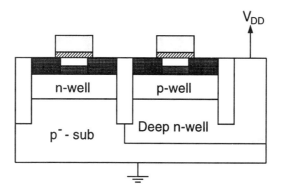

Fig. 3.55 Cross-section of a dynamic-threshold MOS (DTMOS) device in bulk CMOS technology. (Adapted from Wann et al. [33].)

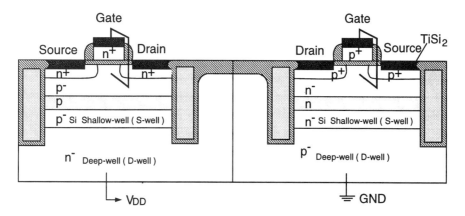

Fig. 3.56 Cross-section of a bulk dynamic-threshold MOS (DTMOS) device. (Adapted from Kotaki et al. [34].)

the requirement for reliability of deep-submicron CMOS devices. At a low supply voltage, low power consumption usually can be reached. On the other hand, for a low-power VLSI circuit, it does not necessarily need low-voltage. Low-power can be reached via techniques including circuit innovation, architecture optimization, in addition to lowering the power supply voltage, which is the most effective way to reduce power consumption. In this section, several low-power CMOS circuit techniques are introduced. Specifically, the low-swing bus architecture approach and the adiabatic logic circuits are described.

The trends on digital VLSI circuits are toward high speed, low power, and large integration. Due to the limited power dissipation capability of packaging, when increasing the size of integration, the power dissipation per transistor of the circuits involved should be reduced. As shown in Eq. (3.25), power dissipation in a digital CMOS VLSI circuit is composed of the power consumed during switching, the power

142 CMOS STATIC LOGIC CIRCUITS

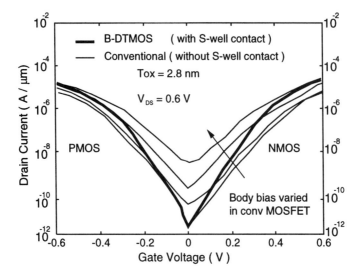

Fig. 3.57 Threshold voltage characteristics of the bulk DTMOS devices. (From Kotaki et al. [34], ©1996 IEEE.)

dissipated under the short-circuited condition, and the power lost in the leakage.

$$P_D = P_{switching} + P_{short-ckt} + P_{leakage}. \qquad (3.25)$$

The power dissipated under the short-circuited condition ($P_{short-ckt}$) is referred to the situation when pull-up and pull-down networks are turned on simultaneously. When the gate input is at ramp up/down, the short-circuited power dissipation exists. As for the switching power dissipation, when a down-step voltage waveform is imposed at its input of an inverter circuit, the energy drawn from the power supply for the positive output transition at the output is $C_L V_{DD}^2$, half of which is stored in the output capacitive load and the other half is dissipated in the PMOS device as the dissipated heat. When the input switches from low to high, at the negative going output transition, no energy is needed from the supply. However, an $\frac{1}{2} C_L V_{DD}^2$ amount of energy stored in the capacitive load is dissipated in the NMOS device. Thus, when the output has an up/down transition, a $C_L V_{DD}^2$ amount of energy is dissipated to the environment to cause the rise in the chip temperature. Conventionally, the requirement on the low-power design is on the performance and the reliability of the circuits. Therefore, for example, in the CPU design, lowering the power dissipation of the circuits has been used to reduce the electromigration problems of the metal interconnects for better reliability. In the DRAM design, lowering the power dissipation has been used to reduce the chip temperature such that the junction leakage current at the DRAM storage node can be shrunk. Recently, due to the booming in mobile telephone and computing, low power dissipation is used to lengthen the battery life—low energy requirement. Thus the low power concept has been evolved quickly.

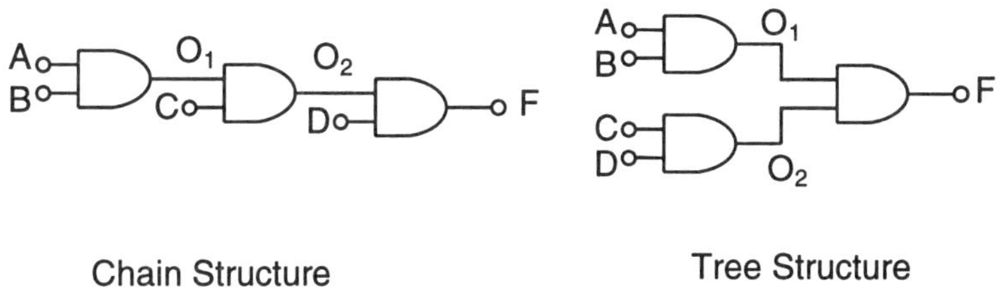

Chain Structure   **Tree Structure**

Fig. 3.58 Logic function $f = A \cdot B \cdot C \cdot D$ realized by various arrangements to demonstrate the influence of circuit topology on switching activity. (Adapted from Chandrakasan and Brodersen [35].)

In a digital CMOS VLSI circuit, the power consumed during switching can be expressed as a function of the probability of the output state changing from logic-0 to logic-1 ($\alpha_{0 \to 1}$), the load capacitance at the output (C_L), the supply voltage (V_{DD}), and the operating frequency (f):

$$P_{switching} = \alpha_{0 \to 1} C_L V_{DD}^2 f. \tag{3.26}$$

In order to raise system performance, the operating clock frequency has been increasing. From the above formula, increasing clock frequency cannot lower power consumption. Instead, it raises it. From Eq. (3.26), in order to reduce power dissipation, a low-power-supply voltage should be used. In addition, the output load capacitance should be minimized. The probability of switching ($\alpha_{0 \to 1}$) can be reduced via selection of the specific circuit design. Fig. 3.58 shows a logic function $f = A \cdot B \cdot C \cdot D$ realized by two arrangements to demonstrate the influence of circuit topology in switching activity[35]. Both chain and tree structures of NAND gates have been used. For an AND gate with two inputs a and b, the probability for the output having a logic-1 is

$$P_1 = P_a P_b, \tag{3.27}$$

where P_a and P_b are the probability for inputs a and b having a logic-1 value. The probability of the AND gate with its output changing from logic-0 to logic-1 is equal to

$$\alpha_{0 \to 1} = P_0 P_1 = (1 - P_a P_b) P_a P_b. \tag{3.28}$$

Therefore, based on the chain structure, the probability of the second AND output (O_2) having a logic-1 value is equal to

$$P_{O_2 c} = P_a P_b P_c = \frac{1}{8}. \tag{3.29}$$

Therefore, based on the chain structure, the probability of the second AND output (O_2) having a transition from logic-0 to logic-1 is

$$\alpha_{0 \to 1} c = (1 - P_{O_2 c}) P_{O_2 c} = \frac{7}{64}. \tag{3.30}$$

144 CMOS STATIC LOGIC CIRCUITS

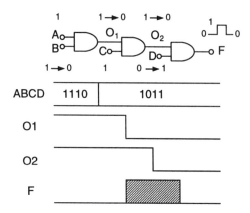

Fig. 3.59 Spurious glitches in a logic function $f = A \cdot B \cdot C \cdot D$ implemented with the chain-structure. (Adapted from Chandrakasan and Brodersen [35].)

Based on the tree structure, the probability of the second AND output (O_2) having a logic-1 value is equal to

$$P_{O_2 t} = P_c P_d = \frac{1}{4}. \tag{3.31}$$

Therefore, based on the tree structure, the probability of the second AND output (O_2) having a transition from logic-0 to logic-1 is

$$\alpha_{0 \to 1} t = (1 - P_{O_2 t})(P_{O_2 t}) = \frac{3}{16}. \tag{3.32}$$

From the above reasoning, the probability of the output of the second AND gate having a transition from logic-0 to logic-1 is different. If tree-structure is used to realize the logic function $f = A \cdot B \cdot C \cdot D$, the chance of having a transition from logic-0 to logic-1 at the output of the second AND is higher. Therefore, higher switching power can be expected at the output of the second AND gate in the tree-structure.

In addition to the probability of switching, glitches in a circuit also affect its power dissipation. Fig. 3.59 shows the spurious glitches in a logic function $f = ABCD$ in chain-structure[35]. For the logic circuit as shown in Fig. 3.59, consider the inputs (ABCD) switching from (1110) to (1011). When the input B switches from logic-1 to logic-0, the output of the first AND gate (O_1) changes from logic-1 to logic-0 with some delay. Since input C maintains its state at logic-1, the output of the second AND gate (O_2) will change from logic-1 to logic-0 with some delay. If input D changes its state from logic-0 to logic-1 at a time earlier than the logic-1 to logic-0 transition of O_2, then a spurious glitch at the output of the circuit may occur. In the chain-structure, due to the propagation delay associated with the output signal of the second AND gate: O_2, if the transition of the input D comes earlier than O_2, a spurious glitch occurs. If the width of a glitch is too short, short circuit may increase power consumption. The glitch problem is only for the static logic circuits. For dynamic logic circuits, there is no glitch problem since any node can have at most

LOW-POWER CMOS CIRCUIT TECHNIQUES 145

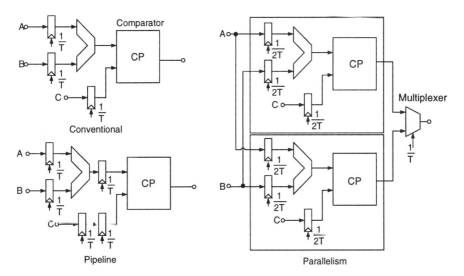

Fig. 3.60 Architecture-driven voltage scaling strategy. (Adapted from Chandrakasan and Brodersen [35].)

one power-consumption transition per clock cycle.

In the power dissipation formula of Eq. (3.26), power dissipation is a square-law function of the power supply voltage. Therefore, reducing power supply voltage is the most effective way in lowering power dissipation of a circuit. Fig. 3.60 shows the architecture-driven voltage scaling strategy[35]. Consider a VLSI system as shown in the figure. When the power supply voltage of a VLSI system is lowered, its power dissipation can be substantially reduced. However, its speed performance may be degraded at this lower power supply voltage. The speed performance can be compensated by using two duplicate systems grouped by a multiplexer at the output. As long as the power consumption of these two duplicate systems operating at a reduced power supply voltage can be much smaller, it's a good voltage scaling strategy to reduce power consumption. Pipeline techniques can also be used to lower power dissipation. Via pipeline, delay can be divided into two segments. When supply voltage is lowered, as long as the delay of each segment is not greater than the total delay when the supply voltage is not lowered, the throughput of the whole system will not be degraded. In addition, when the supply voltage is lowered, power dissipation can be reduced.

3.8.1 Bus Architecture Approach

Reduction of dynamic power dissipation is important for low-power VLSI applications. In order to increase the signal-processing capability, increasing the operating frequency is a straightforward strategy. In order to maintain acceptable power dissipation for the VLSI system with increased complexity and operating speed, low-power

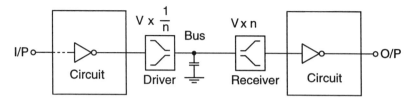

Fig. 3.61 Low-power technique by reducing the internal bus swing. (Adapted from Nakagome et al. [36].)

techniques are important. One general low-power technique is via lowering the supply voltage. The other approach is via new circuit techniques at a less reduced supply voltage. In a VLSI system, a large portion of power is consumed at the data bus. For a VLSI system using 0.1μm CMOS devices, about over 50% of power consumption is associated with the data bus. Fig. 3.61 shows a low-power technique by reducing the internal bus swing[36]. As shown in the figure, reducing its internal voltage swing associated with the internal data bus to below 1V can be used to reduce power dissipation of the overall system. On the other hand, system supply voltage is not changed such that its performance can be maintained. In the architecture with the low-power technique, the logic swing in the circuit block is unchanged. A bus driver is used to convert the output signal of the circuit block, which has a full swing, to a signal with a reduced voltage swing of V_{DD}/n to drive the bus. At the other end of the data bus, a bus receiver is used to recover the data bus signal to the signal with the original full swing such that it can be used in the next circuit block. Under this architecture, the voltage swing of the signal at the data bus is V_{DD}/n. Its corresponding power dissipation at the bus is $fC_bV_{DD}^2/n$, where C_b is the equivalent capacitance of the bus. Compared to the conventional architecture, the internal bus swing reduction architecture provides a $1/n$ times reduction in power dissipation. However, the cost of the internal bus swing reduction architecture is the reduced noise margin in the circuit blocks and the driver/receiver circuits.

Fig. 3.62 shows the bus driver circuit with the internal supply generator and the symmetric level converter[36]. The driver circuit is based on a standard CMOS inverter operating at reduced supply voltages: V_{DL} and V_{SL}. In the driver circuit, the threshold voltages of the CMOS devices have been reduced to $V_{TN} - V_{SL}$ and $|V_{TP}| - (V_{DD} - V_{DL})$, with the values of the reduction equal to the reductions from two supply voltages, respectively. With this arrangement, the driving current of the CMOS devices in the driver circuit is equal to that in the circuit without the reduced supply voltages and the reduced threshold voltages. The reduced power supplies are produced from the internal supply generator, which is based on a resistive voltage divider formed by $R1$–$R3$, where reference voltages can be obtained. A buffer, which is composed of a differential-amplifier $AMP1$, PMOS device $MP2$ and a current source I_{DL}, is used to provide a large charging current to drive the data bus. As shown in the figure, the bus receiver is used to restore the bus signal with a reduced signal swing to the full-swing signal. The receiver functions as a symmetric level

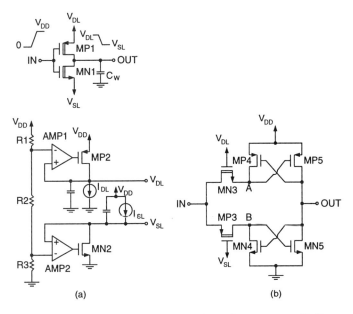

Fig. 3.62 (a) Bus driver circuit with the internal supply generator; (b) Symmetric level converter. (From Nakagome et al. [36], ©1993 IEEE.)

converter, which contains two level conversion circuits. The upper level conversion circuit, which is made of two cross-coupled PMOS devices, is used to transform the intermediate high level (V_{DL}) of the bus signal to the high level (V_{DD}). The lower level conversion circuit, which is made of two cross-coupled NMOS devices, is used to transform the intermediate low level (V_{SL}) to the low level (V_{SS}). When IN rises from V_{SL} to V_{DL}, node A rises from V_{SL} to close to V_{DL} and node B rises from ground to close to V_{DL} with the value determined by the conduction ratio between $MP3$ and $MN4$. Thus $MN5$ turns on to discharge the output to ground. When OUT falls to $V_{DD} - |V_{TP}|$, $MP4$ turns on to charge node A to V_{DD}. Since $MN4$ is off, node B also rises to close to V_{DL}.

Fig. 3.63 shows the architecture of a low-power architecture with the data-dependent logic swing bus[37], which is used to shrink the voltage swing on the data bus such that high speed and low power can be reached. The data-dependent logic swing bus takes advantages of charge sharing between bus wires and an additional bus wire to reduce its voltage swing. As shown in the figure, the architecture of the data-dependent logic includes the driver and the receiver. The data bus is 16-bit wide. In addition, there is an additional bus line called dummy ground line, which is connected to ground via an NMOS transistor. On the dummy ground line, there is also an identical parasitic capacitor connected. Before the cycle, when $\overline{\phi_{PRE}}$ is low, all bus lines are precharged to V_{DD} and the the dummy ground line is pre-discharged to ground. When $\overline{\phi_{PRE}}$ is high, all bus lines are floating. When the driver enable signal ($\overline{\phi_{DE}}$) is low, depending on the logic value of each signal bit,

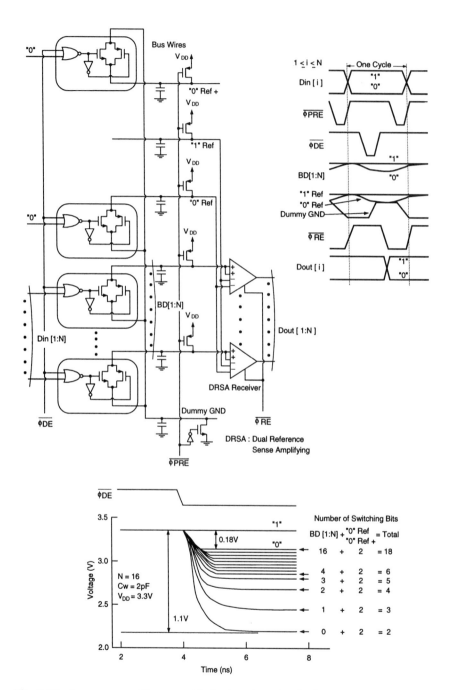

Fig. 3.63 Low-power architecture with the data-dependent logic swing bus. (From Hiraki et al. [37], ©1995 IEEE.)

the driver determines which specific related bus line connected to the dummy ground bus line. If only one signal bit falls from V_{DD} to ground, then the associated bus line is connected to the dummy ground line. Under this situation, the charge in the parasitic capacitor associated with the bus line will redistribute with the capacitor on the dummy ground line. As a result, the voltage on this bus line drops to 1/2 of V_{DD}, which is regarded as the new level for logic 0 instead of ground. If two signal bits fall from V_{DD} to ground, then the two related bus lines are connected to the dummy ground line. Therefore, the voltage on the two bus lines falls to 2/3 of V_{DD}. If n signal bits fall from V_{DD} to ground, after charge redistribution, the voltage on the n bus lines falls to $n/(n+1)$ of V_{DD}. The consumed energy associated with the V_{DD}-to-ground switching on the n bus lines is $\frac{1}{2}nC(\frac{V_{DD}^2}{n+1})$, instead of $\frac{1}{2}nCV_{DD}^2$ for the conventional bus lines. Using this new technique, the consumption of energy has been reduced by $(n+1)$ times. With more bits in switching (n is larger), power reduction is more. When the receiver enable signal is active ($\overline{\psi_{RE}} - 0$), the receiver circuit will restore the original bus signal with a reduced signal swing to the full-swing signal.

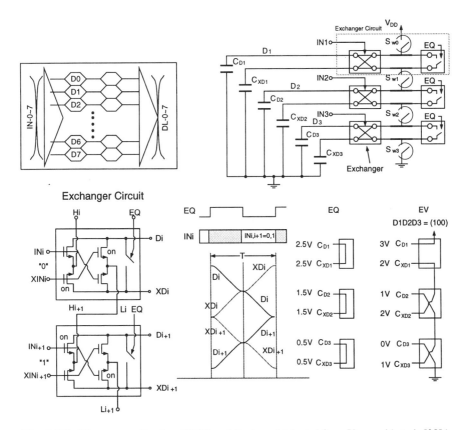

Fig. 3.64 Charge-recycling bus (CRB) architecture. (Adapted from Yamauchi et al. [38].)

150 CMOS STATIC LOGIC CIRCUITS

Fig. 3.64 shows the concept of the charge-recycling bus (CRB) architecture[38], which features virtual stacking of the individual buses connected in parallel into a conceptual series configuration between the power supply and ground. This charge-recycling bus architecture makes it possible to reduce each bus-swing. The extent in charge-up of each bus is given by the recycled charge supplying from the upper adjacent bus capacitance, instead of from the power line. Under this configuration, energy can be transferred from the supply to the ground via each individual bus capacitance sequentially such that for the buses in between, recycling of charge can be utilized.

As shown in Fig. 3.64, there are three pairs of complementary bus lines referred to three signal bits. Each bus line is regarded as a bus capacitance with an equivalent value (C_D). For each signal bit, there are two complementary bus capacitances. One is referred to the bus capacitance C_D of the specific bit. The other is related to the complementary bus capacitance $\overline{C_{XD}}$. The procedure of the charge-recycling bus architecture is described here. During the equalization phase, each pair of the complementary bus lines are connected together via the exchanger to a specific potential. For example, during the equalization phase, the complementary bus lines of the first, second, and third signal bits (D_1, D_2, D_3) are tied to 2.5V, 1.5V, and 0.5V, respectively. During the evaluation period, there are two procedures involved with each pair of the bus capacitances and the exchanger circuit. If the signal is one, then the associated pair of the bus capacitances $(C_D, \overline{C_{XD}})$ maintain their polarity. If the signal is zero, then the associated complementary bus capacitances reverse their polarity to $(\overline{C_{XD}}, C_D)$. Then, in each pair of two complementary bus capacitances, bus capacitance C_D is connected to the adjacent upper bus capacitance and complementary bus capacitance $(\overline{C_{XD}})$ is connected to the adjacent lower complementary bus capacitance. For the first bit, its bus capacitance is connected to the top supply voltage. For the last bit, its complementary bus capacitance is connected to the bottom ground level. For example, during the evaluation period, the signals $(D_1, D_2, D_3) = (1, 0, 0)$ are imposed on the three pairs of the complementary bus lines. The first bit is 1 $(D_1 = 1)$, so the polarity of C_{D1} and $\overline{C_{XD1}}$ is maintained. Since the second and the third bits are 0 $(D_2 = 0, D_3 = 0)$, the corresponding polarities are reversed $(\overline{C_{XD2}}, C_{D2}), (\overline{C_{XD3}}, C_{D3})$. On the first pair of the complementary bus lines, since the first bit D_1 is high, the bus capacitance of the first bit C_{D1} is connected to the top supply voltage, which is 3V. The complementary bus capacitance of the first bit $\overline{C_{XD1}}$ is connected to the complementary bus capacitance of the second bit $\overline{C_{XD2}}$, due to charge redistribution between two capacitances, therefore the voltage of both $\overline{C_{XD1}}$ and $\overline{C_{XD2}}$ becomes 2V. Since the second bit is $D_2 = 0$, the bus capacitance of the second bit C_{D2} is connected to the complementary bus capacitance of the third bit $(\overline{C_{XD3}})$. Since the third bit is $D_3 = 0$, the bus capacitance of the third bit C_{D3} is connected to ground. As shown in the figure, the exchanger circuit is used to facilitate three functions referred to each pair of the complementary bus capacitances of a bit. (1) During the equalization period, equalize each pair of the complementary bus capacitances of each bit. (2) During the evaluation period, reverse the polarity of each pair of the complementary bus capacitances if the signal is 0. (3) During the evaluation period, connect the high-level bus capacitance to its adjacent upper low-level bus capacitance and connect the low-level bus capacitance to its adjacent lower high-level bus capacitance. As shown in the figure, at the end of the evaluation period, the voltages of the first pair of the bus capacitances are $C_{D1} = 3V$ and $\overline{C_{XD1}} = 2V$, which implies logic 1 for the first bit ($D_1 = 1$). The voltages of the second pair of the bus capacitances are $C_{D2} = 1V$ and $\overline{C_{XD2}} = 2V$, which indicates logic 0 for the second bit ($D_2 = 0$). The voltages of the third pair of the bus capacitances are $C_{D3} = 0V$ and $\overline{C_{XD3}} = 1V$, which means logic 0 for the third bit ($D_3 = 0$). As described above, each bus capacitance is referred to a bus line. The evaluation procedure described above is based on the conceptual charge transfer operation involved in bus capacitances connected in series. In reality, during the evaluation procedure, signals on each bus line are being propagated over the bus line. Therefore, after the charge recycling procedure for the bus capacitances in the evaluation period, the voltages on the receiver side of the complementary bus lines have been restored to their respective full-swing values—the propagation of the signals over the entire bus lines has been accomplished. In the charge-recycling bus architecture, during the propagation of the signal over the data buses, charge recycling has been utilized. During the evaluation period, the upper power line supplies the charge needed for the first pair of data buses. During the next evaluation period, the used charge in each bus line between the top power line and ground is transferred to its adjacent lower bus line. In contrast, in a conventional bus architecture, during the propagation of the signal, the upper power line supplies the required charge. At the same time, charge is being transferred to ground. Therefore, the power dissipation of the charge-recycling bus architecture is much smaller than than that of the conventional one since the signal swing on each bus line is V_{DD}/n, where n is the bit number. In addition, only $1/n$ portion of the total data buses draw power from the power supply. Hence, compared to the conventional bus architecture, the total power dissipation of the charge-recycling bus architecture is $1/n^2$ smaller.

In the charge-recycling bus architecture, there are driver and receiver circuits as shown in Fig. 3.64 [38]. As shown in the figure, the driver circuit converts the input signals to all buses, which have a voltage swing of $0-3V$, to their respective reduced signal swings. If the data bus is 3-bit wide, then the voltage swing of the first pair of the complementary data buses (C_{D1}, $\overline{C_{XD1}}$), which is referred to the first bit (D_1), is $3-2V$. The voltage swing of the second pair of the complementary data buses (C_{D2}, $\overline{C_{XD2}}$), which is referred to the second bit (D_2) is $2-1V$. The voltage swing of the third pair (C_{D3}, $\overline{C_{XD3}}$) is $1-0V$. The function of the receiver circuit is just the opposite of the driver function. The transistor-level exchanger circuits of the two adjacent layers are also shown in the figure. As shown in the figure, D_i is connected to C_{D_i} and XD_i is to $\overline{C_{XD_i}}$. When EQ is high, it is the equalization phase. The switch turns on, the voltages of D_i and XD_i are equalized. During the pre-equalize phase, SW_i is open. During the equalization phase, IN_i and XIN_i are low. When EQ is low, it is the evaluation phase. From the values of IN_i and XIN_i, which of D_i and XD_i connected to (H_i, L_i) is determined. When $IN_{i,i+1}$ is '0,1', H_i is connected to XD_i; L_i is connected to D_i; H_{i+1} is to D_{i+1}; L_{i+1} is to XD_{i+1}. Thus,

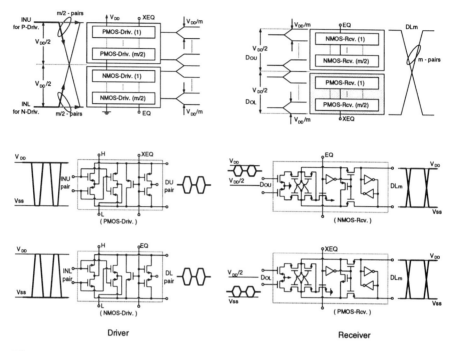

Fig. 3.65 Driver and receiver circuits for the charge-recycling bus architecture. (From Yamauchi et al. [38], ©1995 IEEE.)

charge sharing occurs at C_{D_i} and $C_{D_{i+1}}$. $\overline{C_{XD_i}}$ and the upper layer are connected to the bus capacitance of L_{i-1}. $\overline{C_{XD_{i+1}}}$ and the lower layer are connected to the bus capacitance of H_{i+2}. Thus, virtual stacking and charge recycling are accomplished.

Fig. 3.65 shows the driver and the receiver circuits for the charge-recycling bus architecture[38]. Due to the difference in the bus potential, the gate types in the driver and the receiver circuits are different. As shown in the figure, in the driver circuit, the upper portion is made of PMOS and the lower is NMOS. This arrangement is used to reduce the output bus voltage with a reduced level effectively. During the equalization phase, the input signal pairs in the upper portion are precharged to high such that the switch transistor of the PMOS driver is off. Similarly, the input signal to the lower portion is predischarged to low. In the receiver, the configuration is reversed. In the upper portion, NMOS devices have been used such that a larger driving capability can be obtained for the input device. PMOS devices are used in the lower portion. As shown in the figure, the receiver is made of a dynamic latch-type current sense amp. During the equalization phase ($EQ = 1$), the sense amp is off and the output of the sense amp is equalized. The output latch maintains its state. When $EQ = 0$, during the evaluation phase, the currents of the gate receiver, which serves as a voltage-to-current converter, are different. The difference in the currents

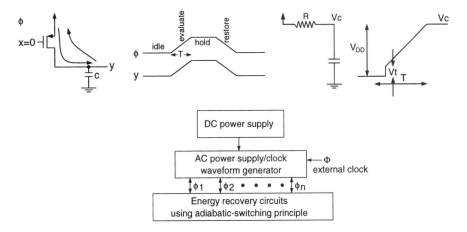

Fig. 3.66 Energy recovery concept. (Adapted from Ye and Roy [46].)

is amplified by the latch sense amp. After the pass transistor, which is controlled by EQ, turns on, the state of the output latch is changed.

3.8.2 Adiabatic Logic

Adiabatic CMOS logic circuits have been reported for their potentials in low-power VLSI applications [39]–[46]. For a standard CMOS circuit, in order to charge its output load to V_{DD}, an amount of energy of CV_{DD}^2 is required to be extracted from the supply, where C is the output load capacitance. Half of the CV_{DD}^2 energy is stored in the output load capacitance and the other half is consumed in the charge process. When the output load is discharged, the output is discharged to ground and a $\frac{1}{2}CV_{DD}^2$ amount of energy is consumed in the transistor. Fig. 3.66 shows the energy recovery concept[46]. As shown in the figure, the power supply voltage is not fixed, it has a ramp up to V_{DD}. The PMOS device can be regarded as a resistor. When the resistance of the resistor is small, the voltage drop on the resistor is small. The energy consumed in the resistor is small. Therefore, most of the energy is saved in charging the output capacitance. When the power supply ϕ ramps down from high to low during the restore mode, the charge in the output capacitance flows to ϕ via the PMOS. Thus the energy in the output capacitance flows back to the power supply. This is close-to-ideal adiabatic switching. Via minimizing the voltage drop over the PMOS, the consumed energy can be minimized. The dissipated energy can be computed for the circuit. Before t_0, the PMOS is off and ϕ is below the threshold voltage of the PMOS. Thus a voltage drop exists between source and drain—energy loss, which is equal to $E_{th} = \frac{1}{2}CV_t^2$. When $t > T$, PMOS turns on, solving the differential equation for the equivalent circuit, one obtains: $RC(\frac{dV_c}{dt}) + V_c = \phi$.

154   CMOS STATIC LOGIC CIRCUITS

Solving this differential equation, the output voltage is[46]:

$$V_c = \begin{cases} 0 & t < t_0, \\ \phi - (\frac{RC}{T})V_{DD}(1 - e^{-(t-t_0)/RC}) + V_t e^{-(t-t_0)/RC} & 0 \le t_0 < t < T, \\ \phi - (\frac{RC}{T})V_{DD}(1 - e^{-(T-t_0)/RC})e^{-(t-T)/RC} & t \ge T. \\ \quad - V_t e^{-(t-t_0)/RC} & \end{cases}$$

The energy loss is:

$$E_{diss} = \int_0^\infty iV_R dt = \int_0^T iV_R dt + \int_T^\infty iV_R dt, \quad (3.33)$$

$$= \left(\frac{RC}{T}\right)CV_{DD}^2 + \frac{1}{2}CV_t^2 + O\left[\left(\frac{RC}{T}\right)^2\right],$$

$$= E_{linear} + E_{th}, \quad (3.34)$$

where $O[(\frac{RC}{T})^2]$ represents other higher-order terms. Without considering the threshold loss, the ratio of the consumed energy of the adiabatic circuit to that of the conventional circuit is $\frac{E_{adia}}{E_{conv}} = \frac{\frac{RC}{T}CV_{DD}^2}{CV_{DD}^2} = \frac{RC}{T}$. From this ratio, the energy loss of the adiabatic circuit is much smaller if $RC \ll T$. Using the framework for the power supply and circuit as shown in the figure, the DC power supplies the total supply current. The external AC power supply/clock generator supplies the power for the adiabatic logic circuit. Via LC resonance, the recycled energy from the output load is stored in the inductance. Thus adiabatic logic circuits with the energy recycling capability can be designed.

Adiabatic logic circuits of early days use diodes or diode-connected devices for the precharge purpose [39][40]. However, the power loss due to the voltage drop over the diode or the diode-connected devices limits their advantages. Although adiabatic logic circuits using a complementary output design have been done [41]–[43], diode-connected devices are still unavoidable. In addition, power supply with complicated waveforms is required. Improved static and ternary logic circuits using the adiabatic concept [44][45] are not suitable for low-voltage operation. In addition, their noise immunity is not good.

Fig. 3.67 shows a 1.5V CMOS energy efficient logic (EEL) circuit[47][49]. This EEL circuit is based on the cascode voltage switch logic (CVSL)[2]. As shown in the figure, four clocks (CK0, CK1, CK2, CK3) connected to the power supply are required. Each clock has four phases. Among four clocks, one clock is different from another. The four phases of each clock are (1) the precharge and evaluate phase, (2) the hold phase, (3) the recover phase, and (4) the wait phase. In the precharge and evaluate phase, the clock voltage gradually increases from 0V to V_{dd}. If the input V_{in} is high and the complementary input $\overline{V_{inb}}$ is low, the NMOS device $MN1$ is on and the NMOS device $MN2$ is off. Therefore, the complementary output node $\overline{V_{outb}}$ is discharged to ground by $MN1$. Consequently, $MP2$ turns on. As a result, the output node V_{out} increases as the supply clock voltage (CK) rises. When the

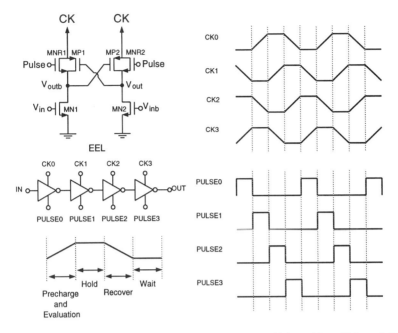

Fig. 3.67 1.5V CMOS energy efficient logic (EEL) circuit. (Adapted from Yeh et al. [49].)

supply clock (CK) enters the hold phase, both the output (V_{out}) and the complementary output ($\overline{V_{out}}$) maintain their previous values. When CK decreases from V_{dd} to ground, it is the recover phase. During the recover phase, the charge previously stored at the V_{out} node flows upward to the CK supply when CK falls. Therefore, the injected energy from the CK supply to the output node is returned to the CK supply. When CK is low, it is the wait phase. During the wait phase, the circuit is idle. From the above analysis, during the precharge and evaluate phase, the energy is supplied by the CK via the PMOS devices to the parasitic load capacitance. During the recover phase, the energy stored in the parasitic load capacitance is returned to CK. The consumed power of the circuit is the product of the PMOS drain current and its drain-to-source voltage. Without inclusion of MNR1 and MNR2, during the recover phase, due to the threshold voltage of the PMOS device, the output voltage cannot drop to 0V as CK falls—the output is not full-swing. As a result, the noise margin of the circuit may not be good. In order to improve the noise margin and the energy consumption for a low supply operation, $MNR1$ and $MNR2$ as shown in Fig. 3.67 are added[49]. As shown in the figure, the EEL circuit has included the NMOS devices $MNR1/MNR2$ between the output node and the CK supply. When $Pulse$ is high, the NMOS devices $MNR1/MNR2$ are on. Therefore, the remaining charge at the output node, which is not able to be discharged via the PMOS device, can be discharged to the CK supply by the NMOS devices $MNR1/MNR2$. As a result, the output voltage can reach 0V—full-swing of the output can be obtained.

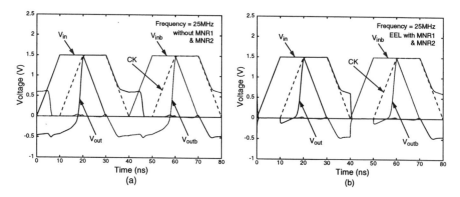

Fig. 3.68 Transient waveforms of the energy efficient logic (EEL) circuit. (a) without MNR1 & MNR2; (b) with MNR1 & MNR2. (Adapted from Yeh et al.[49].)

In addition, more charge can be returned to the CK supply—a higher energy efficiency.

Fig. 3.68 shows the transient waveforms of the energy efficient logic (EEL) circuit (a) without MNR1 and MNR2, (b) with MNR1 and MNR2 using a 1.5V clock. As shown in Fig. 3.68(a), due to the threshold voltage of the PMOS device, during the recover phase, the output voltage V_{out} cannot follow CK to reach the ground level. In addition, because the NMOS device $MN1$ is off during the recover phase and the wait phase, the complementary output node $\overline{V_{outb}}$ is floating. Due to clock feedthrough, the voltage at the output node may drop to below 0V. During the following precharge and evaluate phase, when the PMOS device $MP1$ turns on, the consumed power of $MP1$ is large since the voltage drop between its source and drain is large ($V_{CK} - \overline{V_{outb}}$). In the EEL circuit with $MNR1$ and $MNR2$, the situation is different. During the wait phase, in the EEL circuit with $MNR1$ and $MNR2$, the complementary output node $\overline{V_{outb}}$ is not floating any more. Owing to the NMOS devices $MNR1$ and $MNR2$, both output nodes (V_{out}, $\overline{V_{outb}}$) are pulled to the same voltage level as the CK supply—0V. Therefore, in the following precharge and evaluate phase, the voltage drop between the source and the drain of the PMOS device $MP1$ is smaller than that in the circuit without $MNR1$ and $MNR2$. Consequently, the consumed energy of $MP1$ in the EEL circuit is smaller than that in the circuit without $MNR1$ and $MNR2$. As a result, from the power consumption point of view, a higher operation frequency can be obtained.

Fig. 3.69 shows the consumed energy versus time during the transient of the EEL circuits. As shown in the figure, due to the attachment of the $MNR1$ and $MNR2$, the consumed energy of the EEL circuit is smaller than that without them. Within a time window of 320ns, the EEL circuit consumes $2.5 \times 10^{-13} J$, which is 54% of the ECRL case ($4.66 \times 10^{-13} J$). The EEL circuit described in this section is used in a pipelined system as shown in Fig. 3.67(b). When the EEL circuit of the current stage is in the precharge and evaluate phase, the preceding stage is in the hold phase such that a correct input can be presented at the input of the current stage. At

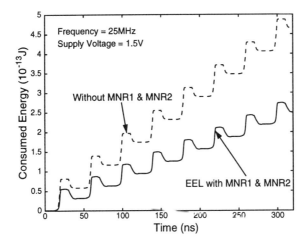

Fig. 3.69 Consumed energy versus time during the transient of the EEL circuits with and without MNR1 and MNR2. (Adapted from Yeh et al.[49].)

the following stage, it is in the wait phase, where it is not influenced by the logic evaluation operation of the current stage. When the current stage enters the hold stage, the preceding stage enters the recover phase and the following stage is in the precharge and evaluate phase. Therefore, the pipeline operation of the whole system is obtained.

3.9 SUMMARY

In this chapter, CMOS static logic circuits have been described. Starting from the basic static CMOS logic circuits, CMOS differential static logic circuits including DCVS, DSL, and DCVSPG have been analyzed. Then, CMOS pass-transistor static logic circuits have been described. After the BiCMOS static logic circuits, low-voltage SOI CMOS static logic circuits have been presented. In the final portion of this chapter, low-voltage CMOS static logic circuit techniques including bootstrap and multi-threshold standby/active techniques have been analyzed, followed by the low-power techniques including the bus architecture approach and the adiabatic logic.

REFERENCES

1. S.-W. Sun and P. G. Y. Tsui, "Limitation of CMOS Supply-Voltage Scaling by MOSFET Threshold-Voltage Variation," *IEEE J. Sol. St. Ckts*, **30**(8), 947–949 (1995).

2. L. G. Heller, J. W. Davis, N. G. Thoma, and W. R. Griffin, "Cascode Voltage Switch Logic: A Differential CMOS Logic Family," *ISSCC Dig.*, 16–17 (1984).

3. K. M. Chu and D. L. Pulfrey, "Design Procedures for Differential Cascode Voltage Switch Circuits," *IEEE J. Sol. St. Ckts*, **21**(6), 1082–1087 (1986).

4. L. C. M. G. Pfennings, W. G. J. Mol, J. J. J. Bastiaens, and J. M. F. Van Dijk, "Differential Split-Level CMOS Logic for Subnanosecond Speeds," *IEEE J. Sol. St. Ckts.*, **20**(5), 1050–1055 (1985).

5. F.-S. Lai and W. Hwang, "Design and Implementation of Differential Cascode Voltage Switch with Pass-Gate (DCVSPG) Logic for High-Performance Digital Systems," *IEEE J. Sol. St. Ckts.*, **32**(4), 563–573 (1997).

6. H. Partovi and D. Draper, "A Regenerative Push-Pull Differential Logic Family," *ISSCC Dig.*, 294–295 (1994).

7. D. Radhakrishnan, S. R. Whitaker, and G. K. Maki, "Formal Design Procedures for Pass Transistor Switching Circuits," *IEEE J. Sol. St. Ckts.*, **20**(2), 531–536 (1985).

8. J. H. Pasternak, A. S. Shubat, and C. A. T. Salama, "CMOS Differential Pass-Transistor Logic Design," *IEEE J. Sol. St. Ckts.*, **22**(2), 216–222 (1987).

9. K. Yano, T. Yamanaka, T. Nishida, M. Saito, K. Shimohigashi, and A. Shimizu, "A 3.8-ns CMOS 16× 16-b Multiplier Using Complementary Pass-Transistor Logic," *IEEE J. Sol. St. Ckts.*, **25**(2), 388–395 (1990).

10. A. Parameswar, H. Hara, and T. Sakurai, "A Swing Restored Pass-Transistor Logic-Based Multiply and Accumulate Circuit for Multimedia Applications," *IEEE J. Sol. St. Ckts*, **31**(6), 804–809 (1996).

11. M. Suzuki, N. Ohkubo, T. Shinbo, T. Yamanaka, A. Shimizu, K. Sasaki, and Y. Nakagome, "A 1.5-ns 32-b CMOS ALU in Double Pass-Transistor Logic," *IEEE J. Sol. St. Ckts.*, **28**(11), 1145–1151 (1993).

12. K. Yano, Y. Sasaki, K. Rikino, and K. Seki, "Top-Down Pass-Transistor Logic Design," *IEEE J. Sol. St. Ckts.*, **31**(6), 792–803 (1996).

13. T. Kuroda and T. Sakurai, "Overview of Low-Power ULSI Circuit Techniques," *IEICE Trans. Elec.*, **E78-C**(4), 334–343 (1995).

14. J. B. Kuo, Y. W. Chen, and K. H. Lou, "Device-Level Analysis of a 1μm BiCMOS Inverter Circuit Operating at 77K Using a Modified PISCES Program," *CICC Dig.*, 23.4.1–3 (1991).

15. J. B. Kuo, G. P. Rosseel, and R. W. Dutton, "Two-Dimensional Analysis of a Merged BiPMOS Device," *IEEE Trans. CAD of ICs*, **8**(8), 929–932 (1989).

16. T. C. Lu and J. B. Kuo, "An Analytical Pull-up Transient Model for a BiCMOS Inverter," *Sol. St. Elec.*, **35**(1), 1–8 (1992).

17. K. Yano, M. Hiraki, S. Shukuri, Y. Onose, M. Hirao, N. Ohki, T. Nishida, K. Seki, and K. Shimohigashi, "Quasi-Complementary BiCMOS for Sub-3-V Digital Circuits," *IEEE J. Sol. St. Ckts*, **26**(11), 1708–1719 (1991).

18. R. B. Ritts, P. A. Raje, J. D. Plummer, K. C. Saraswat, and K. M. Cham, "Merged BiCMOS Logic to Extend the CMOS/BiCMOS Performance Crossover Below 2.5-V Supply," *IEEE J. Sol. St. Ckts*, **26**(11), 1606–1614 (1991).

19. M. Hiraki, K. Yano, M. Minami, K. Sato, N. Matsuzaki, A. Watanabe, T. Nishida, K. Sasaki, and K. Seki, "A 1.5-V Full-Swing BiCMOS Logic Circuit," *IEEE J. Sol. St. Ckts*, **27**(11), 1568–1574 (1992).

20. R. Y. V. Chik, and C. A. T. Salama, "Design of a 1.5V Full-Swing Bootstrapped BiCMOS Logic Circuit," *IEEE J. Sol. St. Ckts*, **30**(9), 972–978 (1995).

21. M. S. Elrabaa, M. S. Obrecht, and M. I. Elmasry, "Novel Low-Voltage Low-Power Full-Swing BiCMOS Circuits," *IEEE J. Sol. St. Ckts*, **29**(2), 86–94 (1994).

22. B. S. Cherkauer and E. G. Friedman, "Design of Tapered Buffers with Local Interconnect Capacitance," *IEEE J. Sol. St. Ckts*, **30**(2), 151–155 (1995).

23. S. H. K. Embabi, A. Bellaouar, and K. Islam, "A Bootstrapped Bipolar CMOS (B^2CMOS) Gate for Low-Voltage Applications," *IEEE J. Sol. St. Ckts*, **30**(1), 47–53 (1995).

24. F. Assaderaghi, D. Sinitsky, S. A. Parke, J. Bokor, P. K. Ko, and C. Hu, "Dynamic Threshold-Voltage MOSFET (DTMOS) for Ultra-Low Voltage VLSI," *IEEE Trans. Elec. Dev.*, **44**(3), 414–422 (1997).

25. I.-Y. Chung, Y. J. Park, and H. S. Min, "A New SOI Inverter Using Dynamic Threshold for Low-Power Applications," *IEEE Elec. Dev. Let.*, **18**(6), 248–250 (1997).

26. T. W. Houston, "A Novel Dynamic Vt Circuit Configuration," *SOI Conf. Dig.*, 154–155 (1997).

27. T. Douseki, S. Shigematsu, J. Yamada, M. Harada, H. Inokawa, and T. Tsuchiya, "A 0.5-V MTCMOS/SIMOX Logic Gate," *IEEE J. Sol. St. Ckts*, **32**(10), 1604–1609 (1997).

28. J. H. Lou and J. B. Kuo, "A 1.5-V Full-Swing Bootstrapped CMOS Large Capacitive-Load Driver Circuit Suitable for Low-Voltage CMOS VLSI," *IEEE J. Sol. St. Ckts*, **32**(1), 119–121 (1997).

29. K. Shimohigashi and K. Seki, "Low-Voltage ULSI Design," *IEEE J. Sol. St. Ckts*, **28**(4), 408–413 (1993).

30. S. Shigematsu, S. Mutoh, Y. Matsuya, Y. Tanabe, and J. Yamada, "A 1-V High-Speed MTCMOS Circuit Scheme for Power-Down Application Circuits," *IEEE J. Sol. St. Ckts*, **32**(6), 861–869 (1997).

31. D. Takashima, S. Watanabe, H. Nakano, Y. Oowaki, K. Ohuchi, and H. Tango, "Standby/Active Mode Logic for Sub-1-V Operating ULSI Memory," *IEEE J. Sol. St. Ckts*, **29**(4), 441–447 (1994).

32. M. Horiguchi, T. Sakata, and K. Itoh, "Switched-Source-Impedance CMOS Circuit for Low Standby Subthreshold Current Giga-Scale LSI's," *IEEE J. Sol. St. Ckts*, **28**(11), 1131–1135 (1993).

33. C. Wann, F. Assaderaghi, R. Dennard, C. Hu, G. Shahidi, and Y. Taur, "Channel Profile Optimization and Device Design for Low-Power High-Performance Dynamic-Threshold MOSFET," *IEDM Dig.*, 113–116 (1996).

34. H. Kotaki, S. Kakimoto, M. Nakano, T. Matsuoka, K. Adachi, K. Sugimoto, T. Fukushima, and Y. Sato, "Novel Bulk Dynamic Threshold Voltage MOSFET (B-DTMOS) with Advanced Isolation (SITOS) and Gate to Shallow-Well Contact (SSS-C) Processes for Ultra Low Power Dual Gate CMOS," *IEDM Dig.*, 459–462 (1996).

35. A. P. Chandrakasan and R. W. Brodersen, "Minimizing Power Consumption in Digital CMOS Circuits," *IEEE Proc.*, **83**(4), 498–523 (1995).

36. Y. Nakagome, K. Itoh, M. Isoda, K. Takeuchi, and M. Aoki, "Sub-1-V Swing Internal Bus Architecture for Future Low-Power ULSI's," *IEEE J. Sol. St. Ckts*, **28**(4), 414–419 (1993).

37. M. Hiraki, H. Kojima, H. Misawa, T. Akazawa, and Y. Hatano, "Data-Dependent Logic Swing Internal Bus Architecture for Ultralow-Power LSI's," *IEEE J. Sol. St. Ckts*, **30**(4), 397–402 (1995).

38. H. Yamauchi, H. Akamatsu, and T. Fujita, "An Asymptotically Zero Power Charge-Recycling Bus Architecture for Battery-Operated Ultrahigh Data Rate ULSI's," *IEEE J. Sol. St. Ckts*, **30**(4), 423–431 (1995).

39. A. G. Dickinson and J. S. Denker, "Adiabatic Dynamic Logic," *IEEE J. Sol. St. Ckts*, **30**(3), 311–315 (1995).

40. R. T. Hinman and M. F. Schlecht, "Power Dissipation Measurements on Recovered Energy Logic," *Symp. VLSI Ckts Dig.*, 19–20 (1994).

41. W. Y. Wang and K. T. Lau, "Adiabatic Pseudo-Domino Logic," *Elec. Let.*, **31**(23), 1982–1983 (1995).

42. K. T. Lau and W. Y. Wang, "Transmission Gate-Interfaced APDL Design," *Elec. Let.*, **32**(4), 317–318 (1996).

43. A. Vetuli, S. D. Pascoli, and L. M. Reyneri, "Positive Feedback in Adiabatic Logic," *Elec. Let.*, **32**(20), 1867–1869 (1996).

44. D. Mateo and A. Rubio, "Quasi-Adiabatic Ternary CMOS Logic," *Elec. Let.*, **32**(2), 99–101 (1996).

45. V. K. De and J. D. Meindl, "Complementary Adiabatic and Fully Adiabatic MOS Logic Families for Gigascale Integration," *ISSCC Dig.*, 298–299 (1996).

46. Y. Ye and K. Roy, "Energy Recovery Circuits Using Reversible and Partially Reversible Logic," *IEEE Trans. Ckts & Sys.–I*, **43**(9), 769–778 (1996).

47. Y. Moon and D. -K. Jeong, "An Efficient Charge Recovery Logic Circuit," *IEEE J. Sol. St. Ckts*, **31**(4), 514–522 (1996).

48. M. Kakumu, M. Kinugawa, and K. Hashimoto, "Choice of Power-Supply Voltage for Half-Micrometer and Lower Submicrometer CMOS Devices," *IEEE Trans. Elec. Dev.*, **37**(5), 1334–1342 (1990).

49. C. C. Yeh, J. H. Lou, and J. B. Kuo, "1.5V CMOS Full-Swing Energy Efficient Logic (EEL) Circuit Suitable for Low-Voltage and Low-Power VLSI Applications," *Elec. Let.*, **33**(16), 1375–1376 (1997).

Problems

1. Derive $V_{OH}, V_{OL}, V_{IH}, V_{IL}$ for the NMOS inverter with a depletion load with its gate connected to source. Suppose $k_i = 25\mu A/V^2$ for the driver NMOS device. $k_l = 6.25\mu A/V^2$ for the depletion load NMOS device. The threshold voltage of the depletion load NMOS device is $-3V$ and $0.8V$ for the enhancement-mode device. $V_{DD} = 5V$.

2. Design a logic function $F = AB + BC + CA$ using 2-input NAND and NOR gates. If an n-well CMOS technology is used, which approach should be used to reduce the body effect on degrading the circuit performance? If p-well CMOS technology is used, which approach should be used?

3. Consider Eqs. (3.13) and (3.14) for t_{PHL} and C_L. How can one lower t_{PHL}? What is the cost?

4. Design $F = AB + BC + CA$ using DCVS and DCVSPG. What are the differences between the DCVS circuit and the DCVSPG circuit? Compare the pull-up and pull-down transients of the DCVS circuit with those of the DCVSPG circuit. Compare the power dissipation of the DCVS with that of the DCVSPG circuit.

5. Use the pass-transistor logic circuits described in this chapter to design $F = AB + \overline{BC} + \overline{A} \cdot \overline{B}$.

6. Design a logic function $F = AB + AC + \overline{AD}\overline{E} + BE$ using CPL and DPTL and LEAP. Use binary decision diagram (BDD) to design the above logic function.

7. Compute the output switching activity $\alpha_{O_1}, \alpha_{O_2}$, and α_F of the chain-type 4-input AND gate as shown in Fig. 3.58. Consider the cases with $P(A = 1) = P(B = 1) = P(C = 1) = P(D = 1) = 0.75$ and 0.25 respectively. Analyze its switching activity.

4
CMOS Dynamic Logic Circuits

In the previous chapter, CMOS static logic circuits were described. Compared to CMOS static logic circuits, CMOS dynamic logic circuits are more concise. However, more problems exist in the CMOS dynamic logic circuits. In this chapter, CMOS dynamic logic circuits are introduced. Starting from the basic concepts of the dynamic logic circuits, charge-sharing, noise, and race problems are described. Then, NORA, Zipper, and domino dynamic logic circuits are analyzed. In the domino dynamic logic circuits, latched domino, skew-tolerant domino, and multiple-output domino logic (MODL) circuits are presented. In the next portion of this chapter, dynamic differential logic circuits are explained. For dynamic logic circuits, two-phase clocks are usually required. In order to simplify the clocking circuits, true-single-phase-clocking (TSPC) dynamic logic circuits are described, followed by the BiCMOS dynamic logic circuits. In the final portion of this chapter, low-voltage techniques for CMOS dynamic logic circuits are described. Specifically, bootstrapped dynamic logic (BDL) circuits, bootstrapped all-N-logic TSPC dynamic logic circuits, and semi-static DCVSPG-domino logic circuits are analyzed.

4.1 BASIC CONCEPTS OF DYNAMIC LOGIC CIRCUITS

In the CMOS static logic circuits, there are two logic networks including the complementary p-network and n-network. Since only one network is active during the logic evaluation period, the other network is redundant. The drawback of the complementary networks in the CMOS static logic circuit is the low packing density, especially for the circuit with multiple-input complex gates. In addition, with the complementary networks, its output load is also larger, thus its speed is slower. To

164 CMOS DYNAMIC LOGIC CIRCUITS

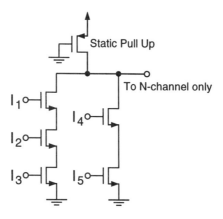

Fig. 4.1 Pseudo-NMOS logic circuits.

improve the CMOS static logic circuit, Fig. 4.1 shows the pseudo-NMOS logic gate circuit. As shown in the figure, the p-network is eliminated. In the pseudo-NMOS logic circuit, PMOS devices have been used to replace the NMOS pull-up load in the NMOS logic circuits. Therefore, as for NMOS logic circuits, pseudo-NMOS logic gate circuits also have ratio logic problems. From this figure, the pull-up and the pull-down structures are not symmetric. In addition, power consumption is high as for NMOS logic circuits. By using the dynamic logic circuit technique, the problems in the pseudo-NMOS logic gate circuits can be solved. Using the dynamic logic circuit technique, only one network—either n-network or p-network—is sufficient. As shown in Fig. 4.2, in n-type and p-type dynamic logic circuits, during the operation, it is composed of two phases—the precharge phase and the evaluation phase. When the clock is low, it is the precharge/predischarge period—the precharge/predischarge transistor turns on. As a result, the output node is precharged to V_{DD} for the n-type dynamic logic circuit; the output node is predischarged to ground for the p-type dynamic logic circuit. During the evaluation phase, the clock is high, hence the precharge/predischarge transistor turns off and the input signals are in effect. For the n-type dynamic logic circuit, if all inputs to the logic gate are high, the related NMOS devices are on. Therefore, the output node is pulled down to ground. Otherwise, if any input NMOS device is off, the pull-down path is off and the output node remains at the high state. For the p-type dynamic logic circuit, it has a complementary mechanism. During the evaluation phase, if all inputs to the logic gate of the charge path are low, the related PMOS devices are on. Therefore, the output node is pulled up to high. Otherwise, if any input is high, the the pull-up path is broken, and the output node remains at the low state. Comparing the dynamic logic circuits with the static logic circuit, the output transition states are different. For the static logic circuits, there is no precharge/predischarge period. The output can change its state any time in accordance with the input state. Therefore, at the output node of the static logic circuit, it can have four transition states: (1) from high to high, (2) from high to low, (3) from low to high, and (4) from low to low. In contrast, at the output node, it

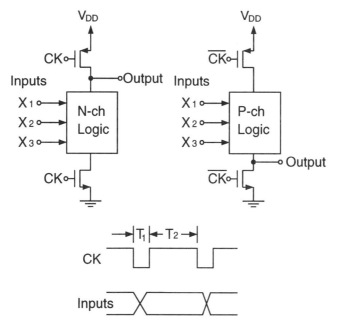

Fig. 4.2 N-type and p-type dynamic logic circuits.

can have only two transition states: (1) from high to high, (2) from high to low, for the n-type dynamic logic circuit; (1) from low to low, (2) from low to high, for the p-type dynamic logic circuit. Due to the deletion of the complementary network, the dynamic logic circuit has less flexibility at the output node. In the dynamic logic circuit, output flexibility has been exchanged for circuit conciseness.

Compared to the static logic circuit, the output load capacitance of the dynamic logic circuit has been reduced since complexity of the circuit has been reduced. Therefore, the speed performance of the dynamic logic circuit is better than that of the static logic circuit. When designing a system using dynamic logic circuits, usually the n-type dynamic logic circuit is preferred. Owing to the higher electron mobility and thus a larger current driving capability, using the n-type dynamic logic circuits, a higher speed performance can be expected. The logic threshold voltage of the dynamic logic circuit is determined by the threshold voltage of the input devices during the evaluation phase. In contrast, the logic threshold voltage of the static logic circuit is usually designed at $V_{DD}/2$—in the middle of the power supply swing. Due to a smaller logic threshold voltage, a higher speed can be expected for the dynamic logic circuit. However, the noise margin of the dynamic logic circuits is much worse than that of the static logic circuits.

Clocks are needed for the dynamic logic circuits. Therefore, the operation of the dynamic logic circuits seems to be a little more complicated. Latches are an

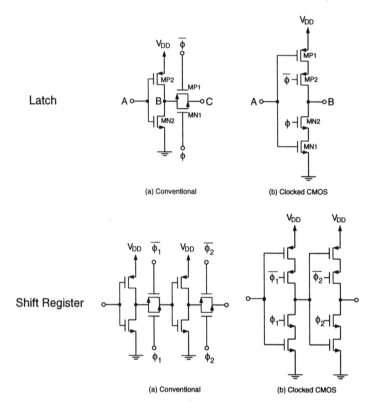

Fig. 4.3 Latch and dynamic shift registers using the conventional CMOS circuit and the clocked CMOS circuit.

important component for CMOS VLSI systems containing sequential machines[4]. Conventionally, controlled by clock, static CMOS transmission-gate latches are used to implement shift registers as shown in Fig. 4.3. In addition to the transmission gate latches, using the clocked CMOS technique, the transmission gates are merged into the CMOS inverter. The operation of a clocked CMOS shift register is composed of the store phase and the pass phase. During the store phase, in the shift register using the conventional CMOS circuit, when the clock is low ($\phi = 0, \overline{\phi} = 1$), the data are stored in the internal node B. When clock is high ($\phi = 1, \overline{\phi} = 0$)—the pass phase, the stored data are passed to the output node C. Using the clocked CMOS circuit, when clock is low, the center inverter is disabled, hence the data are stored at the output node. When clock is high, the center inverter is activated, and the data are present at the output node. As compared to the shift register using the conventional CMOS circuit, no internal node is present in the shift register using the clocked CMOS circuit. In the shift register using the conventional CMOS circuit, both PMOS and NMOS pass transistors are designed to be either off and on at the same time. During the store phase, both NMOS and PMOS pass transistors are supposed to be off simultaneously. If clock signals ϕ and $\overline{\phi}$ are overlapped, during the store phase, the transmission gate,

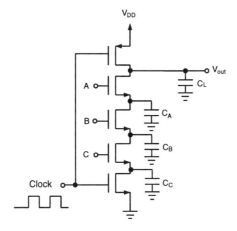

Fig. 4.4 Charge-sharing problems in the CMOS dynamic logic circuit.

which is supposed to be off, may be still on. As a result, the internal output, which may be either 1 or 0, is passed to the output node. This may cause errors in the subsequent stage—the clock race-through problem. On the other hand, for the shift register using the clocked CMOS circuit, during the store phase if the clock signals ϕ and $\overline{\phi}$ are overlapped, only one logic state 1 or 0 can be passed since either pull-up or pull-down path is activated. Compared to the shift register using the conventional CMOS circuit, using the clocked CMOS circuit, the chance of clock race-through problem when clock signals are overlapped is reduced. Using clocked CMOS circuit, the clock race-through problem can be avoided via a proper logic arrangement, as described in the NORA subsection.

4.2 CHARGE-SHARING PROBLEMS

Charge-sharing problems of the CMOS dynamic logic circuit are associated with parasitic capacitances in the logic tree. Consider an n-type CMOS dynamic logic circuit as shown in Fig. 4.4. At the sources of the three input NMOS devices A, B, C, there are three parasitic capacitances C_A, C_B, and C_C. During a logic evaluation period, if three inputs are high, all the capacitances including the output load capacitance (C_L) and the parasitic capacitances (C_A, C_B, C_C) are discharged to a zero voltage. During the following precharge period, the output load capacitance is precharged to V_{DD}. In the following logic evaluation period, if A and B are high and C is low, the output voltage is supposed to stay high. Under this situation, since C_A, C_B and C_L are in parallel, the charge on C_L will redistribute among C_A, C_B, and C_L. As a result, the output voltage may drop to $(C_L/(C_L + C_A + C_B)V_{DD})$. When the parasitic capacitances C_A and C_B are comparable to the output load capacitance C_L, the unwanted drop in the output voltage is substantial. If the output voltage drops to below V_{IH} of the following gate, an erroneous state is produced. This is

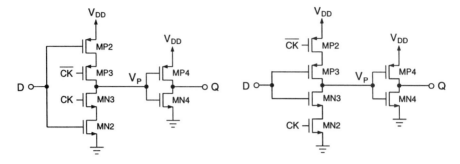

Fig. 4.5 Conventional dynamic latch circuit.

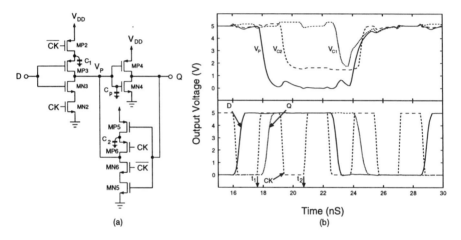

Fig. 4.6 (a) CMOS semi-static latch circuit; and (b) its transient waveform. (Adapted from Lin and Kuo[6].)

the charge-sharing problem, which can be minimized by reducing the number of the inputs to a CMOS dynamic logic circuit such that the parasitic capacitances at the internal nodes can be reduced.

CMOS dynamic latch circuits as shown in Fig. 4.5[4] have advantages in circuit complexity and speed performance. CMOS dynamic latch circuits may also have the charge-sharing problem [7] and the leakage current problem, which are related to the floating internal point (V_p) as shown in Fig. 4.5. The charge-sharing problem may degrade the noise margin of the latch circuit. The leakage current problem limits the lower bound of the operating frequency.

Fig. 4.6 shows a semi-static latch circuit[6], which is composed of the input section ($MP2$, $MP3$, $MN2$, $MN3$), the output section ($MP4$, $MN4$), and the feedback section ($MP5$, $MP6$, $MN5$, $MN6$). As in the conventional dynamic latch circuit, the input section of the semi-static latch circuit is controlled by CK and $\overline{CK}$. Different from the dynamic latch circuit as shown in Fig. 4.5, a feedback section ($MP5$, $MP6$, $MN5$, $MN6$) between the internal node (V_p) and the output node (Q), which is controlled by $\overline{CK}$ and CK, has been added. When CK is high, the output (Q) is equal to the input (D) and the feedback section is disabled. As CK is low, the feedback section starts to function, thus the internal node (V_p) is not floating, which is different from a conventional dynamic latch circuit as shown in Fig. 4.5. Now the internal node V_p is a part of the closed-loop feedback path, which is composed of the output section ($MP4$, $MN4$) and the feedback section ($MP5$, $MP6$, $MN5$, $MN6$). In the conventional dynamic latch circuit as shown in Fig. 4.5(a), when $\overline{CK}$ is high, the internal node (V_p) is floating. Therefore, when $\overline{CK}$ is high, the leakage current of the devices connected to the internal node (V_p) may limit the lower bound of the operating frequency. With the feedback section, the semi-static latch circuit does not suffer from the leakage current problem any more since the internal node (V_p) is not floating during any clock period. Now it is like a static latch circuit, therefore, it is called semi-static latch circuit.

Charge sharing can cause serious operation problems for semi-static circuits[7]. The charge-sharing problem in terms of transient waveforms in a CMOS semi-static latch circuit as shown in Fig. 4.6 is explained below. The charge-sharing problem in a semi-static latch circuit (Fig. 4.6(a)) is divided into two categories: (1) related to the feedback section and (2) related to the input section.

(i) Related to the feedback section

Consider at time ($t = t_1$) for the CMOS semi-static latch as shown in Figs. 4.6(a) with CK switching from low to high and the input (D) is high. When CK is high, the input section is activated and V_p switches from high to low. The output (Q) is turning from low to high. Before the output (Q) becomes high, $MP5$ is on. The parasitic capacitance C_2 at the source of $MP6$ is charged to V_{DD}. When Q becomes high, $MP5$ turns off and C_2 is floating. When CK changes from high to low, $MP6$ turns on, hence the charge on C_2 will redistribute between C_2 and the parasitic capacitance C_p at V_p. Thus V_p may be increased by an amount $\Delta V = \frac{C_2}{C_2+C_p} V_{DD}$. If C_p is not large enough, the output Q may be accidentally pulled low due to an accidental increase in V_p—the charge-sharing problem.

(ii) Related to the input section

Consider at time ($t = t_2$) as indicated in Fig. 4.6(b) with CK switching from low to high and D is high. At this time, the input section is activated and the feedback section is turned off. V_p is low. The parasitic capacitance C_1 is charged to V_{DD} since $MP2$ is on. Then, $MP3$ is off. When CK switches to low, then D turns low; $MP2$ turns off and $MP3$ turns on. V_p will increase by an amount $\Delta V = \frac{C_1}{C_1+C_p} V_{DD}$, which causes the soft charge-sharing problem since V_p is pulled to V_{DD} when CK switches to high again.

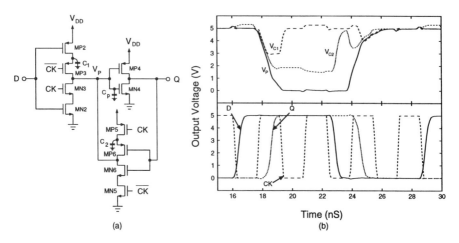

Fig. 4.7 (a) Charge-sharing-problem-free CMOS semi-static latch circuit; and (b) its related transient waveforms. (Adapted from Lin and Kuo[6].)

In both of the input and the feedback sections, charge-sharing problems occur because of the timing sequence in the turn-on of $MP3$ and $MP2$ in the input section and $MP5$ and $MP6$ in the feedback section—$MP2$ turns on earlier than $MP3$ in the input section and $MP5$ turns on earlier than $MP6$ in the feedback section.

Figs. 4.7(a)(b) show the charge-sharing-problem-free CMOS semi-static latch circuit and its related transient waveforms[6]. As shown in Fig. 4.7(a), the charge-sharing-problem-free CMOS semi-static latch circuit is composed of the input section, which is composed of $MN2$, $MN3$ and $MP2$, $MP3$, the output section—$MN4$, $MP4$, and the feedback section—$MN5$, $MN6$, $MP5$, $MP6$. Compared to the semi-static latch circuit as shown in Fig. 4.6(a), the input section in the charge-sharing-problem-free CMOS semi-static latch circuit has a different structure—CK and $\overline{CK}$ are used to control $MP3$ and $MN3$ instead of $MP2$ and $MN2$, and in the feedback section to control $MP5$ and $MN5$ instead of $MP6$ and $MN6$.

The operation of the semi-static latch circuit is divided into two cycles—transfer and store cycles. During the transfer cycle, the input section is activated since when CK is high, $MN3$ and $MP3$ are on. Hence, the logic value of D is transferred to the latch and Q is equal to D. During the transfer cycle, the feedback section is turned off since $MN5$ and $MP5$ are turned off. During the store cycle, CK is low. Therefore, the input section is turned off since $MN3$ and $MP3$ are off. During the store cycle, the feedback section is activated—a latch is formed by the inverter and the feedback section. The charge-sharing problems have been avoided by reversing the turn-on sequence of $MP2$ and $MP3$ in the input section and $MP5$ and $MP6$ in the feedback section as shown in Fig. 4.7(a). As indicated in Fig. 4.7(b), by reversing the turn-on sequence, no accidental change in V_p can be seen during transient.

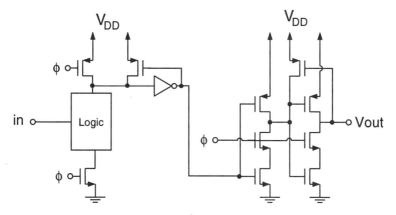

Fig. 4.8 External inverter with feedback to increase the noise immunity for a CMOS dynamic logic circuit. (Adapted from Larsson and Svensson[10].)

4.3 NOISE PROBLEM

Compared to static logic circuits, the major drawback of dynamic logic circuits is their noise immunity. For a static logic circuit, despite the noise, the output can be recovered by itself. In contrast, in the dynamic logic circuit, due to the precharge/predischarge scheme, once the data at the output are destroyed by the noise, it cannot be recovered. For static logic circuits, their noise margin is large. However, in the dynamic logic circuits, the logic noise margin is much smaller—about the threshold voltage. In the dynamic logic circuit system, due to the simultaneous operation of the precharge/predischarge and evaluation schemes, the fluctuation in the current on the power line can be large. Therefore, the sudden change in the current on the ground line (dI/dt) is a major source of noise. In addition, alpha-particles generated charge in a dynamic node can cause noise problems. Electromagnetic radiation from external sources can be another source of noise. Furthermore, capacitively coupled crosstalk from neighboring wires also generates noise[9].

Noise problems in a CMOS dynamic logic circuit can be reduced. Fig. 4.8 shows an external inverter with a pull-up PMOS to increase the noise immunity for a CMOS dynamic logic circuit[10]. As shown in the figure, by adding the inverter with a positive feedback, the noise margin of the CMOS dynamic logic circuit can be improved. During the discharge of the NMOS logic tree, it is like a ratioed logic circuit since the pull-up PMOS device also turns on. The pull-up PMOS device turns off when the inverter output is high. However, the speed is lowered and the power consumption is increased.

172 CMOS DYNAMIC LOGIC CIRCUITS

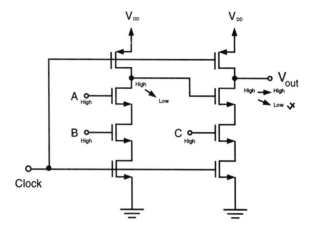

Fig. 4.9 Two cascading n-type dynamic logic circuits.

4.4 RACE PROBLEM

Race problem is a timing problem in a CMOS dynamic logic circuit composed of many cascading dynamic logic gates. Consider two cascading n-type CMOS dynamic logic circuits as shown in Fig. 4.9. During the precharge period, the outputs of the two stages are precharged to V_{DD}. During the logic evaluation period, assume that three inputs A, B, and C turning from low to high. Therefore, during the logic evaluation period, the output of the first stage will turn from high to low. The output of the second stage is supposed to stay high since the output of the first stage is turned low. However, if the propagation delay of the first stage is long, at the very beginning of the logic evaluation period, the output of the second stage may be accidentally switched to low since the input connecting from the output of the first stage cannot turn low instantaneously due to the propagation delay. As a result, an erroneous state at the output of the second stage has been produced. Since this is an n-type CMOS dynamic logic circuit, once the output is pulled low, during this logic evaluation period, it cannot be pulled high again. This is the race problem, which is like a race in timing. Race problem is a major drawback of CMOS dynamic logic circuits. The race problem of the CMOS dynamic logic circuits can be avoided by the domino logic circuit or NORA logic circuit, which will be described in the next section.

4.5 NORA

Clock races are an important problem in a pipelined circuits. To latch the information between two pipelined sections, transmission gates as shown in Fig. 4.10 are necessary. Due to the existence of the clock skew, ϕ and $\overline{\phi}$ may be overlapped. CMOS transmission gates controlled by clocks ϕ and $\overline{\phi}$ may suffer from clock races. As shown in the figure, when $\overline{\phi} = 1$ ($\phi = 0$), the transmission gate TG1 is off and TG2

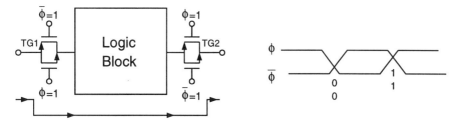

Fig. 4.10 Clock races in CMOS pipelined circuits.

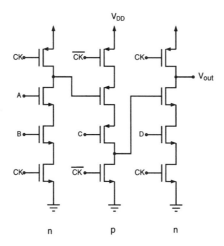

Fig. 4.11 CMOS NORA dynamic logic circuit.

is on. As a result, the data of the previous stage are latched and the output of the present stage is passed via TG2 to the next stage for further processing. Similarly, when $\overline{\phi} = 0$ ($\phi = 1$), transmission gate TG2 turns off and TG1 is on. Via TG1, the output of the previous stage is passed to be processed by the logic block while TG2 is turned off to avoid the pass of the logic block output to the next stage. When TG1 turns on and TG2 is not yet turned off, if the logic block delay is small such that the logic operation can be accomplished before TG2 is turned off, via TG2, the output is erroneously passed to the next stage. Due to the existence of the clock skew, ϕ and $\overline{\phi}$ have overlaps. If ϕ already rises to V_{DD} before $\overline{\phi}$ drops to ground, both TG1 and TG2 are on. The signal of the previous stage may be erroneously transferred via TG1 and TG2 to the next stage—clock signal race-through.

NORA circuit is No Race CMOS dynamic logic circuit[8], which is used to remove the internal signal race problem. In the NORA circuit, n-type and p-type CMOS dynamic logic circuits are connected alternatively as shown in Fig. 4.11. An n-type CMOS dynamic logic circuit is followed by a p-type CMOS dynamic logic circuit. A p-type CMOS dynamic logic circuit is followed by an n-type CMOS dynamic logic circuit. The clock signals for the n-type and p-type CMOS dynamic logic circuits

174 CMOS DYNAMIC LOGIC CIRCUITS

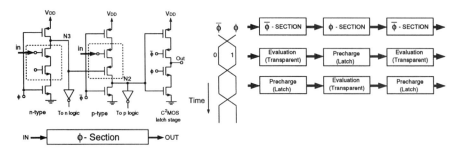

Fig. 4.12 NORA ϕ-section used in a pipelined system. (Adapted from Goncalves and DeMan[8].)

are complementary. For n-type circuits, the clock signal is CK. For p-type circuits, the clock signal is $\overline{CK}$. Therefore, during the precharge/predischarge period, the output of the n-type circuit is precharged to high and the output of the p-type circuit is predischarged to low. Therefore, immediately after the precharge/predischarge period, for a p-type circuit, at the beginning of the logic evaluation period, all inputs are high, hence the input devices are off. No more accidental turn-on of the input devices of the p-type circuits exists at the beginning of the logic evaluation period. Similarly, for an n-type circuit, all inputs are low, hence the input devices are off. No more accidental turn-on of the input devices of the n-type circuits exists at the beginning of the logic evaluation period. Thus race problems have been avoided in the NORA circuit. When implementing the related logic functions using NORA circuits, the n-p-n-p arrangement of the NORA circuits should be kept in mind.

The clock and the internal signal race problems have been described above. To overcome the internal signal race problems, the n-p-n-p arrangement should be used. In order to obtain a clock racefree pipelined system, clocked CMOS (C^2MOS) latch has been used. Fig. 4.12 shows a NORA ϕ-section, which is composed of an n-p-n-p dynamic logic part and a clocked CMOS latch, used in a pipelined system[8]. As shown in the figure, in the pipelined system, a ϕ-section is followed by a $\overline{\phi}$-section, which has a complementary clock timing. Therefore, in the pipelined system, the arrangement of the sections is ϕ-section-$\overline{\phi}$-section-ϕ-section-$\overline{\phi}$-section. When the ϕ-section is active, the dynamic logic part enters the logic evaluation phase and the clocked CMOS latch enters the unlatched—transparent mode. During the logic evaluation phase, the output of the dynamic logic part is passed to the next section via the clocked CMOS latch. At this time, the next section and the previous section are in the precharge/predischarge phase, where their clocked CMOS latches are in the latch mode. When the present section is in the precharge phase, the output of the previous section is transferred as the input to the present section. As shown in the figure, when $\phi = 0$, the ϕ-section is in the precharge/predischarge phase, where its clocked CMOS latch is holding the data. At the same time, the previous section is in the logic evaluation phase. As a result, the output of the dynamic logic part of the previous section is passed via its clocked CMOS latch as the input to the present section. When

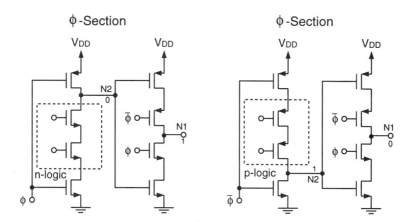

Fig. 4.13 Precharge race-free ϕ-sections. (Adapted from Goncalves and DeMan[8].)

$\phi = 1$, the present section (the ϕ-section) is in the logic evaluation phase and the previous section ($\overline{\phi}$-section) is in the precharge/predischarge phase. At this time, the clocked CMOS latch of the previous section is holding the data, which will be used as the input signals to the ϕ-section for logic evaluation. Therefore, the clocked CMOS latches in the ϕ-section-$\overline{\phi}$-section-ϕ-section-$\overline{\phi}$-section pipelined system are used to relay the signals such that they can be processed in the next sections in the next phase. In order to avoid the internal race problems, there are several guidelines for designing a pipelined system using NORA dynamic logic circuits. Between two consecutive dynamic logic parts in a ϕ-section or a $\overline{\phi}$-section, an even number of static logic circuits can be inserted if the n-p-n-p configuration is used. On the other hand, if instead of the n-p-n-p configuration, n-n or p-p configuration is used, then an odd number of static logic circuits should be inserted between two consecutive dynamic logic blocks. In the following paragraph, methods to avoid the clock race-through problem described in Fig. 4.10 are presented. The key point is to avoid that when clock phases have overlaps the latch is supposed to turn off but it is not turned off. It will be emphasized on two cases—during the logic evaluation phase, (1) the output changes its logic value from its precharged state, and (2) the output maintains its precharged state.

Consider the first case—the output changes its logic value from its precharged state. Fig. 4.13 shows the precharge race-free ϕ- sections[8]. Assuming that in the logic evaluation phase, in the ϕ section, the precharged data has been changed (for the n-logic it changes from high to low; for the p-logic it changes from low to high). When the ϕ-section enters the precharge phase, the output data are changed back to its precharged state. In addition, when considering the phase overlap caused by the clock skew, the input to the following stage, which is supposed to hold the data, may be changed. As shown in the figure, when the ϕ section enters the precharge phase (ϕ is low), the output of the n-type dynamic logic part N2 is precharge to V_{DD}. If ϕ and $\overline{\phi}$ have overlaps such that ϕ and $\overline{\phi}$ are both low, the clocked CMOS latch cannot provide logic 0 at the output. Therefore, during the precharge phase, even when there

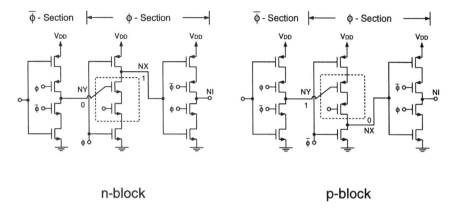

Fig. 4.14 Input variation racefree sequence of dynamic sections with precharge signals kept by the inputs. (Adapted from Goncalves and DeMan[8].)

is a clock overlap, the following stage will not be affected if a clocked CMOS latch is attached to an n-type dynamic logic part in a ϕ-section. A similar reasoning can be done for the p-type dynamic logic part with a clocked CMOS latch. Therefore, in a ϕ-section or $\overline{\phi}$-section, if there is an even number of static inversions inserted between the dynamic logic part and the subsequent clocked CMOS latch, a precharge race-free ϕ- or $\overline{\phi}$- section can be obtained.

Consider the second case. Assuming that during the evaluation phase there is no change from the precharged data in the ϕ-section, thus, for n-block the inputs are 0 and for p-block the inputs are 1. However, before the ϕ-section, in the $\overline{\phi}$-section, during the precharge phase, due to the phase overlap, the output of $\overline{\phi}$-section may lead to a change in the input to the ϕ-section from 0 to 1 for the n-block. The output of the dynamic logic circuit in the ϕ-section, which is supposed to maintain its precharged value, will change its value. Thus, the precharged value in the logic evaluation phase can be altered falsely due to the pass of the clocked CMOS latch of the previous section. Fig. 4.14 shows an input variation race-free sequence of dynamic sections with precharge signals kept by the inputs, which can be used to eliminate this problem[8]. As shown in Fig. 4.14(a), consider the phase overlap caused by the clock skew during the precharge phase ($\overline{\phi} = \phi = 0$), NY changes falsely from logic 0 to logic 1. However, the evaluation phase is at $\phi = 1$. Therefore the above problem will not occur. Similar reasoning can be done for Fig. 4.14(b). Consequently, between the output of a clocked CMOS latch and the following dynamic logic input, if there is an even number of inversions, this output is input variation racefree.

There is another situation—the logic part between C^2MOS latches is static. During the overlapped phase, the output NI will be falsely altered due to NX. If there is an even number of inversions between two clocked CMOS latches as shown in Fig. 4.15, the logic state of N2 is identical to that of NX[8]. At the input transition from low to

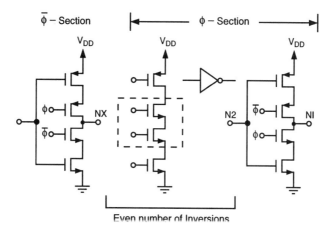

Fig. 4.15 Input variation racefree stages with even number of static inversions inserted between two clocked CMOS latches. (Adapted from Goncalves and DeMan[8].)

high, if N2 and NX switch from high to low, ($\bar{\phi} = \phi = 1$), the output cannot switch from low to high (the PMOS path is blocked). Thus, between two clocked CMOS latches, if there is an even number of static inversions, input variation racefree can be obtained. In summary, by observing the rules for internal delay race-free and clock race-free, using two-phase clock/clocked CMOS latch, by the arrangement of the static, dynamic, and mixed logic circuits, NORA circuits can be obtained.

4.6 ZIPPER

Fig. 4.16 shows a Zipper CMOS dynamic logic circuit[11][12]. As shown in the figure, the Zipper CMOS dynamic logic circuit is made of two parts: (1) n-type and p-type dynamic logic circuits as for NORA, (2) Zipper CMOS driver circuit, which is used to generate the four clock signals (ST, ST', $\overline{ST}$, $\overline{ST'}$) to control the the n-type and p-type dynamic logic circuits.

Fig. 4.17 shows type I and type II driver circuits[11][12]. ST and ST' are used at the CK-controlled NMOS and PMOS devices in the n-type dynamic logic part, respectively. ST and ST' have an identical low level, while the high level of ST' is smaller than that of ST. By this arrangement, when the n-type dynamic logic part enters the logic evaluation phase, the precharge PMOS device is weakly on. Thus, the effect of the leakage current on the output and the charge-sharing problem of the n-type logic part can be reduced. Similarly, $\overline{ST}$ and $\overline{ST'}$ are used to drive the PMOS and the NMOS devices in the p-type dynamic logic part. $\overline{ST}$ and $\overline{ST'}$ have an identical high level, while the low level of $\overline{ST'}$ is higher than that of $\overline{ST}$. As shown in Fig. 4.17, in type I driver, the high level of ST' is $V_{DD} - |V_{TP0}|$ and the low level of $\overline{ST'}$ is V_{TN0}, where V_{TP0} and V_{TN0} are the zero body bias threshold voltage of

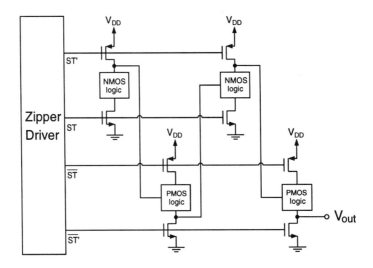

Fig. 4.16 Zipper CMOS dynamic logic circuit. (Adapted from Lee and Szeto[11].)

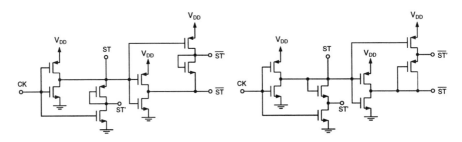

Fig. 4.17 Type I and Type II driver circuits (Adapted from Lee and Szeto[11].)

the PMOS and the NMOS devices, respectively. In type II driver, the high level of ST' is $V_{DD} - V_{TN}(V_{DD} - V_{TN})$ and the low level of $\overline{ST'}$ is $|V_{TP}(|V_{TP}|)|$, where $V_{TN}(V_{DD} - V_{TN})$ and $V_{TP}(|V_{TP}|)$ are the threshold voltage of the NMOS and the PMOS devices considering body effect, respectively. Therefore, using type II driver circuit the conduction of precharge/predischarge device in the logic evaluation phase is more than that using type I driver. Thus, the type II driver circuit is more effective in lowering the leakage current and the charge-sharing problems.

4.7 DOMINO

In order to solve race problems, domino CMOS logic circuit as shown in Fig. 4.18 has been created[1]. 'Domino' is referred to that output stages are pulled down or up sequentially as dominos. As shown in the figure, the domino CMOS logic circuit is composed of the dynamic logic block and an inverter. By adding the inverter, cascading of domino logic blocks does not cause race problems as described in the

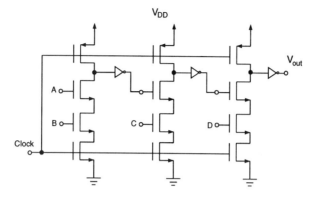

Fig. 4.18 Domino CMOS logic circuit.

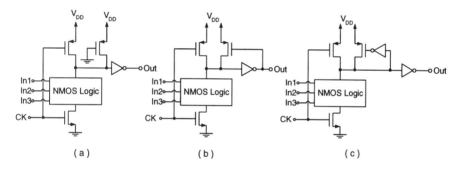

Fig. 4.19 Domino CMOS logic circuit with an additional pull-up device to reduce noise, leakage, and charge-sharing problems. (Adapted from Murabayashi et al. [13].)

previous section. The domino CMOS logic circuit is concise. It can work with simple single-phase clocking for all gates in the circuit. The drawback of the domino CMOS logic circuit is that no inverting capability is available in the circuit. When complementary outputs are needed, in order to avoid race problems, the output of the domino logic circuit cannot be directly connected to an inverter for a complementary output. Instead, another parallel domino gate with a complementary logic output should be implemented. In addition, due to the extra inverter needed, the extra delay may degrade the speed performance. The strong point of the domino CMOS logic circuit is that, owing to the inverter inserted, it can be used to drive a large output load. As for other dynamic logic circuits, leakage current can also cause problems in domino CMOS logic circuits. As shown in Fig. 4.19, by adding a PMOS pull-up load, the noise problem, the leakage current problem and the charge-sharing problem can be lessened[13]. In addition, the input noise margin can be enhanced. The PMOS pull-up load can be realized by grounding the gate or by using an inverter to control the gate. Using a feedback to control the PMOS is a better approach since the PMOS device is turned off when the output switches to high. In addition, the inverter itself is needed by the domino logic circuit. Thus power consumption can be reduced and

180 CMOS DYNAMIC LOGIC CIRCUITS

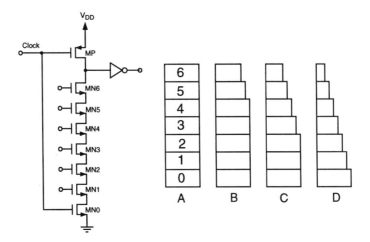

Fig. 4.20 Domino CMOS six-input NAND gate with a nonuniform scaling for a shorter switching delay. (Adapted from Shoji[14].)

it has a smaller problem with the ratioed logic.

The switching performance of a domino CMOS logic circuit can be improved by scaling the size of the input transistors nonuniformly. Fig. 4.20 shows a domino CMOS six-input NAND gate with a nonuniform scaling for a shorter switching delay[14]. As shown in the figure, by arranging the sizes of the input transistors in the n-type logic block nonuniformly, the switching performance can be optimized. For the input transistor closer to ground its aspect ratio is set to be larger, its delay time is smaller. Consider case B as shown in the figure. By shrinking the aspect ratio of the top input transistor, its effective resistance is increased. As a result, its delay increases. On the other hand, its parasitic capacitance at the output is decreased, hence its delay decreases. As shown in the figure, the size of the input transistors is scaled down from bottom to top, at the output node the parasitic resistance increases by a small amount as compared to the case with a uniform size in the input transistors. With the nonuniform scaling, the effective capacitance at the output node has been decreased substantially. As a result, with the nonuniform scaling, the output delay time can be shrunk. This graded scaling technique can be used for NORA and other dynamic logic circuits.

Domino CMOS dynamic logic circuits have advantages in NMOS-like performance and CMOS-like power consumption, and bypass of the race problem in a dynamic logic circuit. However, domino dynamic logic circuits still suffer from charge-sharing and leakage current problems as described before. The charge-sharing problem can be reduced by several techniques: (1) At the important nodes, adding precharge PMOS transistors; (2) selectively increasing the storage capacity of the precharge node in proportion to the number of nodes to which the charge can be redistributed; (3) increasing the size of the inverter such that the output load capacitance

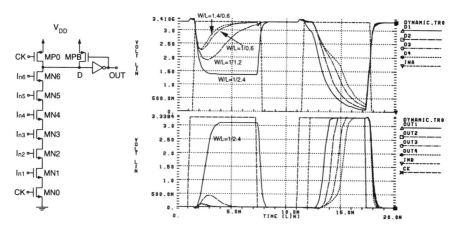

Fig. 4.21 Six-input AND gate with a feedback PMOS to reduce the charge sharing problem and with its transient waveforms.

can be increased to reduce the charge-sharing problem; (4) adding a PMOS pull-up PMOS transistor to compensate for the voltage lowered by charge sharing[13][15]. Fig. 4.21 shows a 6-input AND gate with a feedback PMOS to reduce the charge-sharing problem and its transient waveforms. As shown in the figure, by adding a PMOS pull-up transistor, the leakage current problem can be solved. However, due to charge sharing, the output voltage will fall first. Then, the output voltage will be pulled up by the PMOS pull-up transistor. Therefore, if the pull-up PMOS device is with a larger aspect ratio, charge-sharing problems can be more effectively lessened. In the dynamic logic circuit, the PMOS pull-up transistor and the NMOS logic pull-down devices in the circuit constitute a ratioed logic. Therefore, a larger width in the PMOS device leads to a longer delay. As shown in the figure, the worst-case charge-sharing problem happens when I_{n1}–I_{n6} are high at the previous cycle and I_{n2}–I_{n6} are high during this cycle and I_{n1} is low. The charge in the parasitic capacitance at node D will be redistributed with the parasitic capacitances C_1–C_5—charge-sharing. This charge-sharing problem can be lessened by the PMOS device MPB. As shown in the figure, four cases with four aspect ratios—$W/L = 1\mu m/2.4\mu m$, $1\mu m/1.2\mu m$, $1\mu m/0.6\mu m$, and $1.4\mu m/0.6\mu m$ for MPB have been studied. With an aspect ratio of $1\mu m/2.4\mu m$ for MPB, the charge-sharing problem cannot be solved since before the pull-up of node D, the output is already pulled to high—MPB is turned off. Thus, with a small aspect ratio of MPB, charge-sharing problems cannot be solved. When the aspect ratio of MPB is increased, the charge-sharing problems can be reduced. However, with a larger aspect ratio of MPB, discharge of node D becomes more difficult. With an aspect ratio of $1.4\mu m/0.6\mu m$ for MPB, at the beginning, it is like a ratioed logic. After the output becomes high to turn off MPB, node D is discharged to ground without the interference from the PMOS device. Note that in the circuit simulation the aspect ratio of MP and MN is $4\mu m/0.6\mu m$.

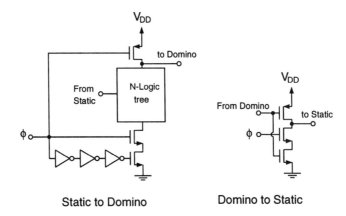

Fig. 4.22 Static-to-domino and domino-to-static interface circuits. (Adapted from Harris and Horowitz[18].)

Fig. 4.22 shows the static-to-domino and the domino-to-static interface circuits[18]. During the logic evaluation phase, the static input to a dynamic gate needs to be stable. In the static-to-domino interface circuit, the 'pulsed latch' technique has been adopted. After the 3-inverter delay in the logic evaluation phase, the static-to-domino interface circuit holds the data—even when the static input is changed, the output will not be destroyed. In the domino-to-static interface circuit, the n-type clocked CMOS latch (N-C^2MOS) is used. During the precharge phase, the input to N-C^2MOS is high and ϕ is low. Thus the output data are latched. When ϕ is high, it is the evaluation phase for the dynamic logic circuit, where the N-C^2MOS latch functions as a static inverter.

Fig. 4.23 shows the domino logic circuit with the clock-and-data precharge dynamic (CDPD) technique[16]. As shown in Fig. 4.23(b), the precharge/predischarge transistor can be controlled by the output of the previous stage, instead of the clock, thus a more concise domino circuit can be obtained. In addition, the load for the clock can be reduced. As shown in the figure, cascading of CDPD domino logic circuits is similar to the n-p-n-p connection of the NORA dynamic logic circuits. As shown in the figure, a p-type dynamic logic circuit is with its input precharged to high and its output predischarge to low. As shown in Fig. 4.23(c), by cascading the domino logic circuits with the CDPD technique, the clock of a stage can be controlled by the system clock, thus the speed in the precharge/predischarge phase of the subsequent dynamic logic circuits can be enhanced. In this domino logic circuit with the CDPD technique, it is internal delay race-free and precharge race-free.

4.7.1 Latched Domino

One drawback of the domino CMOS dynamic logic circuit is the lack of the complementary outputs. In order to obtain a complementary output, another logic gate should be implemented. Fig. 4.24 shows a latched domino CMOS dynamic logic

DOMINO   183

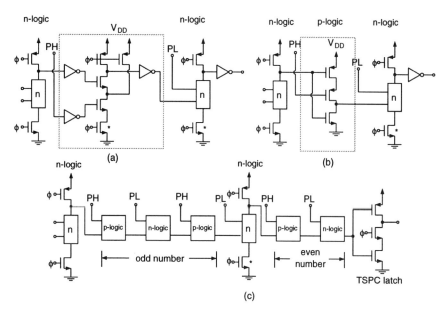

Fig. 4.23 (a) Domino logic circuit; (b) Domino logic circuit with clock-and-data precharged dynamic (CDPD) technique; (c) Cascading of CDPD. (Adapted from Yuan et al. [16].)

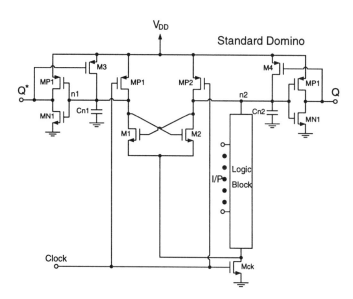

Fig. 4.24 Latched domino CMOS dynamic logic circuit. (Adapted from Pretorius et al. [17].)

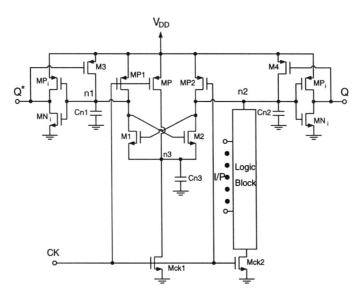

Fig. 4.25 Improved latched domino CMOS dynamic logic circuit. (Adapted from Pretorius et al. [17].)

circuit[17], which has complementary outputs without duplicating the logic circuits. By adding a cross-coupled NMOS latch ($M1$, $M2$) to the domino logic circuit, complementary outputs can be obtained. The inputs to the latched domino logic circuits should be set up during the precharge phase to avoid the charge-sharing problem during the evaluation phase. Therefore, latched domino logic circuits are suitable for use as the first gate in a pipelined logic system or as the gate between static and dynamic logic circuits. The drawback of the latched domino CMOS dynamic logic circuit is that the aspect ratio of $M1$ should be larger than that of $M2$. Therefore, the difference should be considered. The discharge of node $n1$ is by $M1$ and the discharge of node $n2$ is by $M2$ and the n-logic block. Hence, the design procedure is more complicated.

Fig. 4.25 shows the improved latched domino CMOS dynamic logic circuit[17]. As shown in the figure, $Mck1$ has been added to reduce the capacitance load of the logic block. In addition, precharge transistor MP is added such that during the precharge phase the source of $M1$ and $M2$ (n_3) can reach V_{DD}. The operation of the improved latched domino CMOS dynamic logic circuit is described here. When CK is low, it is precharge phase. At this time, $n1$, $n2$, and $n3$ are precharged to V_{DD} and the outputs Q and Q^* are low. The pull-up PMOS devices $M3$ and $M4$ are on. When CK switches from low to high, $Mck1$ turns on. Therefore, $Cn3$ is being discharged until $M1$ and $M2$ turn on to discharge $Cn1$ and $Cn2$. Since $Cn2$ is larger than $Cn1$ and the aspect ratio of $M1$ is designed to be larger than that of $M2$, the potential at $n1$ is discharged faster than $n2$. If there is no conducting path in the n-logic, $M1$ will discharge $n1$ to turn off $M2$. Therefore, $n1$ is low and $n2$ is

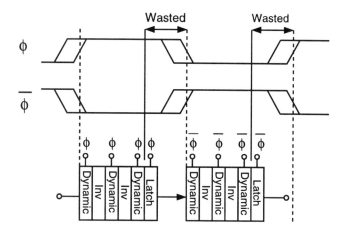

Fig. 4.26 Domino pipeline with clock skew. (Adapted from Harris and Horowitz[18].)

high; output Q is low and Q^* is high. If there is a conducting path in the n-logic, $n2$ is discharged faster than $n1$ and the situation is reversed.

4.7.2 Skew-Tolerant Domino

In a pipelined system as shown in Fig. 4.26, due to the clock skew problem, the clock arrival time in each dynamic gate may be different. The uncertainty in the clock arrival time—clock skew is due to several reasons: (1) phase locked-loop jitter, (2) mismatch in the clock distribution network, (3) data-dependent clock loading, and (4) process variation. The clock skew shortens the time available for logic evaluation. In addition, inclusion of the latches causes extra delays. As a result, the evaluation phase cannot be fully utilized and thus it causes timing waste. When clock frequency is higher, the timing waste problem becomes more serious.

The clock skew problem can be solved by using the phase-overlapping pipelined stage technique instead of the conventional non-overlapping clocking technique. As shown in Fig. 4.27[18], owing to the phase overlap, latches are not needed any more. The input to the domino logic is low initially. Depending on the output of the previous stage, the input will switch high or maintain low. During the overlapped phase, both blocks are in the logic evaluation phase. The dynamic logic result of the ϕ_2-block will not be effective until the conclusion of the final evaluation of the ϕ_1-block and data transfer from the ϕ_1-block to the ϕ_2-block has been accomplished. As long as the last gate of the ϕ_1-block can enter the precharge operation after the first gate of the ϕ_2-block finishes the propagation of the logic result, the operation of the ϕ_2-block will not be affected. Therefore, the clock skew problem can be avoided by the phase overlap technique.

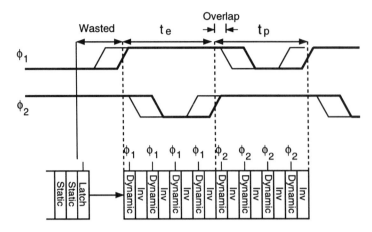

Fig. 4.27 Two-phase overlapping domino clocks. (Adapted from Harris and Horowitz [18].)

With clock skew, how to select the minimum time for the precharge and the evaluation phase is considered here. As shown in Fig. 4.28(a)[18], consider the situation where the clock phase ϕ_{1a} of the previous gate (A and A′) comes late. On the contrary, the clock phase ϕ_{1b} of the next gate (B and B′) comes early. Therefore, when B enters the evaluation phase, we have to make sure that the precharge operation has been accomplished for A. Otherwise, B may have a false discharge. Therefore, the overlap of the low level between the latest and the earliest precharge phases is the minimum time for A and A′ to finish precharge (t_{prech}). By considering this clock skew (t_{skew}), it is the minimum period for the precharge phase:

$$t_p \geq t_{prech} + t_{skew},$$

where t_p is the time of the precharge phase, and t_{prech} is the minimum precharge time. As shown in Fig. 4.28(b), the minimum time for the evaluation phase is considered here. Assume that the clock ϕ_{1b} of the last stage of the ϕ_1-block comes the earliest during the ϕ_1 phase and the clock ϕ_{2a} of the first stage of the ϕ_2-block comes the latest. Therefore, before ϕ_{1b} enters the precharge phase, its output data should have been evaluated by ϕ_{2a} and propagated. When ϕ_{1b} enters the precharge phase, ϕ_{2a} already accomplishes signal propagation. Therefore, the precharge of ϕ_{1b} does cause detrimental damages in ϕ_{2a}. t_{hold} is defined as the overlap time when ϕ_{2a} enters the evaluation phase and ϕ_{1b} enters the precharge phase. Considering an N-overlap systems, the delay between two logic blocks of different phases is T/N, the total precharge time (t_e) should be:

$$t_e \geq \frac{T}{n} + t_{skew} + t_{hold},$$

where t_{hold} is the hold time, which is generally negative or zero. Considering the clock period, which is the sum of the precharge time and the evaluation time

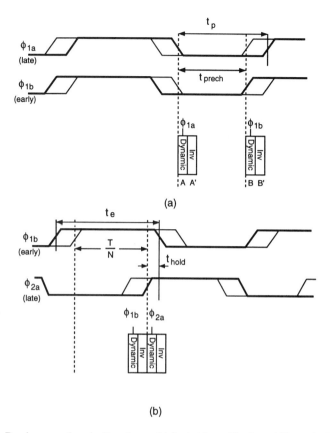

Fig. 4.28 Precharge and evaluation times. (Adapted from Harris and Horowitz [18].)

$T = t_p + t_e$, the maximum skew time is

$$t_{skew-max} = \left(\frac{N-1}{N}T - t_{hold} - t_{prech}\right)/2.$$

In a well-distributed pipelined system with local clock generation to reduce local clock skew, the overlapped phase can be larger than the actual clock skew. The extra overlap phase can be used for time borrowing—the logic evaluation of phase ϕ_1 can be extended to the logic evaluation phase of phase ϕ_2. After the completion of the logic evaluation for the domino logic gate, a change in the logic input from high to low does not change its gate output. In addition, when the logic evaluation of the previous stage has not been finished, the output of its logic gate will not change from low to high. Thus the first gate of phase ϕ_2 cannot enter the logic evaluation period until the last gate of phase ϕ_1 has accomplished logic evaluation. When the last gate of phase ϕ_1 enters the precharge phase, the logic output of the first gate of phase ϕ_2 does not change. Phase ϕ_1 and phase ϕ_2 can be regarded as another valid phase—

188 CMOS DYNAMIC LOGIC CIRCUITS

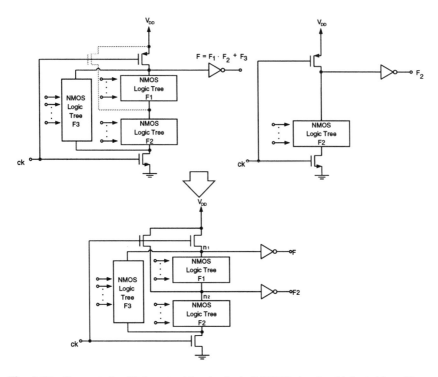

Fig. 4.29 Concept of multiple-output domino logic (MODL) circuits. (Adapted from Hwang and Fisher[19].)

the logic evaluation phase has been lengthened. Therefore, in addition to being used to increase the clock skew tolerance, the phase overlap can also be used for logic evaluation purposes.

4.7.3 Multiple-Output Domino Logic (MODL)

In the conventional domino logic circuits, if a logic function is made of serial connection of two logic functions, when one of the two functions needs to be used as another output, the logic functions need to be implemented twice. The concept of multiple-output domino logic (MODL) circuits[19] as shown in Fig. 4.29 is used to take advantages of the subfunction already existing in the logic tree such that no duplicate logic circuits are needed. As shown in the figure, in the MODL circuit, by adding an inverter and a pull-up PMOS device, the subfunction in the logic tree can be used. MODL circuits are especially convenient for a complex logic system. In order to solve the charge-sharing problems in a domino logic circuit, at important nodes, pull-up PMOS devices are added. Therefore, to implement an MODL circuit, only an extra inverter is required. However, there is a situation worth pointing out. As shown in the figure, consider a case when logic trees F_1 and F_3 are on and F_2 is off. Since F_1 is on, there is a conducting path from $n2$ to $n1$, but $n1$ is connected to

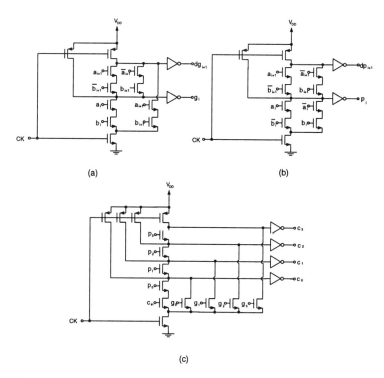

Fig. 4.30 Adder designed using MODL circuits. (Adapted from Hwang and Fisher [19].)

ground via F_3. $n2$ is erroneously connected to ground via F_1 and F_3. As a result, an erroneous situation occurs at $n2$. When doing designs, in order to avoid the above sneak path problem, F_1 and F_3 need to be mutually disjoint, which implies that F_1 and F_3 cannot have a conducting path simultaneously.

Fig. 4.30 shows the adder designed using MODL circuits[19]. In an adder, the ith-bit carry signal is produced by $c_i = a_i \cdot b_i + c_{i-1} \cdot (a_i + b_i)$, where a_i and b_i are the two inputs of ith-bit. By defining the generate signal as $g_i = a_i \cdot b_i$ and the propagate signal as $p_i = a_i + b_i$, the carry signal becomes $c_i = g_i + p_i \cdot c_{i-1}$, which can be expressed as function of c_{i-2}: $c_i = g_i + p_i \cdot g_{i-1} + p_i \cdot p_{i-1} \cdot c_{i-2}$. By defining $dg_i = g_i + p_i \cdot g_{i-1}$ and $dp_i = p_i \cdot p_{i-1}$, one may obtain the relationship between c_i and c_{i-2}: $c_i = dg_i + dp_i \cdot c_{i-2}$. Based on $c_i = g_i + p_i \cdot c_{i-1}$, c_1-c_4 can be realized using MODL circuits as shown in Fig. 4.30(c), which is the well-known dynamic Manchester carry chain. Consider the following case: $p_3 = g_3 = 1$, $p_2 - p_0$, and $g_2 - g_0 = 0$ in the figure. At this time, the output is $c_3 = 1$, $c_2 = c_1 = c_0 = 0$. Since $p_3 = 1$, a sneak path leads to $c_3 = 1$, $c_2 = 1$, $c_1 = c_0 = 0$, which is a wrong output state. Therefore, p_i needs to be rearranged as $p_i = a_i \oplus b_i$, such that p_i and g_i are mutually disjoint to avoid the sneak path problem.

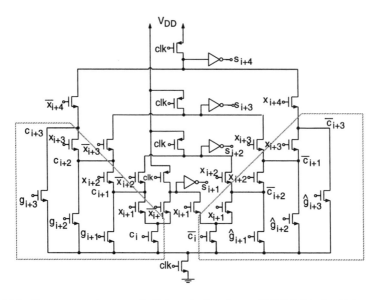

Fig. 4.31 Four consecutive sums built in a single enhanced MODL block. (Adapted from Wang et al. [20].)

The output of the MODL circuit is non-inverting. Therefore, the complementary carry signal $\overline{c_i}$ should be regenerated by: $\overline{c_i} = \overline{g_i + p_i \cdot c_{i-1}} = \overline{g_i} \cdot \overline{p_i} + \overline{g_i} \cdot \overline{c_{i+1}} = \overline{a_i} \cdot \overline{b_i} + (\overline{a_i} + \overline{b_i}) \cdot \overline{c_{i-1}}$. From above, the logic expression for $\overline{c_i}$ is similar to c_i except all inputs are inverted. Hence, c_i, $\overline{c_i}$, and s_i are implemented in a single enhanced MODL block as shown in Fig. 4.31[20]. The sum signal is expressed as $s_i = x_i \oplus c_{i-1} = x_i \overline{c_{i-1}} + \overline{x_i} c_{i-1}$, where $x_i = a_i \oplus b_i$. As marked by the dashed line in Fig. 4.31, using the carry chain circuit, s_i has been implemented. Fig. 4.31 looks like a MODL, but in fact, it is not. The basic concept of MODL circuits is to take advantages of the subfunctions, which are also outputs. In Fig. 4.31, s_{i+1} is not related to s_i. As shown in the figure, s_{i+1}, s_{i+2}, s_{i+3}, s_{i+4} are four independent outputs with inputs from the same carry chain circuit block as marked in the dashed line. The MODL concept has been utilized to provide subfunction at the output for use in the next stage without inverters—it implies from cascade to cascode. Taking advantage of the c_i logic tree, s_{i+1} is formed—enhanced MODL circuits.

4.8 DYNAMIC DIFFERENTIAL

In a domino logic circuit, complementary output signals cannot be produced by the same gate. Therefore, sometimes to realize a complicated logic function using domino logic circuits can be cumbersome. In this section, the dynamic differential logic, which has a complementary logic tree, can be used to produce complementary output signals simultaneously. In a dynamic differential logic circuit as shown in Fig. 4.32[21], during the precharge phase, its output nodes are pulled up to high.

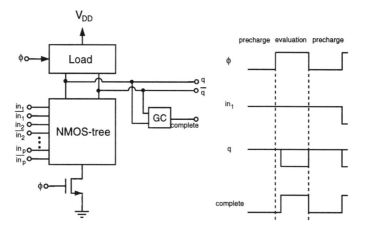

Fig. 4.32 Concept of a differential circuit. (Adapted from Acosta et al. [21].)

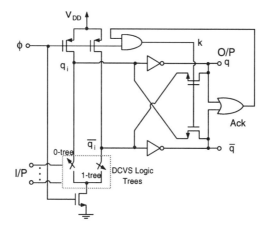

Fig. 4.33 Self-timed circuit. (Adapted from Lau [22].)

During the evaluation phase, depending on the inputs of the logic tree (1-tree or 0-tree), an output node is discharged to ground. Therefore, the dynamic differential logic circuit is suitable to be incorporated in the domino logic circuit to produce complementary outputs. In addition, during the logic evaluation period, since one of two complementary outputs q and $\bar{q}$ must be discharged from high to low, it can be used in the self-timed circuit—using q and $\bar{q}$ to generate a complete signal to indicate the conclusion of the logic evaluation for other circuits. Fig. 4.33 shows a self-timed circuit using the DCVS dynamic logic circuit technique[22]. As shown in the figure, it has a DCVS dynamic logic circuit and a self-timed signal generating circuit. When ϕ is low, the DCVS dynamic logic circuit enters the precharge period. q_i and $\overline{q_i}$ are precharged to V_{DD} by a precharge PMOS device. Thus, both q and $\bar{q}$ are low and $Ack = q + \bar{q}$ is also low. When ϕ is high, one of q_i and $\overline{q_i}$ becomes low. Thus, Ack changes to high and k also becomes high. Two NMOS transistors turn on to form

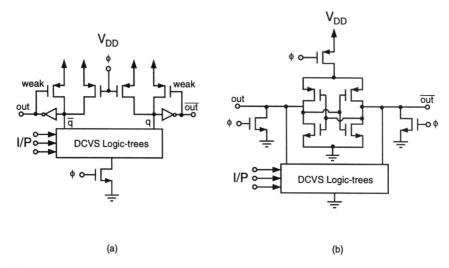

Fig. 4.34 (a) Differential cascode voltage switch (DCVS) domino logic circuit (after Heller et al. [23]); and (b) enable-disabled CMOS differential logic (ECDL) circuit. (Adapted from Lu and Ercegovac [24][25].)

a latch with two inverters. Therefore, q_i and $\overline{q_i}$ maintain a stable state. The Ack signal can be used as the ϕ signal for the next stage. Though its operation is similar to a dynamic logic circuit, its precharge-evaluation periods are not controlled by a common clock. Instead, they are controlled by the signals produced in the previous stage. Thus, self-timed logic operation can be reached and clock loading can be substantially reduced.

Fig. 4.34 shows the differential cascode voltage switch (DCVS) domino logic circuit[23] and the enable-disabled CMOS differential logic (ECDL) circuit[24][25]. DCVS-domino is a differential domino logic circuit with two complementary outputs. A pull-up PMOS device controlled by an output feedback is used to decrease leakage and charge-sharing problems. The operation is basically identical to that of a domino logic except that only one of the two logic trees has a conducting path. The operation of ECDL is different from that of DCVS-domino. When ϕ is high, it's the predischarge phase, where the back-to-back CMOS inverter latch is turned off by a PMOS device. When ϕ is low, it's the evaluation phase, where the activated CMOS latch pulls up the outputs q and $\overline{q}$. If a conduction path exists in any logic tree, due to the regenerative feedback of the CMOS latch, a final stable value will be reached for the outputs. Different from DCVS-domino, after the evaluation phase, the ECDL functions as a static logic circuit. Only when ϕ is high, it functions as a dynamic logic circuit in the precharge phase, where the CMOS latch is disabled.

Fig. 4.35 shows the sample-set clocked differential logic (SSDL) circuit[26]. As shown in the figure, it has two PMOS pull-up transistors, a DCVS logic tree, and a latching sense amp. The operation of the sample-set clocked differential logic circuit

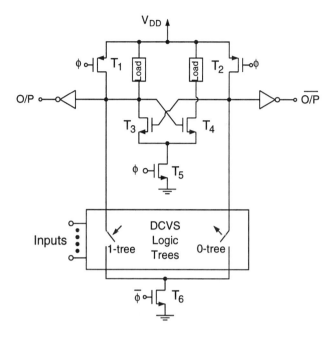

Fig. 4.35 Sample-set clocked differential logic (SSDL) circuit. (Adapted from Grotjohn and Hoefflinger [26].)

is different from the circuit described before. When ϕ is low, it is the sample phase, where PMOS T1 and T2 and NMOS T3 turn on and NMOS T6 is off. Thus the latching sense amp does not have any action. Since PMOS T1 and T2 are turned on, in the DCVS logic tree, the node without a conduction path will be pulled up to V_{DD}. The voltage at the other node with a conduction path is determined by conduction ratio of the pull-up PMOS and pull-down NMOS logic trees—ratioed logic. Thus its voltage is lower than V_{DD}. When ϕ changes from low to high, it is the set phase, where PMOS T1, T2, and NMOS T3 are off. Thus the logic tree is not active. At this time, the turn-on of T6 activates the latching sense amp. As a result, the node with a voltage lower than V_{DD} is discharged to ground. By this arrangement, as compared to a standard DCVS domino logic, the discharge capability of the SSDL latching sense amp is much larger especially when the serial path is long for a large number of inputs to the logic tree. However, during the sample phase, since it is a ratioed logic, power dissipation is large.

Fig. 4.36 shows switched output differential structure (SODS) dynamic logic[21]. As shown in the figure, SODS looks like a differential split-level logic (DSL). Only T4 and T5 are controlled by clock instead of a reference voltage. In addition, an equalizing PMOS device T3 is added for equalization. When ϕ is low, it is the precharge phase, where NMOS T4 and T5 are off and PMOS T3 is on. Since one of q and $\bar{q}$ must be low, hence, one of PMOS T1 and T2 must turn on. In addi-

194   CMOS DYNAMIC LOGIC CIRCUITS

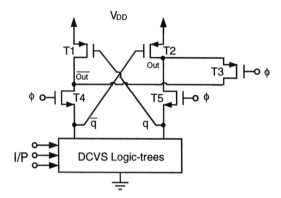

Fig. 4.36 Switched output differential structure (SODS) dynamic logic. (Adapted from Acosta et al. [21].)

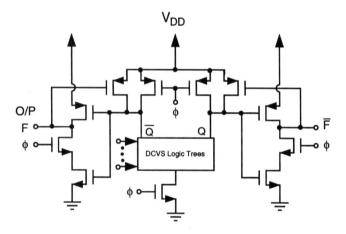

Fig. 4.37 Differential NORA dynamic logic circuit. (Adapted from Ng et al. [27].)

tion, T3 is also turned on. Therefore, in the precharge phase, both *out* and $\overline{out}$ are precharged to high. When ϕ is high, T3 turns off and T4 and T5 are on. The function of the whole circuit is identical to DSL. Therefore, during the evaluation phase, SODS is in the static mode. Only when ϕ is low, it functions as a dynamic logic circuit. Therefore, SODS combines the reliability and the noise margin of the static logic circuits with the n-logic tree advantages of the dynamic logic circuits. During the precharge phase, one of q and $\overline{q}$ is low. When ϕ is high, one of both *out* and $\overline{out}$ has the charge-sharing problem. When ϕ is high, SODS is under the static operation. Thus the charge-sharing problem does not cause an error in logic.

Fig. 4.37 shows the differential NORA dynamic logic circuit[27]. Differential NORA circuit, which is used in a pipelined system, is similar to DCVS-domino logic circuit. The output inverter in the DCVS-domino logic circuit has been replaced

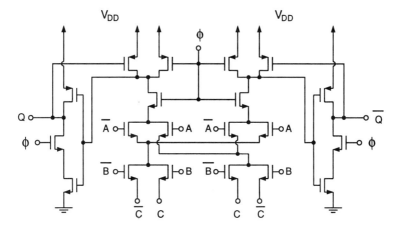

Fig. 4.38 Dynamic DCVSPG NORA-like sum circuit. (Adapted from Lai and Hwang [28].)

by an n-latch. During the precharge phase, both Q and $\overline{Q}$ are precharged to V_{DD}, therefore the clocked PMOS device in the clocked CMOS (C^2MOS) latch is not necessary. During the precharge phase, the output is latched. Therefore, it is suitable for high-speed pipelined systems. When CK is high, it is the evaluation phase, where the n-latch functions like an inverter. During the evaluation phase, it is identical to DCVS domino logic.

DCVSPG logic circuit described in the previous chapter can be used for dynamic operation. Fig. 4.38 shows the dynamic DCVSPG NORA-like sum circuit[28]. The operation of the DCVSPG NORA-like circuit is similar to DCVS-NORA. In the DCVS-NORA, there is only one logic tree with a conducting path. In DCVSPG-NORA, via the pass logic tree, both logic trees have conduction paths to transfer the complementary pass variables to the outputs. Therefore, no similar leakage problems as in dynamic or DCVS domino exist in the DCVSPG-NORA. The noise problem can be smaller than in the dynamic logic or DCVS domino logic. The lowest level of the logic-1 output is at $V_{DD} - V_{TN}$ despite the leakage current.

Fig. 4.39 shows the implementation of a logic function $\overline{A \cdot B}$ using the dynamic DCVS logic, and the regenerative push-pull differential logic (RPPDL)[29]. In the regenerative push-pull differential logic (RPPDL) circuit, it has a precharge-high/evaluate-low LO_{NET} and a predischarge-low/evaluate-high HI_{NET}, and a cross-coupled NMOS/PMOS, where the NMOS device is used to help discharge of LO_{NET} and the PMOS device is used to help charge of the HI_{NET}. From the structure, it is a regenerative feedback path formed by two dynamic logic circuits—precharge high/evaluate low and precharge low/evaluate high, with the cross-coupled PMOS/NMOS transistors. The operation of RPPDL is similar to a dynamic logic circuit. When ϕ is high, it is the precharge/predischarge phase, where the output is precharged/predischarged to V_{DD}/ground for $A \cdot B / \overline{A \cdot B}$. When ϕ is low, it

196 CMOS DYNAMIC LOGIC CIRCUITS

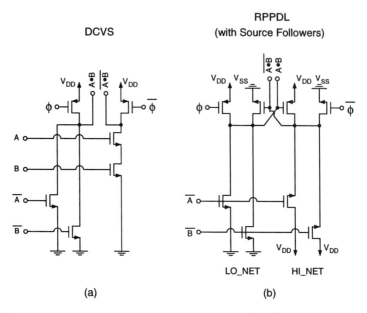

Fig. 4.39 Implementation of the logic function $A \cdot B$ and $\overline{A \cdot B}$ using (a) dynamic DCVS logic; and (b) regenerative push-pull differential logic (RPPDL). (Adapted from Partovi and Draper [29].)

is the evaluate phase. If an input— $\overline{A}$ or $\overline{B}$ is high, one side is discharged and the other side is charged. In addition, the cross-coupled PMOS/NMOS turns on to form a regenerative positive feedback to speed up the charge/discharge process. Thus the speed of RPPDL is highest among all dynamic logic circuits. However, due to the regenerative positive feedback formed by the cross-coupled PMOS/NMOS, its noise immunity is the worst. If the noise at the output node triggers the formation of the positive feedback, the output may be falsely discharged/charged to a wrong unrecoverable state. Fig. 4.40 shows the regenerative push-pull differential logic (RPPDL) implementation of a generalized parallel/serial logic network[29]. As shown in the figure, an 8-bit carry chain has been designed. From this figure, the circuit configuration of the RPPDL is systematic. Owing to the regenerative positive feedback, its speed is faster than any other dynamic logic circuit.

4.9 TRUE-SINGLE-PHASE CLOCKING (TSPC)

For a VLSI system using dynamic logic circuits, clock timing, which provides a safe operation, can be very complicated. Non-overlapping two-phase clock was used in the early years. Pseudo-two-phase clocking, which actually provides four clock phases, was used[4]. Using an n-p dynamic logic circuit technique, a two-phase clock has been developed[8]. Fig. 4.41 shows a pipeline of precharged logic circuits using

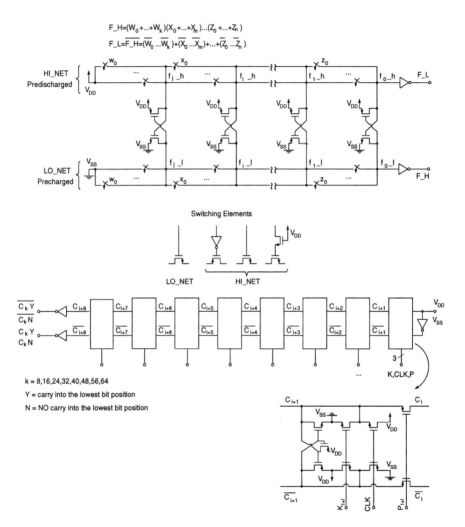

Fig. 4.40 Regenerative push-pull differential logic (RPPDL) implementation of a generalized parallel/serial logic network. (From Partovi and Draper [29], ©1994 IEEE.)

198 CMOS DYNAMIC LOGIC CIRCUITS

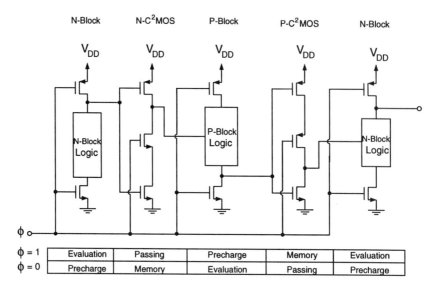

Fig. 4.41 Pipeline of precharged logic circuits using single-phase clock. (Adapted from Yuan et al. [31].)

single-phase clocking[31]. As shown in the figure, an n-dynamic logic block with an n-clocked CMOS (n-C^2MOS) latch forms the ϕ-block. A p-dynamic logic block with a p-clocked CMOS latch forms the $\overline{\phi}$-block. As shown in the figure, only one clock phase is needed to realize the pipeline operation. When $\phi = 1$, the n-dynamic logic block of the ϕ-block is in the logic evaluation phase. The n-C^2MOS latch passes the output of the n-dynamic logic block to the next stage. At this time, the p-dynamic logic block of the $\overline{\phi}$-block is in the predischarge phase and the p-C^2MOS latch holds the output such that the predischarged output of the p-dynamic logic block does not affect the evaluation operation of the n-dynamic logic block in the ϕ-block.

In the conventional clocked CMOS (C^2MOS) latch, four clock phases (quasi-two phase) are needed. In NORA dynamic logic circuit as shown in Fig. 4.42, only two clock phases (ϕ, $\overline{\phi}$) are required[8]. However, as described in the NORA section, there are some restraints on the logic arrangements. From the figure, in the ϕ-section, the output of the n-dynamic logic is precharged to high, hence $M2$ can be eliminated. Similarly, in the $\overline{\phi}$-section, the output of the p-dynamic logic is predischarged to ground, hence $M7$ can be eliminated. After eliminating $M2$ and $M7$, it has become the true-single-phase clocking (TSPC) circuit. Fig. 4.43 shows two cascaded n-clocked and p-clocked latch stages[5]. As shown in the figure, the circuits before the latches can be used for logic function. In the first stage circuit in the TSPC latch, as long as they are not clock-controlled transistors, they can be used to implement logic evaluation blocks.

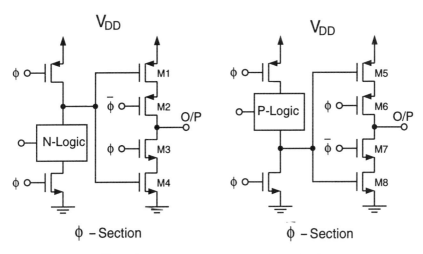

Fig. 4.42 NORA CMOS dynamic logic circuit.

Fig. 4.44 shows the split-output latch and the improved TSPC circuits[5]. As shown in the figure, when the clock-controlled transistor turns on, the split outputs can be used to drive the next stage like an inverter. Since one input is affected by the body effect of the clock-controlled transistor, the speed of the output is degraded. As shown in the figure, the TSPC-1 and the TSPC-2 circuits are alike. In the TSPC-1 n-block, when ϕ is high, the output is pulled down first. Then it may be pulled up depending on the input. Therefore, glitches may exist. In the TSPC-2 n-block, the glitch problem has been lessened at the cost of a weaker output discharge capability.

Fig. 4.45 shows the positive and negative-edge triggered D flip-flops (ETDFF) realized by p/n-blocks, and p/n-C^2MOS stages and the circuits realized by split-output latch stages[5]. As shown in the figure, the positive and the negative ETDFFs are made by cascading two TSPC-1 circuits without the input stage of the first TSPC-1 circuit—only the N-C^2MOS latch is left in the first TSPC-1 circuit. As shown in the figure, the positive and the negative ETDFFs can be also realized by the split-output latch, which has half the clock load as compared to the other two circuits. Thus the clock load can be reduced. However, it is not suitable for low-voltage operation.

Fig. 4.46 shows the single-transistor-clocked TSPC dynamic and static differential latches[32]. As shown in the figure, DSTC(P) and DSTC(N) are DCVS latches. One thing regarding the operation is worth pointing out. When DSTC1(P) and DSTC1(N) enter the latch/hold phase, if there is an overlap at the inputs (In, $\overline{In}$), both input transistors turn on. Its common source voltage changes and its output voltage may be damaged. For example, in DSTC1(P), when ϕ is high and both In and $\overline{In}$ are low, charge sharing may occur at the two output nodes via the two PMOS devices. As a result, the data that are supposed to be held are destroyed. In SSTC1(P) and SSTC1(N), due to the cross-coupled inverter latch, the charge-sharing problem in

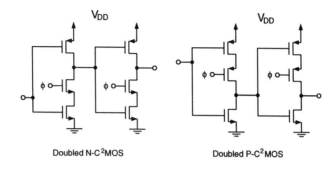

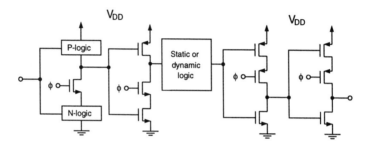

Fig. 4.43 True single-phase-clocking (TSPC) latch stages. (Adapted from Yuan and Svensson[5].)

DSTC does not exist. The transistors marked '*' in the figure are only used to the hold data level—thus, minimum-size transistors are sufficient.

Fig. 4.47 shows a high-speed dynamic differential flip-flop and a termination stage used in a pipelined system[32]. As shown in the figure, DSTC2(P) is derived from DSTC1(P). The dynamic differential flip-flop is made of DSTC2(P) and DSTC1(N). The input transistors in DSTC2(P) are NMOS instead of PMOS in DSTC1(P). The operation principle of DSTC2(P) latch is described here. When ϕ is low, the PMOS devices controlled by ϕ turn on. It operates as a DCVS inverter, which is identical to the operation of DSTC1(N) when ϕ is high. The PMOS device of DSTC2(P) turns off. When ϕ is high, DSTC2(P) is supposed to hold the output data. However, assuming that when ϕ is high, I_n and $\overline{I_n}$ reverse their states ($I_n = 0 \rightarrow 1$, $\overline{I_n} = 1 \rightarrow 0$). Therefore, both outputs of DSTC2(P) become low, which may affect the function of the logic evaluation in the next stage DSTC1(N). As long as the output of DSTC2(P) changes after DSTC1(N) has accomplished logic evaluation, the result of the logic evaluation of DSTC1(N) is not affected. When the above situation occurs, two NMOS devices in DSTC1(N) are turned off, therefore its output will not change. In a pipelined system, it may be ended with DSTC2(P). In order to avoid the problem mentioned above. By adding a DCVS inverter, which is similar to DSTC1(N), used

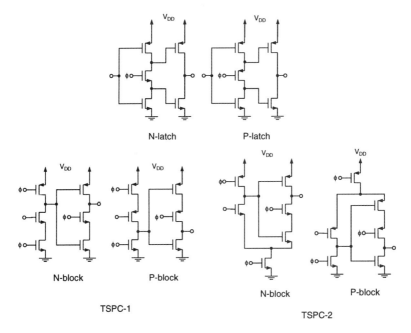

Fig. 4.44 Split-output latch and the improved TSPC circuits. (Adapted from Yuan and Svensson [5].)

as a termination stage for DSTC2(P), the problem mentioned above can be avoided.

Fig. 4.48 shows the fully static and semi-static differential flip-flops[32]. As shown in the figure, they are based on the static and dynamic latches described in this section. The advantages of the fully static and semi-static operation shown in Fig. 4.48 are good noise immunity, high speed, and low parasitic load.

Fig. 4.49 shows a finite-state machine structure using a single-phase clock[33]. As shown in the figure, via combinational logic, the present state signal and the input signal generate the output and the next state signals. At the next positive clock edge, a new state signal combined with the new input signal is computed via the combinational logic again. For a proper operation of the single-phase finite state machine, within a clock cycle, no two outputs can be generated in the positive-edge-triggered D flip-flop. Also shown in Fig. 4.49 is the circuit and the timing diagram of the single-phase positive edge triggered flip-flop. As shown in the figure, the input should come in at a time period of U before the positive edge of the clock (setup time), otherwise the p-clocked CMOS latch will not pass the signal. In addition, after a time period of H (hold time), the input D can change from low to high, otherwise $\overline{Q}$ may be affected. As shown in the figure, D_{CQ} is the output delay time of the slave latch when the clock phase is high. Fig. 4.50 shows the selection mechanism of the minimum clock cycle when clock skew exists in a single-phase system[33]. As shown in the figure, clock skew exists between CLK A and CLK B for the positive edge-triggered D

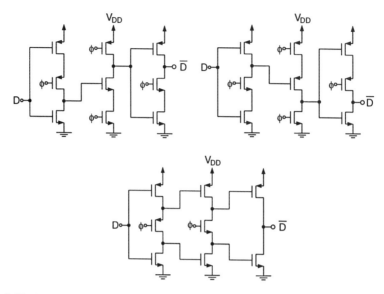

Fig. 4.45 Positive and negative-edge triggered D flip-flops (ETDFF) realized by p/n-blocks and p/n-C^2MOS stages, and the circuit realized by split-output latch stages. (Adapted from Yuan and Svensson [5].)

flip-flops (ETDFF) A and B. Here, CLK A is used as a referenced clock signal. U is the setup time of the positive ETDFF. H is the hold time. D_Q is the output delay of ETDFF. D_L is the logic propagation delay. D_{QM} and D_{Qm} represent the maximum and minimum values of D_Q, respectively. When CLKA switches from low to high, the output data of ETDFF need to pass through the worst delay $D_{QM} + Q_{LM}$ and reach A before the setup time U prior to the next positive clock edge of CLKA, and reach B before the setup time U prior to the next positive clock edge of CLKB. As shown in the figure, the combinational logic output should arrive at B at the hold time U prior to the next positive clock edge of the CLKB. Then from the case as shown in Fig. 4.50(b) (CLKA is earlier), $P + S_{ABM} \geq D_{QM} + D_{LM} + U$, where P is the period of the clock. However, as shown in Fig. 4.50(c), since CLKB arrives earlier than CLKA, the above equation becomes $P - S_{ABM} \geq D_{QM} + D_{LM} + U$. On the other hand, if the delay of the logic and the flip-flop is too small, then the signal may arrive at B within the hold time of B, thus the output is damaged. In order to avoid this problem in Fig. 4.50(b), $D_{Qm} + D_{Lm} \geq S_{ABM} + H$, which is not necessary for the case in Fig. 4.50(c). Therefore, the minimum safe value for the clock period P should be $P \geq D_{QM} + D_{LM} + D_{Qm} + D_{Lm} + U - H$. From the above analysis, the minimum delay ($D_{Qm} + D_{Lm}$) cannot be too small. On the other hand, $D_{QM} + D_{LM}$ should be as small as possible. This will cause difficulties in circuit designs. By adding a CLKA-controlled ETDFF before ETDFF B and CLKA, the problem can be resolved.

TRUE-SINGLE-PHASE CLOCKING (TSPC) 203

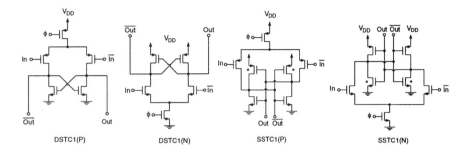

Fig. 4.46 Single-transistor-clocked TSPC dynamic and static differential latches. (Adapted from Yuan and Svensson [32].)

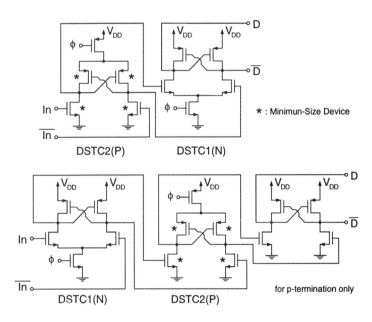

Fig. 4.47 High-speed dynamic differential flip-flop and termination stage used in a pipelined system. (Adapted from Yuan and Svensson [32].)

204 CMOS DYNAMIC LOGIC CIRCUITS

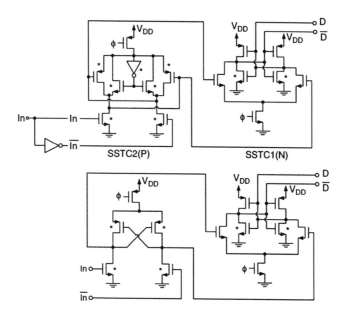

Fig. 4.48 Fully static and semi-static differential flip-flops. (From Yuan and Svensson [32], ©1997 IEEE.)

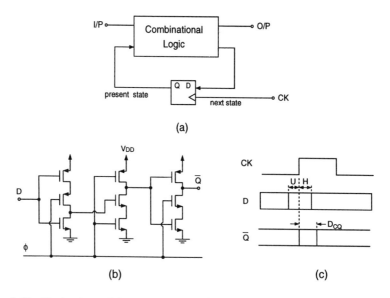

Fig. 4.49 Single-phase finite-state machine structure and single-phase positive-edge-triggered D flip-flop. (Adapted from Afghahi and Svensson [33].)

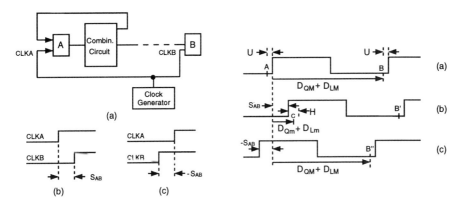

Fig. 4.50 Selection mechanism of the minimum clock cycle time when clock skew exists in a single-phase finite-state machine system. (Adapted from Afghahi and Svensson [33].)

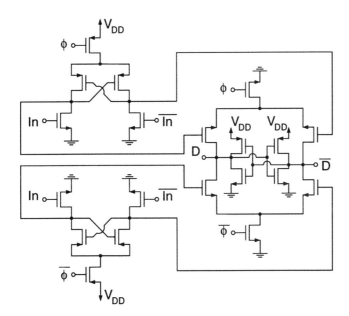

Fig. 4.51 High-speed differential double-edge triggered flip-flop. (Adapted from Blair [34].)

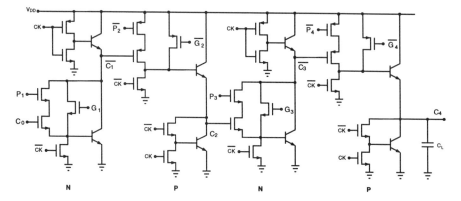

Fig. 4.52 Four-bit BiCMOS dynamic carry chain circuit. (Adapted from Kuo et al. [35].)

Fig. 4.51 shows the double-edge triggered flip-flop based on the semi-static differential flip-flop[34]. As shown in the figure, the double-edge triggered D flip-flop is composed of the positive-edge and the negative-edge triggered flip-flops. The cross-coupled inverter is shared. When ϕ is low, the master latch in the upper left is transferring the data and the slave latch in the upper right does not function. On the other hand, the master latch in the lower left is holding the output data and the slave latch in the lower right is transferring the output data. When ϕ is high, the operation is reversed. Therefore, when ϕ switches from low to high and from high to low, the output node outputs data.

4.10 BiCMOS DYNAMIC LOGIC CIRCUITS

BiCMOS circuit techniques have been used to enhance the speed performance of CMOS static logic circuits. BiCMOS circuit techniques can also be used to improve the speed performance of CMOS dynamic logic circuits. Fig. 4.52 shows the 4-bit BiCMOS dynamic carry chain circuit using two pairs of N and P-type BiCMOS dynamic carry cells as shown in Fig. 4.53[35]. The function of the conventional Manchester carry chain circuit is: $C_i = G_i + P_i \cdot C_{i-1}, i = 1 \sim n$, where n is the bit number, G_i and P_i are the generate and propagate signals ($G_i = a_i \cdot b_i$, $P_i = a_i \oplus b_i$) produced from two inputs (a_i, b_i) to the half adder. As shown in Fig. 4.52[35], all pass transistors in the conventional dynamic Manchester carry chain circuit are replaced by the cascading BiCMOS dynamic logic gates in the BiCMOS carry chain circuit. Each BiCMOS dynamic Manchester carry chain cell's output—the carry signal—is taken as an input to the following cell. In order to shorten the precharge/predischarge time, BiCMOS precharge/predischarge circuits have been used in every cell.

As in a pipelined system, cascading dynamic logic gates may have race problems[8]. In the BiCMOS dynamic carry chain circuit, race problems have been avoided by placing the "complementary" BiCMOS dynamic carry cells as shown in Fig. 4.52

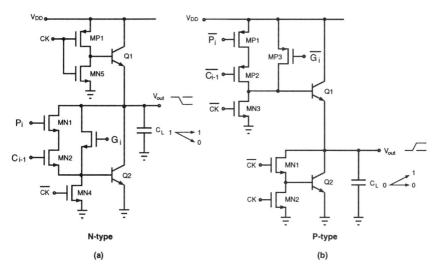

Fig. 4.53 BiCMOS dynamic carry chain cell circuit. (a) N-type; (b) P-type. (Adapted from Kuo et al.[35].)

alternatively in the BiCMOS dynamic carry chain circuit. For example, in the N-type cell as shown in Fig. 4.53(a), the pull-up BiCMOS device, which is composed of the bipolar transistor $Q1$ and MOS transistors $MP1$ and $MN5$, is used as the precharge circuit. Initially, during the precharge period when CK is low, V_{out} is precharged to a high voltage close to V_{DD}. At this time, the pull-down bipolar transistor $Q2$ is turned off by the MOS transistor $MN4$. During the logic evaluation period, the precharge circuit is turned off and V_{out} is determined by the n-type BiCMOS logic block including the MOS transistors $MN1$, $MN2$, $MN3$ and the bipolar transistor $Q2$. As shown in Fig. 4.53(b), the P-type cell has a complementary scheme. During the predischarge period (CK is low), its output node is predischarged to close to ground by $MN1$ and $Q2$. During the logic evaluation period (CK is high), V_{out} is determined by the p-type BiCMOS logic block. In addition, in order to avoid charge-sharing problems[2], in the N-type BiCMOS logic cell in Fig. 4.52, the carry signal of the previous bit output has been arranged to control the NMOS device below the NMOS device where the propagate signal is connected. In order to avoid race problems, in each N or P cell, one transition state is prohibited at the input. Specifically, in the N-type cell, inputs cannot have a transition state from V_{DD} to ground since the output may be accidentally switched to an incorrect state. Similarly, in the P cell, inputs cannot have a transition state from ground to V_{DD}. In order to avoid race problems, N- and P-type cells are placed alternatively in the BiCMOS dynamic carry chain circuit such that the output of an(a) N(P)-cell is also the input to a(an) P(N)- cell. After the precharge/predischarge period, in the BiCMOS dynamic carry chain, internal output nodes are set to high and low alternatively. With this arrangement, race problems are avoided.

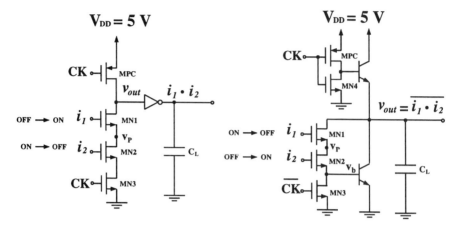

Fig. 4.54 5V CMOS and BiCMOS dynamic logic circuits. (Adapted from Kuo and Chiang[38].)

The BiCMOS dynamic circuit techniques described in this subsection are applicable not just to the Manchester carry chain circuits. It can be used in any large-scale pipelined system implemented by the CMOS dynamic circuits to enhance the speed performance at a cost of two extra bipolar transistors per stage—one bipolar transistor is for the precharge/predischarge and the other one is for the logic gate. For example, the BiCMOS dynamic logic circuit has been used in a parallel multiplier with Wallace tree reduction architecture[4] to shrink the propagation delay time associated with the complex irregular routing.

BiCMOS dynamic logic circuits also have charge-sharing problems. Before the charge-sharing problem is addressed, the charge-sharing problem in a CMOS dynamic logic circuit is reviewed. In the CMOS dynamic logic circuit as shown in Fig. 4.54[38], during the precharge period, the v_{out} node is precharged to 5V by MPC. During the logic evaluation period, if the input i_2 is high ($MN2$ is on) and the input i_1 is low ($MN1$ is off), the v_p node is predischarged to 0V. During the next logic evaluation period, if the input i_2 switches to low ($MN2$ turns off) and the input i_1 switches to high ($MN2$ turns on), the v_{out} node is supposed to stay high but it may be pulled low since the charge at the v_{out} node will redistribute with the parasitic capacitance at the v_p node. If the parasitic capacitance at the v_p node is comparable to that at the v_{out} node, the noise margin of the 5V CMOS dynamic logic circuit may be degraded due to charge sharing. Therefore, in the 5V CMOS dynamic logic circuit, the charge-sharing problem is due to charge redistribution between the parasitic capacitances at the v_{out} node and the v_p node—the load capacitance at the output node is the dominant factor in determining charge sharing. In other words, owing to the inverter as shown in Fig. 4.54, after charge redistribution between the v_p node and the v_{out} node, if v_{out} falls below 2.5V, charge sharing triggers noise immunity problems. The charge-sharing problem in the 5V BiCMOS dynamic logic circuits is quite different. In the n-type BiCMOS dynamic logic circuit as shown in the figure, during the precharge period, the v_{out} node is also precharged to high. During the logic evaluation period, if input i_1 is high ($MN1$ turns on) and input i_2 is low ($MN2$ is off), v_p is charged to high. In the next logic evaluation period, if input i_1 turns low ($MN1$ turns off) and input i_2 turns high ($MN2$ turns on), v_{out} is supposed to stay high since the bipolar device is off but it may be pulled low since the bipolar device may turn on as a result of the rise of the voltage at the v_b node due to charge redistribution with the parasitic capacitance at the v_p node. Therefore, different from the situation in the CMOS dynamic logic circuit, the charge-sharing problem in the BiCMOS dynamic logic circuit is due to turn-on of the bipolar device triggered by the rising of the v_b node voltage as a result of charge redistribution with the parasitic capacitance at the v_p node. Therefore, in the 5V BiCMOS dynamic logic circuit, the charge-sharing problem is also due to charge redistribution between the parasitic capacitances at the v_b node and the v_p node—the load capacitance at the output node is not the dominant factor in determining charge sharing because of the large driving capability of the bipolar device. Instead, the charge-sharing problem is determined by charge redistribution between the v_p node and the v_b node. After charge redistribution between the v_p node and the v_b node, if the v_b node voltage rises to above 0.7V, charge sharing triggers noise immunity problems. Therefore, compared to the CMOS case, the 5V BiCMOS dynamic logic gate may have a much worse charge-sharing problem.

4.10.1   1.5V BiCMOS Dynamic Logic Circuit

For advanced BiCMOS technologies, scaling power supply voltage is unavoidable [39][40]. For an advanced deep-sub-quarter-micron BiCMOS technology, a 1.5V supply may be necessary. With a 1.5V supply, the BiCMOS dynamic circuit introduced in the previous subsection cannot be used. Full-swing BiCMOS static logic circuits for a small power supply voltage have been reported [41][42]. The difficulties in designing BiCMOS dynamic circuits using an advanced BiCMOS technology with a small power supply voltage can be understood by considering the BiCMOS static inverter circuit as shown in Fig. 4.55, where BiPMOS and BiNMOS structures have been used. The BiPMOS structure is used in the pull-up operation. The emitter of the BiPMOS pull-up structure is connected to the output. Due to the V_{be} voltage drop, the output cannot go beyond $V_{DD} - V_{be}$. On the other hand, the BiNMOS structure is used for the pull-down operation. As a result of the V_{be} voltage drop, the output has difficulties to go below V_{be}. The non-full-swing of the output limits the application of the BiCMOS circuits for low-voltage applications.

Fig. 4.56 shows the BiCMOS dynamic logic 'OR' circuit suitable for 1.5V operation[43]. This BiCMOS dynamic logic 'OR' circuit is made of the PMOS precharge transistor MPC, two control transistors $MN1$, $MP1$, the CMOS NAND gate as feedback, and the BiPMOS pulldown structure[43]. The operation of the BiCMOS dynamic logic circuit is divided into two periods—the precharge period and the logic evaluation period. During the precharge period, CK is low and the

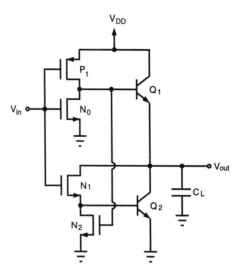

Fig. 4.55 Standard BiCMOS inverter circuit

output is pulled up to V_{DD}. In addition, the bipolar device is turned off since the base is pulled down to ground via $MN1$. During the logic evaluation period, both MPC and $MN1$ are turned off by $MN1$ and $MP1$ is on. If both inputs (I_1, I_2) are low (0V), the bipolar device (Q_1) will be turned on. As a result, the output is pulled low. After the output is pulled down, FB will switch to high. Then, $MN1$ is turned on and $MP1$ is turned off. As a result, the bipolar device is turned off and no current flows in $MP1$, $MP2$, and $MP3$. During operation, the bipolar device is on only during switching, where the transient saturation technique described in Chapter 3 has been used[41]. In addition, a full swing of 1.5V at the output of the BiCMOS dynamic logic gate can be obtained.

In the BiCMOS dynamic logic circuit, the output can go down to close to 0V owing to the BiPMOS pulldown structure—$MP1$, $MP2$, $MP3$ and the bipolar device. In the BiNMOS pull-down structure as shown in Fig. 4.55, the drain and the collector are connected to the output. Due to the V_{be} drop, the output voltage has difficulties to go down to close to 0V because of V_{be}. Different from the BiNMOS pull-down structure as shown in Fig. 4.55, the collector and the drain in the BiPMOS pulldown structure are not connected together any more. As a result, during the pulldown transient, the output can go down to close to 0V although the bipolar device may even be in saturation. Once the output comes down, the bipolar device is turned off by setting the base to ground using $MN1$ and $NAND$. Consequently, no DC power dissipation exists.

Fig. 4.57[45] shows a 16-bit full adder, which is composed of half adders and a 4-bit Manchester carry chain circuit, using the 1.5V BiCMOS dynamic logic circuit. As shown in Fig. 4.57[45], in the 4-bit adder circuit, BiCMOS dynamic logic circuit

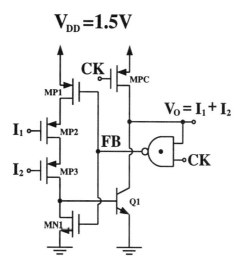

Fig. 4.56 1.5V BiCMOS logic OR circuit. (Adapted from Kuo et al. [43].)

with the BiPMOS pulldown structure has been utilized. The Manchester carry chain circuit is used to process the propagate and generate signals produced by the half adders to generate the carry signals – $C_i = G_i + P_i \cdot C_{i-1}$, for $i = 1 \sim n$ where n is the bit number. G_i and P_i are the generate and propagate signals ($G_i = a_i \cdot b_i$, $P_i = a_i \oplus b_i$) produced from two inputs (a_i, b_i) to the half adder. As described before, in a pipelined system, cascading dynamic logic gates may have race problems[8]. In the BiCMOS dynamic circuits with the BiPMOS pull-up structure shown in Fig. 4.53, N-type and P-type BiCMOS dynamic cells have been placed alternatively to avoid race problems [35]. In the full adder circuit using the BiCMOS dynamic logic circuit with the BiPMOS pull-down structure, no race problems exist since the operation of the BiPMOS pull-down structure is a 'buffer' instead of an 'inverter'. The operation of it is similar to that of domino logic.

With a larger power supply voltage, the BiCMOS dynamic circuit can be even more advantageous. Fig. 4.58 shows the propagation delay of a 16-bit Manchester carry look-ahead adder using the BiCMOS dynamic circuit and the CMOS static one versus the power supply voltage based on a $0.8\mu m$ BiCMOS technology. Also shown in the figure is the propagation delay of the 16-bit Manchester carry look-ahead adder using the BiCMOS dynamic circuit described in the previous subsection [35] (BiCMOS*). For a supply voltage of less than 2.3V, the speed of the 16-bit carry look-ahead adder using the previous BiCMOS dynamic circuit is worse than that of the CMOS one. For a supply voltage of less than 2.3V, the previous BiCMOS dynamic circuit is not practical for use. As shown in the figure, for a supply voltage of less than 3.3V, the 16-bit carry look-ahead adder using the new BiCMOS dynamic circuit is better than that using the previous BiCMOS dynamic circuit. Generally speaking, for a larger power supply voltage, the improvement of the 16-bit carry

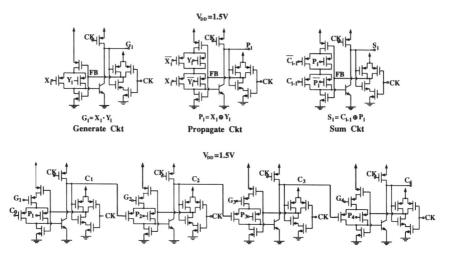

Fig. 4.57 Adder circuit using BiCMOS dynamic logic circuit. (Adapted from Kuo et al.[45].)

look-ahead adder using the new BiCMOS dynamic circuit over that using the CMOS circuit is more than that for the 1.5V case.

In the 1.5V BiCMOS dynamic logic circuit, when the bipolar device is on, the output voltage may go down to close to 0V, where the bipolar device is saturated. Usually, in a typical bipolar circuit, when the bipolar device is saturated, its base is filled with lots of minority carriers and its turn-off time may be long since it takes a long time for the minority carriers in the base to be removed. In the BiCMOS dynamic circuit described in this subsection, saturation of the bipolar device does not affect the speed performance owing to the feedback CMOS NAND gate as shown in Fig. 4.56. When the output is pulled low, the FB signal produced by the NAND is turned high. Therefore, $MN1$ is turned on and the bipolar device is turning off. Consequently, saturation of the bipolar device does not affect the switching time.

Compared to the CMOS dynamic circuits [1][5], BiCMOS dynamic logic circuits also provide speed advantages owing to their unique circuit structure. Generally speaking, N/P-type CMOS dynamic circuits may suffer from speed degradation due to the N/P-channel device under/above the logic block. For the BiCMOS dynamic circuits, no such bottleneck exists since the BJT device becomes the major pull-up/pull-down device during the logic evaluation period. In addition, compared to the CMOS dynamic circuits, BiCMOS dynamic circuits may be more suitable for advanced VLSI systems with down-scaled power supply voltages as a result of its BJT related structure—less voltage drop in the pull-down BJT device as compared to the N-channel device under the logic block in the CMOS dynamic circuits.

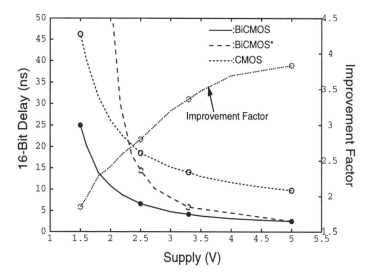

Fig. 4.58 Propagation delay time versus supply voltage of the 16-bit carry look-ahead adder circuit using BiCMOS dynamic and CMOS static techniques and the propagation delay improvement factor. The propagation delay improvement factor is defined as the ratio of the switching time of the CMOS static circuit to that of the BiCMOS dynamic one. Also shown in the figure is the propagation delay of the 16-bit carry look-ahead adder circuit using the BiCMOS dynamic circuit described in the previous subsection. (Adapted from Kuo et al.[45].)

During the precharge period, the PMOS device (MPC) has been used. As a result, during the precharge period, the BiCMOS dynamic logic circuit functions like a CMOS circuit. Since in a BiCMOS dynamic logic system, all cascading stages are precharged at the same time. As a result, the speed performance during the precharge period is not critical in determining the overall speed performance. The speed performance of the BiCMOS dynamic logic system is mainly determined by the propagation delay during the logic evaluation period.

4.10.2   1.5V BiCMOS Latch

As in a clocked CMOS dynamic logic using a low supply voltage, for a 1.5V BiCMOS dynamic pipelined digital logic VLSI system as shown in Fig. 4.59, in addition to the 1.5V BiCMOS dynamic logic gates, 1.5V clocked BiCMOS dynamic latches are also required to enhance speed. The switching speed of 1.5V clocked CMOS dynamic latches may be low when the capacitive load is large. Fig. 4.60 shows the 1.5V clocked BiCMOS dynamic latch circuit. As shown in the figure, the 1.5V clocked BiCMOS dynamic latch is composed of the active pull-up and pull-down circuits. In the active pull-up circuit, a BiPMOS pull-up structure (Q_{n1}, M_{p1}, M_{p2}, M_{p3}) is used. In the active pull-down circuit, a BiPMOS pull-down structure (Q_{n2}, M_{p5}, M_{p6}, M_{p7}) is used. In addition, a Schmitt trigger (I_1) with a CMOS inverter (I_2) in the feedback path is used to implement dynamic operation of the circuit—BJTs are

214 CMOS DYNAMIC LOGIC CIRCUITS

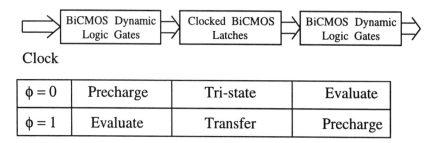

Fig. 4.59 1.5V BiCMOS dynamic pipelined digital logic VLSI system.

only on during transient. M_{p3}, M_{p6} are used to implement the tristate/logic transfer operation. As $\overline{CK}$ is high, both BJTs are off—the latch is in tri-state. When $\overline{CK}$ is low, depending on the input (V_{in}), only one BJT is on only during switching. In order to enhance the output swing for 1.5V operation and thus the switching time, a bootstrapped capacitor (C_b) is placed between the output node and the source node of M_{p2}. When the output is charged to high by Q_{n1}, due to C_b, V_b is bootstrapped to surpass V_{DD}. Thus the base of Q_{n1} also rises to over V_{DD}. Therefore, V_{out} can be pulled up to close to V_{DD} by Q_{n1}, which is different from the result in the conventional BiCMOS logic—in the conventional BiCMOS logic, due to the V_{be} drop of Q_{n1}, it is difficult for V_{out} to be greater than $V_{DD} - V_{be}$.

4.10.3 Charge-Sharing Problems

Using the 'BiPMOS pull-down' structure, charge sharing in the 1.5V BiCMOS dynamic logic circuit is a little bit different from the 5V case as described in Fig. 4.53. As shown in Fig. 4.61[43], during the precharge period, the output is precharged to high by MPC and the two input transistors $MP2$ and $MP3$ are off ($i_1 = 1.5V$, $i_2 = 1.5V$). During the logic evaluation period, if i_2 remains high such that $MP3$ stays off and i_1 switches from 1.5V to 0V, the output is high since the bipolar transistor is off. During this period, the parasitic capacitance at the internal node is charged to V_p=1.5V. In the next logic evaluation period, if i_1 switches to high and i_2 switches to low, the charge at the parasitic node V_p will redistribute with the parasitic capacitance at V_b. If the parasitic capacitances at these two nodes are comparable, the base voltage may rise to 0.7V, thus the bipolar device may turn on. Therefore, the output voltage will be pulled down accidentally. As in the 5V BiCMOS dynamic logic circuit as shown in Fig. 4.53, the charge-sharing problem in the 1.5V BiCMOS dynamic logic circuit is also due to charge redistribution between the v_b node and the v_p node. Fig. 4.62 shows the output voltage, the v_p node voltage, the v_b node voltage during the transient[38]. When CK is low, i_1 is low ($MP2$ is on) and i_2 is high ($MP3$ is off), the v_p node voltage is 1.5V and the v_b node voltage is set to 0V. When CK switches to high, $MP2$ turns off and $MP3$ is on. Due to charge redistribution, the v_p node voltage slews downward and the v_b node voltage goes

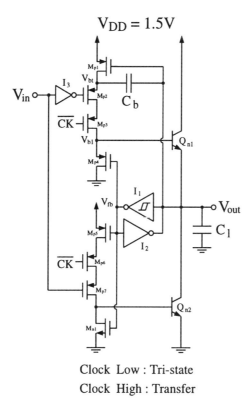

Fig. 4.60 1.5V clocked BiCMOS dynamic latch circuit.

up. Without adding a parasitic capacitance at the base node, the base-node voltage goes up quickly over 0.8V and the output voltage quickly comes down. After the output voltage comes down, the bipolar device is turned off by the feedback signal v_{fb}, and the based voltage comes down again. Adding a parasitic capacitance of 0.2pF at the base node, the base voltage, which goes up slowly, never exceeds 0.7V. Consequently, the bipolar device never turns on and the output voltage stays high. Therefore, a large parasitic capacitance at the base node is helpful for reducing the charge-sharing problems.

Imitating the CMOS solution to the charge-sharing problems, a predischarge PMOS device $MP4$ has been added as shown in Fig. 4.63[38] such that during the precharge period, the internal node V_p will go down to $0.7V$. (Note that V_p will not go below $0.7V$ during the precharge period.) Thus the charge-sharing problems can be removed. However, when the input chain is long, many predischarge PMOS devices are required. In addition, a voltage of 0.7V at the internal node may slow down the pull-down process in the following logic evaluation period if both i_1 and i_2 switch to low. To overcome the drawbacks of the 1.5V BiCMOS dynamic logic circuit with the predischarge scheme, Fig. 4.64 shows the 1.5V BiCMOS dynamic

216 CMOS DYNAMIC LOGIC CIRCUITS

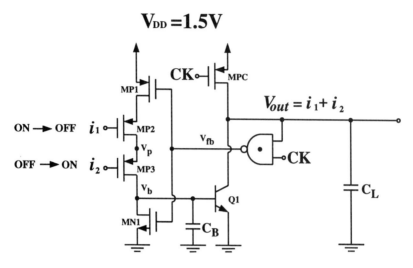

Fig. 4.61 1.5V BiCMOS dynamic logic circuit. (Adapted from Kuo et al. [43].)

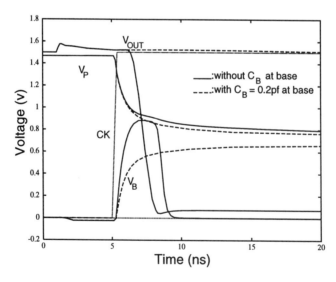

Fig. 4.62 Output voltage, the v_p node voltage, the v_b voltage of the 1.5V BiCMOS dynamic logic circuit with and without a parasitic capacitance of 0.2pF at the base node during transient. (Adapted from Kuo and Chiang[38].)

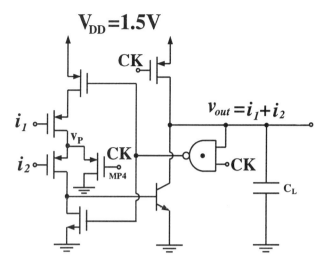

Fig. 4.63 1.5V BiCMOS dynamic logic circuit with a PMOS predischarge scheme. (Adapted from Kuo and Chiang[38].)

logic circuit, which is free from charge-sharing problems. Instead of the predischarge PMOS devices as shown in Fig. 4.63, an NMOS predischarge device is connected to the base node. If i_1 is high, $MN2$ is on and the base node is pulled down to ground. In addition, V_p stays at $1.5V$ instead of $0.7V$. Therefore, a higher switching speed can be expected in the following logic evaluation period.

4.11 LOW-VOLTAGE DYNAMIC LOGIC TECHNIQUES

For next-generation deep-submicron CMOS VLSI technology, low supply voltage is the trend[46]. In the past, for $1\mu m$ CMOS technology, a 5V supply voltage is used. The power supply voltage has been scaled down to 3.3V for $0.5\mu m$ CMOS technology and 2.5V for $0.3\mu m$ CMOS technology. The down-scaled supply voltage reduces power dissipation but brings in difficulties. Since threshold voltage of a deep-submicron MOS device cannot be scaled down accordingly with the supply voltage, the propagation delay of a logic gate using deep-submicron CMOS devices increases when the supply voltage is scaled down. For sub-$0.1\mu m$ CMOS technology, a 1.5V supply voltage may be used. At a supply voltage of 1.5V, the current driving capability of CMOS devices may be small. How to enhance the speed performance of 1.5V CMOS logic gates using deep-submicron CMOS devices has become an important task. At a low supply voltage, the speed performance of CMOS dynamic logic circuits such as NORA, domino, and Zipper is better than that of CMOS static ones as a result of reduced internal parasitic capacitances. However, when the serial fan-in is large, its associated propagation delay may increase drastically[14], which is especially serious at a low supply voltage. How to enhance the speed performance of 1.5V CMOS dynamic logic circuits is an important issue.

218 CMOS DYNAMIC LOGIC CIRCUITS

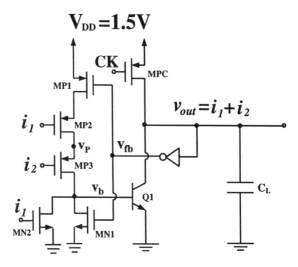

Fig. 4.64 1.5V BiCMOS dynamic logic circuit free from charge-sharing problems. (Adapted from Kuo and Chiang[38].)

4.11.1 Bootstrapped Dynamic Logic (BDL) Circuit

Fig. 4.65 shows a two-stage 1.5V CMOS bootstrapped dynamic logic (BDL) circuit [50], which is composed of the CMOS dynamic logic circuit and the CMOS bootstrapper circuit[47][48]. As shown in the figure, $MPD, MND, MN1$, and $MN2$ comprise the dynamic logic circuit. The CMOS inverter—Inv, MP, MPB, MN and the bootstrap capacitor C_b form the bootstrapper circuit. The function of the bootstrapper circuit is explained as follows: When CK is low, it is the precharge period. During the precharge period, the internal node (V_{do}) is precharged to V_{DD}. At the same time, the output voltage V_{out} is discharged to ground via MN. The bootstrap capacitor (C_b) is charged to V_{DD}—the left side is grounded and the right side is at $V_b = V_{DD}$. During the precharge period, the right side of the bootstrap capacitor is separated from the output since MPB is off. When CK turns high, it's the logic evaluation period. During the logic evaluation period, MPD turns off. During the logic evaluation period, the internal node voltage V_{do} is determined by inputs V_{ina} and V_{inb}. If both V_{ina} and V_{inb} are high, V_{do} is pulled low and V_I is high. Thus MP and MN turn off. Owing to the charge in the bootstrap capacitor, V_b will be bootstrapped to over V_{DD}—the internal voltage overshoot. In addition, when MPB turns on, V_{out} is pulled high to over V_{DD}.

Fig. 4.66 shows the application of the bootstrapper circuit to the Multiple Output Domino Logic (MODL) circuit[19]. MODL circuit is used for avoid cascading logic gates such that the device count can be reduced, hence the propagation delay time can be reduced. As shown in the figure, applying the bootstrapper circuits to the output nodes, the speed performance can be enhanced. Using the bootstrapper circuit, the MODL circuit is called BMODL (bootstrapped MODL) circuit. MODL

LOW-VOLTAGE DYNAMIC LOGIC TECHNIQUES   **219**

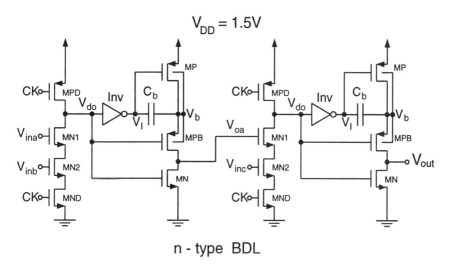

Fig. 4.65 1.5V CMOS bootstrapped dynamic logic (BDL) 2-input AND circuit including the CMOS bootstrapper circuit. (Adapted from Lou and Kuo[50].)

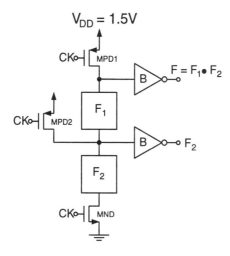

Fig. 4.66 Bootstrapped 1.5V CMOS multiple-output dynamic logic (MODL) circuit.

circuit is especially useful for recursive logic such as carry-generating circuit for the adder circuit [19][49]. Specifically, MODL is useful for Manchester carry chain [19][49] used in the adder circuit. In Manchester carry chain, the ith carry is expressed as: $c_i = g_{i-1} + \{p_{i-1} \cdot [g_{i-2} + p_{i-2} \cdot (......p_0 \cdot c_0)]\}$, where p_i and g_i are propagate and generate signals: $g_i = a_i \cdot b_i$, $p_i = a_i \oplus b_i$, where a_i and b_i are the two inputs to the ith half adder. Using BMODL circuit, Fig. 4.67 shows the group

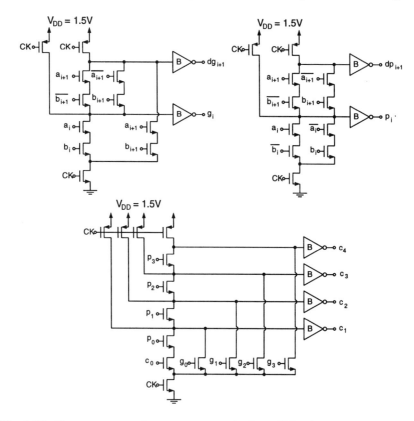

Fig. 4.67 Group generate and group propagate circuits and Manchester carry chain circuit in the 4-bit adder circuit using the 1.5V CMOS BMODL circuit.

generate and group propagate circuits[19], where "generate" and "propagate" are: $dg_{i+1} = g_{i+1} + p_i \cdot g_i$, $dp_{i+1} = p_{i+1} \cdot p_i$, respectively. Carry signal can be generated by $C_i = dg_i + dp_i \cdot C_{i-2}$. The CMOS bootstrapper circuit is especially useful when the Manchester carry chain is long. As shown in Fig. 4.67(c), the critical path involves a 5-input AND. Therefore, for a supply voltage of 1.5V, the delay associated with the serial fan-in of the circuit can be critical. Adding the CMOS bootstrapper circuits to it helps shrink propagation delays of the P_i circuit.

Fig. 4.68 shows the propagation delay time of the CMOS BDL circuit versus the supply voltage[50]. Also shown in the figure are the results for the CMOS NORA

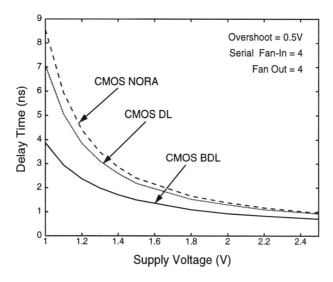

Fig. 4.68 Propagation delay time versus supply voltage of the CMOS BDL circuit. (Adapted from Lou and Kuo[50].)

circuit and the CMOS domino logic circuit without the CMOS bootstrapper circuit. As shown in the figure, when the supply voltage is scaled down from 2.5V to 1.0V, the propagation delay time increases. Among three circuits, the BDL circuit is the most insensitive to the shrinkage in the supply voltage. Therefore, the CMOS bootstrapper circuit is especially useful for low-voltage applications. In the bootstrapper circuit, the extent of V_b surpassing V_{DD}—the internal voltage overshoot is dependent on C_b and the equivalent load capacitance at the output node. In this design, considering a fan-out of four, the equivalent load capacitance is 50fF at the output node. Using a C_b of 120fF, during the bootstrap operation, V_b surpasses 1.5V by 0.5V. A higher internal voltage overshoot may bring in a greater improvement in speed performance. In the CMOS bootstrapper circuit, the body of MP and MPB is connected to V_b, instead of V_{DD}. Therefore, the internal voltage overshoot at V_b does not cause the forward bias of the source/drain-substrate junction.

The CMOS bootstrapper circuit is not just for CMOS dynamic logic circuits. It can also be useful for 1.5V BiCMOS dynamic logic circuits. BiCMOS circuits have been advantageous for driving a large capacitive load. Using the CMOS bootstrapper circuit, the speed performance of the 1.5V BiCMOS bootstrapped dynamic logic (BiCMOS BDL) circuit as shown in Fig. 4.69[50] can be improved further. As shown in Fig. 4.69, the 1.5V BiCMOS BDL circuit is composed of the 1.5V CMOS BDL circuit and the 1.5V BiCMOS dynamic logic circuit described in the previous section, In order to be compatible with the BiPMOS pull-down structure in the 1.5V BiCMOS dynamic logic circuit, the CMOS BDL circuit used in the 1.5V BiCMOS BDL circuit has a complementary configuration (p-type BDL) as compared to the one as shown in Fig. 4.65. (Note that the BDL circuit in Fig. 4.65 is called

222 CMOS DYNAMIC LOGIC CIRCUITS

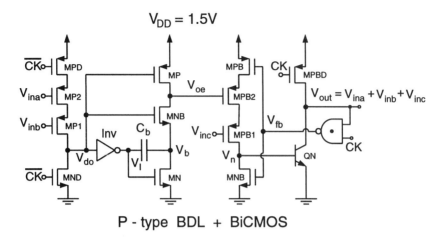

Fig. 4.69 1.5V BiCMOS dynamic logic circuit with the bootstrapper circuit (BiCMOS BDL)—p-type CMOS BDL with a BiCMOS dynamic logic circuit. (Adapted from Lou and Kuo[50])

n-type BDL.) Here, serial inputs are connected to the PMOS devices instead of NMOS devices. During the predischarge cycle, V_{do} is predischarged to 0V instead of 1.5V. The CMOS bootstrapper circuit also has a complementary configuration (p-type CMOS bootstrapper circuit) as compared to that in Fig. 4.65. (Note that the CMOS bootstrapper circuit in Fig. 4.65 is called n-type CMOS bootstrapper circuit.) During the operation, V_b is bootstrapped to below zero—the internal voltage undershoot. As shown in Fig. 4.69, cascading of the p-type CMOS BDL circuit with the 1.5V BiCMOS dynamic logic circuit can be helpful to boost the switching speed of the system—the 1.5V BiCMOS dynamic logic circuit can be used to speed up the switching speed associated with a large capacitive load.

4.11.2 Bootstrapped All-N-Logic TSP Bootstrapped Dynamic Logic

As described before, for CMOS dynamic logic circuits, clocks with complicated timing may be required. A single-phase CMOS dynamic circuit is convenient[5]. Using the bootstrapper circuit technique described in the previous subsection, Fig. 4.70 shows the 1.5V CMOS true-single-phase-clocking (TSPC) dynamic logic circuits: n-block and p-block as described before. Compared to NORA and domino logic circuits, only a single-phase clock is needed. No clocked CMOS latches are required in the CMOS TSPC dynamic logic circuit. Instead, in the n- or the p-block, it is composed of the input logic branch followed by the output latch branch. In addition, the circuit configuration of the n- and the p-blocks is complementary. In the n-block, the internal node output V_{dn} is between the PMOS device MNP and the n-logic input branch. In the p-block, the internal node output V_{dp} is between the NMOS device MPN and the p-logic input branch. In the CMOS TSPC dynamic logic circuits, the n-p-n-p block arrangement is necessary. The operation of the CMOS TSPC dynamic

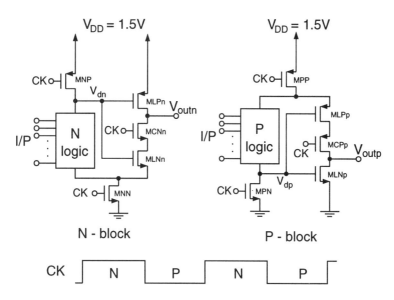

Fig. 4.70 True-single-phase-clocking (TSPC) CMOS dynamic logic circuit using the n-p-n-p configuration.

logic circuit is also divided into two periods: the precharge period and the logic evaluation period. In a CMOS TSPC dynamic logic circuit, if the n-blocks (p-blocks) are in the precharge period, the p-blocks (n-blocks) must be in the logic evaluation period. During the precharge period, the internal node voltages V_{dn} and V_{dp} of the n-block and the p-block are precharged to high and low, respectively. During the precharge period, the output of the n- and the p-blocks is floating—the previous states are maintained at the output nodes. During the logic evaluation period, depending on the input signals, the internal node V_{dn} of the n-block may be pulled low. If there exists a conduction path in the n-block, the internal node V_{dn} will be pulled low, therefore, the output V_{outn} is pulled high. P-block has a complementary scheme. During the logic evaluation period, the internal node V_{dp} of the p-block is pulled high if there is a conduction path in the p-block. Under this situation, the output V_{outp} of the p-block will be pulled low. In the TSPC dynamic logic, its slower speed of the p-block due to the smaller hole mobility degrades the overall speed performance.

Fig. 4.71 shows the 1.5V CMOS all-N-logic TSP BDL circuits[51]: the n1-block and the n2-block, where the p-block as described in the previous subsection has been replaced by the n2-block using the all-N-logic true-single-phase technique[52]. The n1-block is identical to the n-block as described in Fig. 4.70 except that the bootstrapper circuit containing the bootstrap capacitor C_b has been added. In the n1-block, when CK is low, MNN is off, and V_{dn} node is separated from the ground. Under this situation, the V_{dn} node is precharged to V_{DD} since MNP is on. Therefore, the internal node V_I is low, hence $MLP1$ is on and V_b is charged to high. At this time, in the bootstrap capacitor C_b an amount of charge $C_b V_{DD}$ is stored. Since

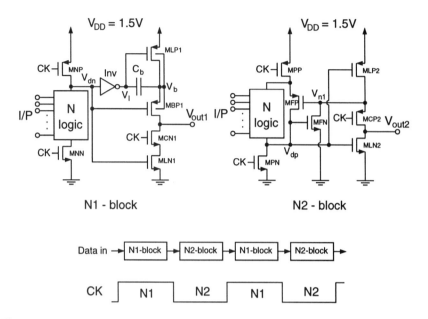

Fig. 4.71 CMOS all-N-logic true-single-phase (TSP) bootstrapped dynamic logic (BDL) circuit using the 1.5V supply voltage. (Adapted from Lou and Kuo[51].)

both $MBP1$ and $MCN1$ are off, the output of n1-block keeps its previous state. When CK is high, MNP is off and MNN is on. Therefore, when there exists a conduction path in the n1-block, the internal node voltage V_{dn} is pulled low, thus V_I switches from low to high and $MLP1$ is off and $MBP1$ is on. Due to the charge stored in the bootstrap capacitor, V_b is bootstrapped to exceed V_{DD}. Since $MBP1$ is on, the output voltage of the n1-block will be pulled up to a value V_{out1}, which will exceed V_{DD}. Owing to the n1-n2-n1-n2 block arrangement, the bootstrapped output of the n1-block may help drive the following n2-block. Note that the n2-block is used to replace the p-block in Fig. 4.70. When CK is high, MPP is off, therefore the n-logic is separated from V_{DD}. At the same time, the V_{dp} node is predischarged to ground since MPN is on. Under this situation, the output of n2-block V_{out2} keeps its previous state since both $MCP2$ and $MLN2$ are off. When CK is low, MPP turns on and MPN is off—the n2-block enters the logic evaluation period. If there exists a conduction path in the n2-block, V_{dp} is pulled high to $V_{DD} - V_{tn}(V_{dp})$, where $V_{tn}(V_{dp})$ is the threshold voltage of the NMOS device at its source voltage of V_{dp} and considering the body effect of the NMOS device. Therefore, $MLN2$ turns on, hence, the output voltage V_{out2} is pulled low. The switching speed of V_{out2} and the stability of the V_{dp} level can be improved by raising the V_{dp} level by MFN and MFP. When $MLN2$ turns on, MFN also turns on, which pulls the node voltage V_{n1} to ground. Thus MFP turns on and V_{dp} is pulled high to V_{DD}.

The n1-n2-n1-n2 block arrangement is required for the CMOS all-N-logic TSP BDL circuit. When the n1-blocks are in the precharge period, the n2-blocks are in the

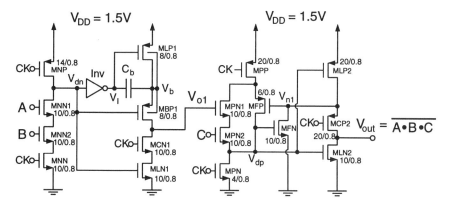

Fig. 4.72 Three input NAND circuit using the 1.5V CMOS all-N-logic TSP BDL circuit. (Adapted from Lou and Kuo[51].)

logic evaluation period. With the bootstrapper circuit in the n1-block, the output of the n1-block may exceed V_{DD}. Therefore, the internal voltage V_{dp} of the following n2-block can be raised to $V_{out1} - V_{tn}(V_{dp})$. As a result, the output voltage can be discharged faster by an enhanced current drive of the $MLN2$.

Fig. 4.72 shows a 3-input NAND circuit using the 1.5V CMOS all-N-logic TSP BDL circuit[51]. It's worth pointing out that the n1-block is non-inverting and the n2-block is inverting. At the output of the n1-block, it is $V_{o1} = A \cdot B$. At the output of the n2-block, it is $V_{out} = \overline{V_{o1} \cdot C} = \overline{A \cdot B \cdot C}$.

4.11.3 Semi-Dynamic DCVSPG-Domino Logic Circuits

Is dynamic logic suitable for low-voltage operation? Maybe the answer is pessimistic. Lowering the threshold voltage to increase the gate overdrive voltage is a required approach for enhancing the performance of the circuit operating at a low supply voltage. In addition, noise immunity of dynamic logic is inferior. For dynamic logic, when the noise voltage at the input is greater than the threshold voltage of MOS devices, the precharge node may be falsely discharged. Although the threshold voltage can be lowered, the drop in the threshold voltage may lead to a more serious leakage current problem due to the increased subthreshold current. Therefore, pull-up PMOS devices are needed to reduce these problems and the charge-sharing problems. However, increasing the pull-up strength of the PMOS as described in Fig. 4.22, may bring in a more serious ratioed logic problem of the NMOS logic tree during pull-down and thus a degradation in the speed performance. The principle of the dynamic logic is based on precharging the output to high and selectively discharging to low during the logic evaluation phase. Thus the precharged output voltage should not be destroyed. Once the output node is accidentally discharged to low, there is no way to recover. The input logic threshold voltage of dynamic logic is much smaller than that of the static logic. (Note that by adding the pull-up PMOS,

226 CMOS DYNAMIC LOGIC CIRCUITS

the input logic threshold voltage can be increased.) From the above consideration, dynamic logic is less suitable for low-voltage operation when compared with static logic. Notwithstanding, if the n-logic tree is changed to the pass-transistor logic tree, this problem can be effectively resolved. In the conventional dynamic logic tree, there is a discharging path only—no charging path. With the pass-transistor logic tree, one of both discharging and charging conduction paths can exist in the circuit, depending on the logic value of the pass variable. Thus, in the conventional dynamic logic circuit without a charging path, the accidental destruction of the precharged data can be eliminated when pass-transistor logic trees are added. In addition, the disadvantage of the inferior pull-up capability in the static pass-transistor logic can be improved by the initial precharge of the output to high. During the logic evaluation, it is based on the discharge scheme. During the logic evaluation phase, if the output maintains high, its output voltage will only drop to $V_{DD} - V_t$, which will not change the output value of the subsequent inverter. By adding the CPL technique with the cross-coupled PMOS latch, its operation is identical to DCVSPG described in Chapter 3, except that at the complementary output a precharge-to-high scheme has been added. During the logic evaluation period, the stronger driving capability of the NMOS logic block has been used for discharge. The cross-coupled PMOS latch has been used to maintain the stability of the output. Note that when the logic evaluation is completed, it functions as a static DCVSPG logic gate. The simplicity and the high speed of the dynamic logic and the stability of the static DCVSPG have been combined. Thus combining the static DCVSPG with the dynamic logic technique is suitable for realizing low-voltage dynamic logic circuits. During precharge and at the beginning of the logic evaluation period, it is a dynamic logic circuit. After the conclusion of the logic evaluation period, it becomes a static logic circuit. Thus it is called the semi-dynamic DCVSPG-domino logic circuit. Fig. 4.73 shows the semi-dynamic DCVSPG-domino logic circuit and the sum circuit in a 1-bit full adder[53]. The operation of the circuit is described here. When CK is low, it is the precharge period. At this time, $MN1$ and $MN2$ are off and $MP3$ are on. Thus, q and $\bar{q}$ are equalized. Since one of the devices $MP1$ and $MP2$ must be on, during the precharge phase, both q and $\bar{q}$ are equalized to high. When CK is high, it is the logic evaluation phase, where $MP3$ is off and $MN1$ and $MN2$ are on. In the DCVSPG logic tree, the logic-0 tree is identical to the logic-1 tree. Only the pass variables are complementary. Therefore, there is always a conducting path—either q or $\bar{q}$ is discharged to ground and the other one stays high. At this time, the circuit functions as a static DCVSPG. Unlike DCVS, any of the logic trees has a conducting path—the leakage current problem does not occur.

Due to the SODS type logic circuit involved, in the semi-static DCVSPG-domino logic circuit as shown in Fig. 4.73[53], when CK is high, one of n and $\bar{n}$ must be low—there is a charge-sharing problem existing at q or $\bar{q}$. Fig. 4.74 shows the improved semi-dynamic DCVSPG-domino logic circuit without the charge-sharing problem. As shown in the figure, during the precharge phase, the n-type clocked CMOS latches at the output of the previous stage, which are connected to the input pass variable $p_a,\ldots p_j$ nodes of the current stage, are turned off by CK. Therefore,

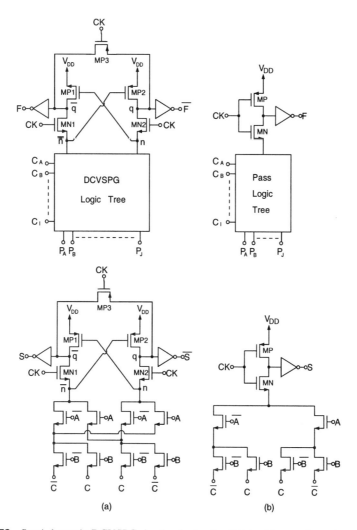

Fig. 4.73 Semi-dynamic DCVSPG-domino logic circuit (a) with complementary outputs; (b) without complementary outputs. (Adapt from Lou and Kuo[53].)

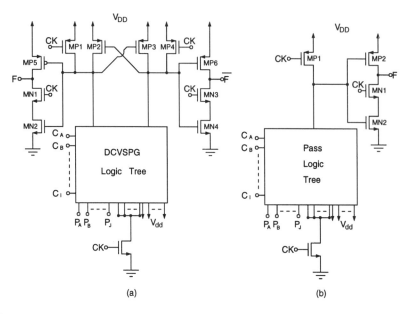

Fig. 4.74 Improved semi-dynamic DCVSPG-domino logic circuit without the charge-sharing problem. (a) with complementary outputs; (b) without complementary outputs. (Adapt from Lou and Kuo[53].)

the conduction path to ground is blocked. In the semi-dynamic DCVSPG-domino logic circuit shown in Fig. 4.73, CK controls $MN1$ and $MN2$. In the improved semi-dynamic DCVSPG-domino logic circuit, CK-controlled $MN1$ and $MN2$ of Fig. 4.73 have been moved to the n-type clocked CMOS latch of the proceeding stage. Therefore, compared to the semi-dynamic DCVSPG-domino logic circuit as shown in Fig. 4.73, the improved semi-dynamic DCVSPG-domino logic circuit is more concise and without the charge-sharing problem. In addition, with the clocked CMOS latches, the improved semi-static DCVSPG-domino logic circuits can be used in a pipelined stage without an additional latch. Note that the $P_A, P_B,..., P_J$ signals should be valid earlier than the $C_A, C_B,..., C_I$ signals. If any of the inputs $C_A, C_B,..., C_J$ is not one of $P_A, P_B,...,P_J$, for example C_B, then no clocked NMOS latch is required for the output of the stage driving C_B—an inverter can be used directly. Only for the output, which is the pass input to the next-stage gate, a clocked NMOS latch is needed. For the output, which is used as the control input to the next-stage gate, an inverter is sufficient. Note that the semi-dynamic DCVSPG circuit mentioned in this section is like a dynamic logic circuit during the precharge period and at the beginning of the logic evaluation period. When logic evaluation is completed in the logic evaluation period, it functions as a static circuit. Thus it has good noise immunity as for the static logic and the high speed as for the dynamic logic.

4.12 SUMMARY

In this chapter, CMOS dynamic logic circuits have been described. Starting from the fundamental concepts of the dynamic logic circuits, charge-sharing, noise, and race problems have been analyzed. Then, NORA circuit to resolve the race problem has been introduced, followed by Zipper and domino logic. In the domino logic, latched domino, skew-tolerant domino, and multiple-output domino logic (MODL) circuits were described. Then, a dynamic differential logic circuit was introduced, followed by the true-single-phase-clocking (TSPC) dynamic logic circuits and the BiCMOS dynamic logic circuits. Low-voltage circuit techniques for the dynamic logic were also presented. Specifically, low-voltage bootstrapped dynamic logic (BDL), bootstrapped all-n-logic TSP dynamic logic, and semi-dynamic DCVSPG-domino logic have been described.

REFERENCES

1. R. H. Krambeck, C. M. Lee, and H.-F. S. Law, "High-Speed Compact Circuits with CMOS," *IEEE J. Sol. St. Ckts.*, **17**(3), 614–619 (1982).

2. V. Friedman and S. Liu, "Dynamic Logic CMOS Circuits," *IEEE J. Sol. St. Ckts.*, **19**(2), 263–266 (1984).

3. Y. Suzuki, K. Odagawa, and T. Abe, "Clocked CMOS Calculator Circuitry," *IEEE J. Sol. St. Ckts.*, **8**(6), 462–469 (1973).

4. N. H. E. Weste and K. Eshraghian, "Principles of CMOS VLSI Design: A Systems Perspective," Addison-Wesley, Reading, MA, 1985.

5. J. Yuan and C. Svensson, "High-Speed CMOS Circuit Technique," *IEEE J. Sol. St. Ckts.*, **24**(1), 62–70 (1989).

6. Taiwan Pat. 087194(1997), P. F. Lin and J. B. Kuo, "A CMOS Semi-Static Latch Circuit without Charge Sharing and Leakage Current Problems."

7. J. A. Pretorius, A. S. Shubat, and C. A. T. Salama, "Charge Redistribution and Noise Margins in Domino CMOS Logic," *IEEE Tran. Ckts. & Sys.*, **33**(8), 786–793 (1986).

8. N. F. Goncalves and H. J. DeMan, "NORA:A Racefree Dynamic CMOS Technique for Pipeline Logic Structures," *IEEE J. Sol. St. Ckts.*, **18**(3), 261–266 (1983).

9. E. T. Lewis, "An Analysis of Interconnect Line Capacitance and Coupling for VLSI Circuits," *Sol. St. Elec.*, **27**(8/9), 741–749 (1984).

10. P. Larsson and C. Svensson, "Noise in Digital Dynamic CMOS Circuits," *IEEE J. Sol. St. Ckts.*, **29**(6), 655–662 (1994).

11. C. M. Lee and E. W. Szeto, "Zipper CMOS," *IEEE Ckts. and Dev. Mag.*, (5), 10–17 (1986).

12. Q. Tong and N. K. Jha, "Testing of Zipper CMOS Logic Circuits," *IEEE J. Sol. St. Ckts.*, **25**(3), 877–880 (1990).

13. F. Murabayashi, T. Yamauchi, H. Yamada, T. Nishiyama, K. Shimamura, S. Tanaka, T. Hotta, T. Shimizu, and H. Sawamoto, "2.5V CMOS Circuit Techniques for a 200MHz Superscalar RISC Processor," *IEEE J. Sol. St. Ckts.*, **31**(7), 972–980 (1996).

14. M. Shoji, "FET Scaling in Domino CMOS Gates," *IEEE J. Sol. St. Ckts.*, **20**(5), 1067–1071 (1985).

15. V. G. Oklobdzija and R. K. Montoye, "Design-Performance Trade-Offs in CMOS-Domino Logic," *IEEE J. Sol. St. Ckts.*, **21**(2), 304–306 (1986).

16. J.-R. Yuan, C. Svensson, and P. Larsson, "New Domino Logic Precharged by Clock and Data," *Elec. Let.*, **29**(25), 2188–2189 (1993).

17. J. A. Pretorius, A. S. Shubat, and C. A. T. Salama, "Latched Domino CMOS Logic," *IEEE J. Sol. St. Ckts.*, **21**(4), 514–522 (1986).

18. D. Harris and M. A. Horowitz, "Skew-Tolerant Domino Circuits," *IEEE J. Sol. St. Ckts.*, **32**(11), 1702–1711 (1997).

19. I. S. Hwang and A. L. Fisher, "Ultrafast Compact 32-bit CMOS Adders in Multiple-Output Domino Logic," *IEEE J. Sol. St. Ckts.*, **24**(2), 358–369 (1989).

20. Z. Wang, G. A. Jullien, W. C. Miller, J. Wang, and S. S. Bizzan, "Fast Adders Using Enhanced Multiple-Output Domino Logic," *IEEE J. Sol. St. Ckts.*, **32**(2), 206–214 (1997).

21. A. J. Acosta, M. Valencia, A. Barriga, M. J. Bellido, and J. L. Huertas, "SODS: A New CMOS Differential-Type Structure," *IEEE J. Sol. St. Ckts.*, **30**(7), 835–838 (1995).

22. C. H. Lau, "SELF: A Self-Timed Systems Design Technique," *Elec. Let.*, **23**(6), 269–270 (1987).

23. L. G. Heller, W. R. Griffin, J. W. Davis, and N. G. Thoma, "Cascode Voltage Switch Logic: A Differential CMOS Logic Family," *ISSCC Dig.*, 16–17 (1984).

24. S.-L. Lu, "Implementation of Iterative Networks with CMOS Differential Logic," *IEEE J. Sol. St. Ckts.*, **23**(4), 1013–1017 (1988).

25. S.-L. L. Lu and M. D. Ercegovac, "Evaluation of Two-Summand Adders Implemented in ECDL CMOS Differential Logic," *IEEE J. Sol. St. Ckts.*, **26**(8), 1152–1160 (1991).

26. T. A. Grotjohn and B. Hoefflinger, "Sample-Set Differential Logic (SSDL) for Complex High-Speed VLSI," *IEEE J. Sol. St. Ckts.*, **21**(2), 367–369 (1986).

27. P. Ng, P. T. Balsara, and D. Steiss, "Performance of CMOS Differential Circuits," *IEEE J. Sol. St. Ckts.*, **31**(6), 841–846 (1996).

28. F.-S. Lai and W. Hwang, "Design and Implementation of Differential Cascode Voltage Switch with Pass-Gate (DCVSPG) Logic for High-Performance Digital Systems," *IEEE J. Sol. St. Ckts.*, **32**(4), 563–573 (1997).

29. H. Partovi and D. Draper, "A Regenerative Push-Pull Differential Logic Family," *ISSCC Dig.*, 294–295 (1994).

30. D. J. Myers and P. A. Ivey, "A Design Style for VLSI CMOS," *IEEE J. Sol. St. Ckts.*, **20**(3), 741–745 (1985).

31. J. R. Yuan, I. Karlsson, and C. Svensson, "A True Single-Phase-Clock Dynamic CMOS Circuit Technique," *IEEE J. Sol. St. Ckts.*, **22**(5), 899–901 (1987).

32. J. Yuan and C. Svensson, "New Single-Clock CMOS Latches and Flipflops with Improved Speed and Power Savings," *IEEE J. Sol. St. Ckts.*, **32**(1), 62–69 (1997).

33. M. Afghahi and C. Svensson, "A Unified Single-Phase Clocking Scheme for VLSI Systems," *IEEE J. Sol. St. Ckts.*, **25**(1), 225–233 (1990).

34. G. M. Blair, "Comments on 'New Single-Clock CMOS Latches and Flip-Flops with Improved Speed and Power Savings'," *IEEE J. Sol. St. Ckts.*, **32**(10), 1610–1611 (1997).

35. J. B. Kuo, H. J. Liao, and H. P. Chen, "A BiCMOS Dynamic Carry Lookahead Adder Circuit for VLSI Implementation of High-Speed Arithmetic Unit," *IEEE J. Sol. St. Ckts.*, **28**(3), 375–378 (1993).

36. P. K. Chan and M. D. F. Schlag, "Analysis and Design of CMOS Manchester Adders with Variable Carry-Skip," *IEEE Trans. Comp.*, **39**(8), 983–992 (1990).

37. R. F. Hobson, "Refinements to the Regenerative Carry Scheme," *IEEE J. Sol. St. Ckts.*, **24**(6), 1759–1762 (1989).

38. J. B. Kuo and C. S. Chiang, "Charge Sharing Problems in the Dynamic Logic Circuits: BiCMOS versus CMOS and a 1.5V BiCMOS Dynamic Logic Circuit Free from Charge Sharing Problems," *IEEE Trans. Ckts. and Sys.–I*, **42**(11), 974–977 (1995).

39. D. L. Harame, E. F. Crabbe, J. D. Cressler, J. H. Comfort, and J. Y.-C. Sun, S. R. Stiffler, E. Kobeda, J. N. Burghartz, M. M. Gilbert, J. C. Malinowski, A. J. Dally, S. Ratanaphanyarat, M. J. Saccamango, W. Rausch, J. Cotte, C. Chu, and J. M. C. Stork, "A High Performance Epitaxial SiGe-Base ECL BiCMOS Technology," *IEDM Dig.*, 19–22 (1992).

40. T. M. Liu, G. M. Chin, D. Y. Jeon, M. D. Morris, V. D. Archer, H. H. Kim, M. Cerullo, K. F. Lee, J. M. Sung, K. Lau, T. Y. Chiu, A. M. Voshchenkov, and R. G. Swartz, "A Half-micron Super Self-aligned BiCMOS Technology for High Speed Applications," *IEDM Dig.*, 23–26 (1992).

41. M. Hiraki, K. Yano, M. Minami, K. Sato, N. Matsuzaki, A. Watanabe, T. Nishida, K. Sasaki, and K. Seki, "A 1.5-V Full-Swing BiCMOS Logic Circuit," *IEEE J. Sol. St. Ckts.*, **27**(11), 1568–1574 (1992).

42. C.-L. Chen, "2.5-V Bipolar/CMOS Circuits for 0.25-μm BiCMOS Technology," *IEEE J. Sol. St. Ckts.*, **27**(4), 485–491 (1992).

43. J. B. Kuo, K. W. Su, J. H. Lou, S. S. Chen, and C. S. Chiang, "A 1.5V Full-Swing BiCMOS Dynamic Logic Gate Circuit Suitable for VLSI using Low-Voltage BiCMOS Technology," *IEEE J. Sol. St. Ckts.*, **30**(1), 73–75 (1995).

44. S.-S. Lee and M. Ismail, "1.5V Full-Swing BiCMOS Dynamic Logic Circuits," *IEEE Trans. Ckts. and Sys.–I*, **43**(9), 760–768 (1996).

45. J. B. Kuo, S. S. Chen, C. S. Chiang, K. W. Su, and J. H. Lou, "A 1.5V BiCMOS Dynamic Logic Circuit Uisng a "BiPMOS Pull-Down" Structure for VLSI Implementation of Full Adders," *IEEE Trans. Ckts. and Sys.—I*, **41**(4), 329–332 (1994)

46. M. Kakumu, M. Kinugawa, and K. Hashimoto, "Choice of Power-Supply Voltage for Half-Micrometer and Lower Submicrometer CMOS Devices," *IEEE Trans. Elec. Dev.*, **37**(5), 1334–1342 (1990).

47. S. H. K. Embabi, A. Bellaouar, and K. Islam, "A Bootstrapped Bipolar CMOS (B^2CMOS) Gate for Low-Voltage Applications," *IEEE J. Sol. St. Ckts.*, **30**(1), 47–53 (1995).

48. R. Y. V. Chik and C. A. T. Salama, "Design of a 1.5V Full-Swing Bootstrapped BiCMOS Logic Circuit," *IEEE J. Sol. St. Ckts.*, **30**(9), 972–978 (1995).

49. J. Kernhof, M. A. Beunder, B. Hoefflinger, and W. Haas, "High-Speed CMOS Adder and Multiplier Modules for Digital Signal Processing in a Semicustom Environment," *IEEE J. Sol. St. Ckts.*, **24**(3), 570–575 (1989).

50. J. H. Lou and J. B. Kuo, "1.5V CMOS and BiCMOS Bootstrapped Dynamic Logic Circuits Suitable for Low-Voltage VLSI," *Symp. VLSI Tech., Sys., Appl.*, 279–282 (1997).

51. J. H. Lou and J. B. Kuo, "A 1.5V CMOS All-N-Logic True-Single-Phase (TSP) Bootstrapped Dynamic Logic (BDL) Circuit Suitable for Low Supply Voltage and High Speed Pipelined System Operation," *IEEE Trans. Ckts. and Sys.*, to be published.

52. R. X. Gu and M. I. Elmasry, "All-N-Logic High-Speed True-Single-Phase Dynamic CMOS Logic," *IEEE J. Sol. St. Ckts.*, **31**(2), 221–229 (1996).

53. J. H. Lou and J. B. Kuo, "A Low-Voltage Semi-Dynamic DCVSPG-domino logic circuit," private communication (1998).

Problems

1. Why is the noise immunity of CMOS dynamic logic circuits worse as compared to the static logic circuit? Is there any method to raise the noise immunity? At what cost?

2. Using NORA and domino dynamic logic circuits, design a logic function $F = A \oplus B \oplus C$ in one stage and two cascading stages. Simulate and discuss the transient performance of the circuit for load capacitances of 0.01pF, 0.1pF, and 0.5pF, and at supply voltages of 5V, 3.3V, 2.5V, and 1.5V.

3. What are the advantages of multiple output domino logic (MODL)? What are the possible problems? How can they be avoided?

4. In the skew-tolerant domino logic, from the restraints on t_p and t_n in Fig. 4.28, the $t_{skew-max}$ formula has been obtained. Assume that, by an appropriate clock distribution and an independent clock driver in the local block, clock skew can be divided into $t_{skew-local}$ and $t_{skew-global}$. Assume that, for the block in the same clock phase such as ϕ_1 of Fig. 4.28(a), its clock skew is from the local clock driver. Between two different clock phases such as ϕ_1 and ϕ_2, its clock skew is global clock skew. Assume that the clock skew of the local clock driver can be controlled within a certain value. Derive the $t_{skew-global}$ for a known $t_{skew-local}$.

5. The above figure shows a finite-state machine using two latches and combinational logic. Follow the procedure in Fig. 4.50 to derive the safe value of the clock period.

6. Discuss the mechanisms of the charge-sharing problems in CMOS and BiCMOS dynamic logic circuits.

7. Use the semi-dynamic DCVSPG-domino logic circuit to design a logic function $F = AB + BC + CA$. Compare it with the DCVS-domino logic circuit.

8. Consider the 6-input AND gate in Fig. 4.20. $W(MN3) = 5\mu m$. $W(MN_i) = W(MN3)(1 - \alpha(i - 3))$, $i = 0 - 6$. Compare the differences in the propagation delay for $\alpha = -0.05, 0, 0.05, 0.1, 0.2$. Compare the differences in influence of the worst-case charge-sharing problem for $\alpha = -0.05, 0, 0.05, 0.1, 0.2$.

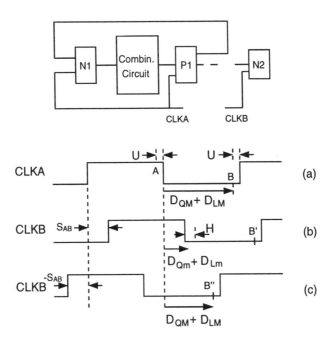

Fig. 4.75 For Problem 5.

5
CMOS Memory

Memory ICs are usually used in a computer system. As shown in Fig. 5.1, memory ICs are used as buffers between hard disks and CPUs to form a hierarchical architecture. Memory ICs are provided to store a large amount of data temporarily such that the speed of the CPU can be maintained—during each data access, it does not need to fetch data from hard disk or other I/O ports. In an advanced computer system, there is an embedded cache SRAM unit near the CPU. Main memory based on DRAM unit is placed at a location far from the CPU close to hard disk. Thus SRAM can be regarded as the cache for DRAM unit, which can be regarded as the cache for the hard disk and other I/O ports. The key parameter in an SRAM-based cache is speed. The key parameters in DRAM are size and cost.

Memory ICs can be divided into two categories: volatile and non-volatile. The data in a volatile memory will disappear when power supply is turned off. Volatile memories include SRAM and DRAM. A latch structure is used in the SRAM memory cell to store data. Data storage in a DRAM memory cell is based on dynamic storage of charge. Therefore, more transistors are needed in the SRAM memory cell. The area of a DRAM memory cell is much smaller than that of an SRAM memory cell. Therefore, the integration size of DRAM is larger—a larger storage capability at a lower cost. The specialty of SRAM is its high speed for read and write operations. The data in a non-volatile memory will not disappear when power supply is turned off. Nonvolatile memories include ROM (read only memory), PROM (program ROM), EPROM (erasable PROM), and EEPROM (electrically EPROM), flash, and FRAM (ferroelectric RAM).

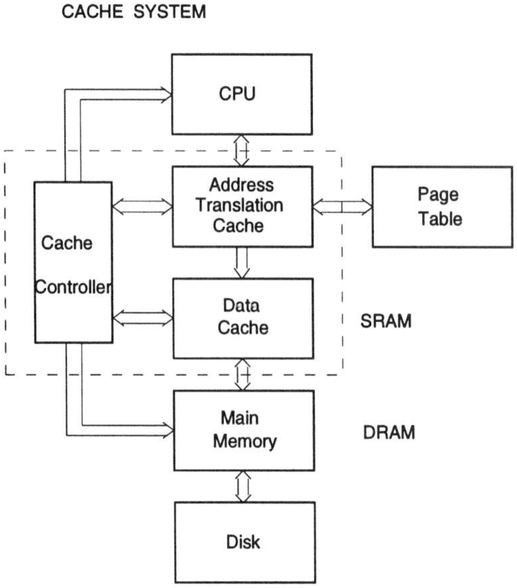

236 CMOS MEMORY

Fig. 5.1 Role of memory in a computer system.

DRAM, SRAM, and ROM have various requirements. DRAM emphasizes on cost, therefore requirements on the integration density of processing technology are high. SRAM emphasizes on speed, therefore circuit design skill is required. ROM is focused on performance, programmability, and especially cost. In the past, DRAM, SRAM, and ROM may have their preferences on the processing technology. Due to the rapid progress in CMOS technology, nowadays DRAM, SRAM, ROM, and other nonvolatile memories are based on CMOS technology. In this chapter, these memory circuits are described. In the final portion of this chapter, ferroelectric RAM (FRAM) is also introduced.

5.1 SRAM

Static random access memory (SRAM) itself is a complicated circuit system. In this section, operation and characteristics of CMOS SRAM are described. Specifically, structures of SRAM, SRAM memory cells, word line related circuits, bit line related circuits, critical path analysis, application specific SRAMs, and low-voltage SRAMS are described.

5.1.1 Basics

For an SRAM memory chip, there are address input pins. For example, for a 16Mbit SRAM, 24 address pins are required. In addition, there are data in (DIN), data out ($DOUT$), chip select ($\overline{CS}$), write enable ($\overline{WE}$), supply voltage (V_{DD}), and ground

I/P		O/P	
$\overline{CS}$	$\overline{WE}$	D_{IN}	D_{OUT}
H	X	X	L
L	L	L	L
L	L	H	L
L	H	X	D_{OUT}

Fig. 5.2 Operation of an SRAM.

(GND). When the chip select $(\overline{CS})$ signal is high, the chip is not accessed and dataout (D_{out}) is set to low. When the chip select signal is low, the chip is accessed. If the write enable $(\overline{WE})$ signal is low, it's the write cycle. The data on datain (D_{in}) are written to the memory cell, whose location is defined by the address bits. If the write enable $(\overline{WE})$ signal is high, it is the read cycle. At D_{out}, there are data from the memory cell, whose location is determined by the address bits. As shown in Fig. 5.2, when both $\overline{CS}$ and $\overline{WE}$ are low, it's the write cycle. At this time, D_{in} contains the data to be written into the designated memory cell. Under this situation, D_{out} is pulled low. When $\overline{CS}$ is low and $\overline{WE}$ is high, it is the read cycle. At this time, D_{out} contains the data, which are fetched from the designated memory cell.

Fig. 5.3 shows the typical timing waveforms of an SRAM chip during the write operation. During the write access, $\overline{CS}$ is low. The address information needs to be available at the address bits $A_0 - A_{23}$. Data to be written into the memory should be available at D_{in}. During the low state of $\overline{WE}$, the data provided by D_{in} are written into the specified location of the memory cell. Before write enable $(\overline{WE})$ turns low, data should be presented at D_{in} for some time, which is the data setup time prior write. Before $\overline{WE}$ turns low, address signals should be existent at the address bits for some time, which is the address setup time prior to write. In addition, before $\overline{WE}$ turns low, $\overline{CS}$ should also exist for a certain time—the chip select setup time prior to write. To accomplish the write operation, $\overline{WE}$ should be low for a certain time—the minimum write pulse width. After the write operation, $\overline{WE}$ will turn high. Before D_{in} can change, the signal on D_{in} should already stay at least for a certain time, which is called the data hold time after write. After turn-high of $\overline{WE}$, before the address signal can change, signal on the address bits should stay at least for a certain time, which is called the address hold time after write. When $\overline{WE}$ is low, D_{out} is pulled low. When $\overline{WE}$ turns from low to high, the time for D_{out} not to be restricted at low is called the write recovery time, since normally an SRAM cannot be accessed for read and write simultaneously. During the write access, there is no data output, therefore D_{out} is pulled low. The write recovery time defines the time needed after write before the read access can begin. From the minimum width of the write pulse needed for the SRAM, the speed of the write access can be known.

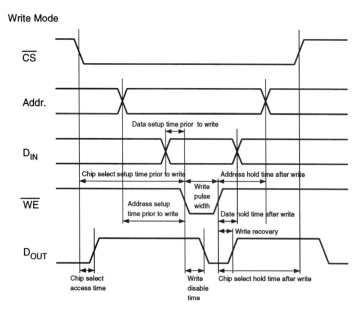

Fig. 5.3 Timing waveforms during the write operation of SRAM.

During the read access of an SRAM chip, its associated timing waveforms are much simpler as compared to the write access. As shown in Fig. 5.4, during the read access, $\overline{CS}$ should be low first. When the address signals are available, the data stored at the designated location are transferred to D_{out}. The time from the availability of the address signals to D_{out} is called the address access time.

Fig. 5.5 shows the block diagram of a typical 16M-bit SRAM memory chip. The address signals ($A_0 - A_{23}$) are divided into two groups. One group contains the row address bits ($A_0 - A_{11}$)—AXI. The other group contains the column address bits ($A_{12} - A_{23}$)—AYI. Based on the twelve row address bits, the row decoder produces $2^{12} = 4096$ horizontal word lines. With twelve column address bits, the column decoder generates 4096 vertical bit lines. The array produced by the intersections of the 4096 horizontal word lines and the 4096 vertical bit lines is the 4096 × 4096 memory cell array.

For an asynchronous SRAM, no clock is required. Instead, the address transition detection (ATD) circuit is required, which is needed to control the operation. Fig. 5.6 shows the address transition detection (ATD) circuit for the x address[1]. As shown in the figure, $ATDXi$ is the address transition detection signal of bit i, which is generated by XORing the address signal x_i with its delay. When the address signal x_i has a change, $ATDX_i$ outputs a pulse with a width equal to its delay. Then, all $ATDXi$ signals are ORed together. When $\overline{CS}$ is low, it is an OR gate. If any $ATDXi$ is high, signal ATD is high, which represents an address transition. After

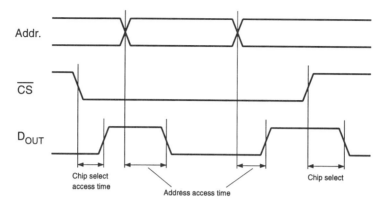

Fig. 5.4 Timing waveforms during the read operation of SRAM.

the signal ATD is produced, it can be used to generate other timing control signals, such as the signal to start sensing of the sense amplifier.

5.1.2 Memory Cells

Fig. 5.7 shows SRAM memory cells, which are based on the latch structure with two back-to-back connected inverters. As shown in the figure, there are two kinds of SRAM memory cells—the 6T cell and the 4T cell. The 6T memory cell is composed of two CMOS inverters. In addition, two NMOS pass transistors are used as the access transistors to connect two internal nodes to the bit lines. The access pass transistors to the memory cell are controlled by the horizontal word line. When the voltage of the word line is high, the access transistors turn on. Therefore, the internal nodes of the memory cell are connected to the bit lines. The 6T memory cell has advantages in noise margin, soft error immunity, and speed. The size of an SRAM cell, which determines the size of the whole chip, is determined by the layout area of the internal latch, two access transistors, and the interconnect. Since PMOS devices may occupy a large area due to latchup consideration, the two PMOS devices in the 6T cell can be replaced by polysilicon resistors—the 4T memory cell. Polysilicon resistors in the 4T memory cell are located above the four NMOS devices to save layout area. Two polysilicon resistors of $1G\Omega$ generated by ion implantation are used in the 4T memory cell to reduce the standby current. Using the polysilicon resistors and gates structure, double-poly CMOS technology is required to implement SRAMs. The problems with the 4T cell are its smaller noise margin and its smaller soft error immunity. In addition, pull-up of the internal node is via the high resistive load. In order to raise the speed performance, the resistive load should be as small as possible. With the reduced resistive load, the consumed current may rise—power consumption may increase. In order to reduce power dissipation, load resistance should be as large as possible. Recently, for high-density SRAMs, a memory cell with polysilicon

240 CMOS MEMORY

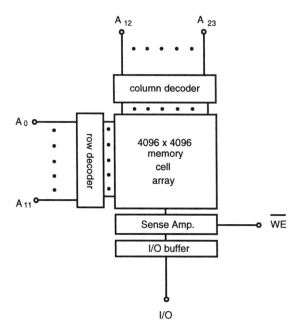

Fig. 5.5 Block diagram of an SRAM memory chip.

thin-film transistor (TFT) load has been adopted. The top layer polysilicon resistors are replaced by the polysilicon thin-film transistors (TFT), which share the same gate with the bulk NMOS device, to form a 3D 6T memory cell structure[2][3]. In the fabrication process for the polysilicon TFT load 6T memory cell, the high-resistive polysilicon load is annealed to become the polysilicon TFT used as a PMOS device. When the polysilicon TFT is turned off, its property is similar to a high-resistive load. When the polysilicon TFT turns on, its on conductance is much larger than the resistive load—pull-up speed and soft error immunity can be improved substantially.

Fig. 5.8 shows the equivalent circuit of SRAM cells during the read access[4]. During the read operation, bit lines are precharged to V_{DD}, therefore the equivalent circuit is as shown. There are two cases associated with the equivalent circuit. (Case A) at the side of the memory cell with a high at the internal node, the access pass transistor is off. As a result, it is equivalent to a CMOS inverter (6T) or a resistive-load inverter (4T). (Case B) at the side of the memory cell with a low at the internal node, the access pass transistor is on. Under this situation, it is regarded as a ratioed NMOS enhancement-load (saturation-load) inverter with the access pass transistor as the load. The transfer curve of the memory cell is also shown in the figure. At the side of the memory cell with a high at the internal node (Case A), its voltage is V_{DD}. At the side of the memory cell with an output low at the internal node (Case B), its voltage is not zero due to the ratioed NMOS enhancement-load inverter structure. From the transfer curve of the memory cell, noise margin of the memory

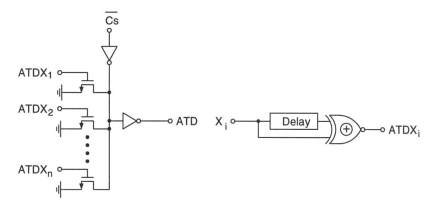

Fig. 5.6 Address transition detection (ATD) circuit for the x address.

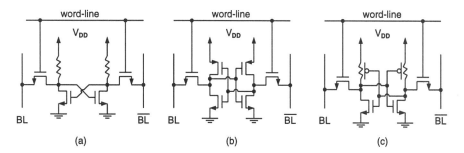

Fig. 5.7 SRAM memory cells

cell can be obtained. The output low voltage (V_{OL}) at the internal node is determined by the β ratio of the NMOS driver in the inverter to the enhancement-load inverter ($\beta_{driver}/\beta_{access}$, where $\beta = C_{ox}\mu_n \frac{W}{L}$) in Case B. In order to increase the noise margin, V_{OL} should be as low as possible—at least the output low voltage should be lower than the threshold voltage of the NMOS device ($V_{OL} < V_{thn}$). Especially for the resistive-load memory cell, V_{OL} should be low enough such that the output high voltage (V_{OH}) of the inverter at the other side is not affected. For the 6T memory cell, V_{OL} can be a little bit higher. Generally speaking, for a 6T memory cell, a β ratio of 2–2.5 is designed. For a 4T memory cell, a β ratio of 3.5 is designed.

The noise margin of the memory cell can be estimated by considering the transfer curve and its 45-degree mirror. By judging the size of the maximum square in the area bounded by the transfer curve of the memory cell and its 45-degree mirror, noise margin can be obtained. For a larger maximum square, its noise margin is better. Thus the noise margin of the CMOS memory cell is better than that of the resistive-load one as shown in the figure. Here, a method to measure the noise margin of the SRAM memory cell is described. Fig. 5.9 shows the noise margin estimation method based on the maximum square in the area bounded by the transfer curve of

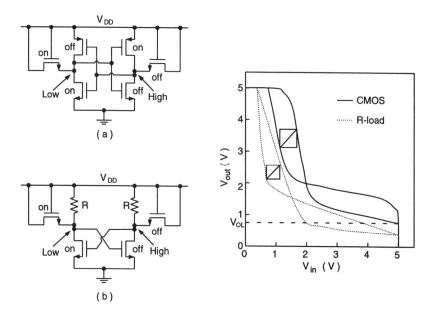

Fig. 5.8 Equivalent circuit of SRAM cells during read access.

the memory cell and its 45-degrees mirror[4][5]. By rotating the (x,y) coordinate system 45 degrees, a new coordinate system in terms of (u,v) can be obtained. By measuring the difference between the normal transfer curve of the memory cell (F_1) and its 45-degree mirror (F_2') in the (u,v) coordinate system, then multiplying the difference by $\frac{1}{\sqrt{2}}$, one obtains the length of the maximum square. The worst-case noise margin is determined by the smaller of the lengths of the two maximum squares. After coordinate transformation, the (x,y) coordinates can be expressed as:

$$x = \frac{1}{\sqrt{2}}u + \frac{1}{\sqrt{2}}v, \qquad (5.1)$$
$$y = \frac{-1}{\sqrt{2}}u + \frac{1}{\sqrt{2}}v.$$

The normal transfer curve and its 45-degree mirror can be expressed as:

$$y = F_1(x), \qquad (5.2)$$
$$y = F_2'(x).$$

In the (u,v) coordinate, the normal transfer curve and its 45-degree mirror become:

$$y = F_1(x) \quad \rightarrow \quad v = u + \sqrt{2}F_1\left(\frac{1}{\sqrt{2}}u + \frac{1}{\sqrt{2}}v\right),$$
$$y = F_2'(x) \quad \rightarrow \quad x = F_2(y) \quad \rightarrow \quad v = -u + \sqrt{2}F_2\left(-\frac{1}{\sqrt{2}}u + \frac{1}{\sqrt{2}}v\right).$$
(5.3)

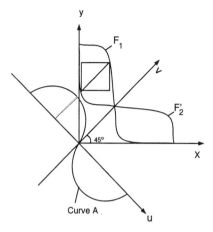

Fig. 5.9 Noise margin estimation based on the maximum square in the area bounded by the transfer curve of the memory cell and its 45-degree mirror.

The equivalent circuits for the above equations are shown in Fig. 5.10[4]. As shown in the figure, F_1 and F_2 can be simulated using the actual circuits used in the memory cell. By extracting the maximum values in the difference between v_1 and v_2, the noise margin of the memory cell can be determined.

5.1.3 Decoders

The function of a row decoder in an SRAM is to decode the row address to produce the word line signals. When the number of the row address bits is not large, row decoders can be implemented using CMOS and pseudo-NMOS gates as shown in Fig. 5.11. In order to raise speed, the pseudo-NMOS structure is more widely used to implement row decoders. Since usually the word line load is large, an inverter is added to provide the buffer function. Word lines are usually implemented in polysilicon or polycide. Polysilicon word lines may cross the whole memory cell array. As shown in Fig. 5.12, when the integration size is large, the polysilicon word line may be very long, hence the associated RC delay may be long. The RC delay of the polysilicon word line can be shortened by rearranging the structure of the memory cell array. As shown in Fig. 5.12, if the row decoder is placed adjacent to one end of the memory cell array, from the output of the decoder to the memory cell located at the other end of the chip, the polysilicon word line is the longest—its RC delay is the longest, which can be shortened by rearranging the structure. If the memory cell array is divided into two halves with the row decoder placed in the center, thus the length of the longest polysilicon word line is cut in half. As a result, the longest RC delay associated with the longest polysilicon word line is reduced by four times. In an advanced architecture, the word line can be equally divided into several sub-word lines by the hierarchical word line decoder to reduce the length of the poly/polycide word line. Global word lines, which are implemented in metal lines, are connected to

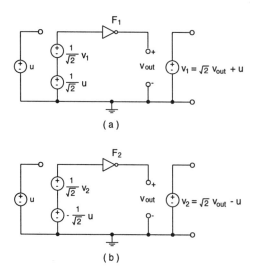

Fig. 5.10 Circuit implementation of Eq. (5.3). (Adapted from Seevinck et al. [4].)

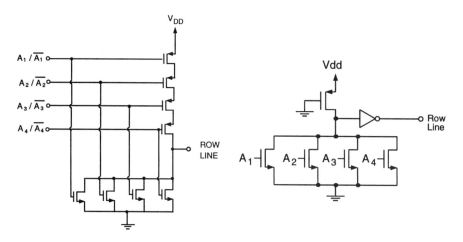

Fig. 5.11 Row decoders in CMOS and pseudo-NMOS.

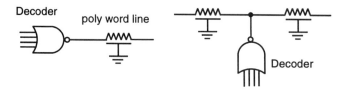

Fig. 5.12 Structures of the polysilicon word line.

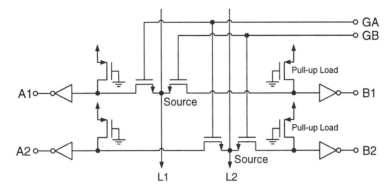

Fig. 5.13 Transfer word decoder. (Adapted from Matsumiya et al. [7].)

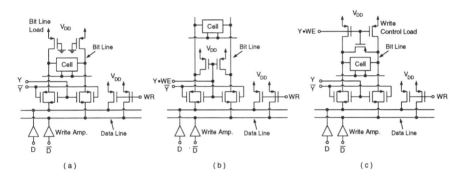

Fig. 5.14 Bit-line load circuits. (Adapted from Matsumiya et al. [7].)

the sub-wordlines to reduce RC delays. One other method to shorten the propagation delay associated with the polysilicon word line can be accomplished by using the BiCMOS word line driver[6]. Taking advantage of the strong driving capability of the bipolar device in the driver, the delay time associated with the word line can be reduced.

Fig. 5.13 shows the transfer word decoder[7] with a 2-input AND gate. As shown in the figure, the transfer word decoder is used in the last stage of the high-capacity hierarchical architecture to drive the cell word line. Its structure combines the pseudo-NMOS and pass transistor circuit techniques[7]. When the row select signal L1 is low and the global row select signal GA is high, the left word line A1 is high ($A = \overline{L1} \cdot GA$), which implies selected. Similarly, when GB is high (GA is low), B1 is high. Using this arrangement, the decoder circuit is concise. In addition, GA and GB-controlled transistors have a shared source area to minimize the layout area.

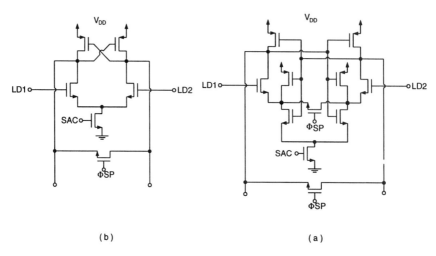

Fig. 5.15 (a) Latched sense amp; and (b) NMOS source-controlled latched sense amp. (Adapted from Seki et al. [8].)

5.1.4 Bit-Line Related Architecture and Sense Amps

In a large SRAM, several bit lines may be connected to a data line, where the sense amp is located. During the write access, the bit line signals are at full swing. Fig. 5.14 shows bit-line load circuits[7] with various write control structures. At the beginning of read, bit lines are set to V_{DD}. WR (write recovery) is used to equalize and precharge the data lines after the write cycle to improve the read access time—the worst-case read operation occurs when there is a read operation immediately after write for accessing the same bit line. As shown in the figure, three types of bit-line load circuits are listed—(A) the common bit-line load circuit, (B) the write recovery load placed at the same end of the column select gate, and (C) the write recovery load placed at the opposite end of the column select gate. In the common bit-line load circuit (A), during the write operation, a substantial amount of current in the bit line load still flows into the bit line, which brings in short-circuited power waste and a reduction in the write speed. In (B), during write recovery, the precharge current can only flow from one end of the bit line—when the bit line is long, due to the RC delay the charge time is long. In (C), the load-line currents flow from two ends of the bit lines—the most efficient.

In addition to the memory cell and the load line, the other important circuit at the bit line is the sense amp. Fig. 5.15 shows the latched sense amp and the NMOS source-controlled latched sense amp[8]. As shown in the figure, when SAC is high, the operation of the sense amp is activated. The latched sense amp has advantages in large charge/discharge currents. However, when the input signal swing is small, gain may not be sufficient. In the NMOS source-controlled latched sense amp, two feedback inverters are added to control the source voltage of the pull-down NMOS driver. As a result, when the voltage of the input at one end (LD1) is decreased,

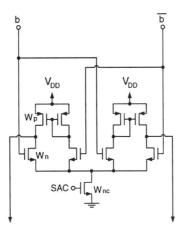

Fig. 5.16 Double-ended current-mirror differential amplifier. (Adapted from Sasaki et al. [9].)

the elevated output voltage will lower the source voltage of the pull-down NMOS driver (LD2). Consequently, the pull-down capability of the LD2 NMOS can be enhanced. Therefore, the source voltage of the LD1 NMOS device is raised, hence the pull-down current of LD1 is reduced and the gain is increased.

Fig. 5.16 shows the double-ended current-mirror differential amplifier[9]. As shown in the figure, via the current mirror, from the difference in the input voltages, the difference in the relative currents is used to form the voltage difference at the output node. As shown in the figure, two single-ended differential amplifiers are combined to provide two single-ended outputs for producing the differential outputs. Fig. 5.17 shows the current sense amp (CSA)[7]. As shown in the figure, in the current sense amp, the voltage swing at the input end can be very small. Hence, when the loads at the inputs to data bus DB, $\overline{DB}$ are large, a small voltage swing leads to a high switching speed. In addition, due to a small voltage swing, the current sense amp is suitable for low-voltage operation. In the current sense amp, the input data current I_{DB} is amplified by PMOS and NMOS cascoded current mirrors (via W/L). As shown in the figure, the output currents are:

$$\begin{aligned} I_{out1} &= m_1 I_{DB} - m_1 m_2 I_{DBX}, \\ I_{out2} &= m_1 I_{DBX} - m_1 m_2 I_{DB}. \end{aligned} \quad (5.4)$$

If the complementary input current (I_{DBX}) is close to zero, then

$$\begin{aligned} I_{out1} &= m_1 I_{DB}, \\ I_{out2} &= -m_1 m_2 I_{DB}. \end{aligned} \quad (5.5)$$

Thus current amplification can be accomplished in the current sense amp.

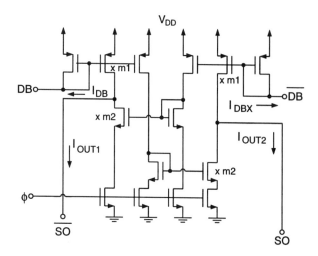

Fig. 5.17 Current sense amp (CSA). (Adapted from Matsumiya et al. [7].)

Fig. 5.18 shows the stabilized feedback current sense amp (SFCA)[10]. As shown in the figure, the fundamental structure is based on the small input resistance (R_{in}) by the feedback scheme made of M1–M4. As shown in the figure, the loop gain of the feedback scheme is:

$$A = \left(\frac{g_{m3}}{g_{m1} + g_{ds1} + g_{ds2}}\right)\left(\frac{g_{m2}}{g_{m4} + g_{ds3} + g_{ds4}}\right). \quad (5.6)$$

If there is a small voltage variation (δV) at the input, considering the feedback effect, the variation in the current is:

$$\delta I + A\delta I + A^2 \delta I + \ldots = \frac{\delta I}{1 - A}. \quad (5.7)$$

By increasing transconductances g_{m1}–g_{m4} or decreasing output conductances g_{ds1}–g_{ds4}, $A \to 1$. Therefore, the input resistance becomes:

$$R_{in} = \frac{\delta V}{\delta I} = \frac{\delta V}{\frac{\delta I}{1-A}} = \frac{1 - A}{g_{m3}}. \quad (5.8)$$

Thus a very small R_{in} can be obtained—the voltage swing at the input can be effectively reduced. As shown in Fig. 5.18(c), a current sense amp followed by a current-to-voltage converter to convert the current output to the voltage output is included.

5.1.5 SRAM Architecture

In the conventional architecture, when the selected word line is high, all cells connected to the word line in the row are active. When the size of the SRAM increases,

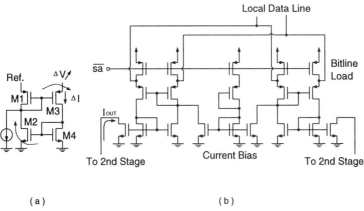

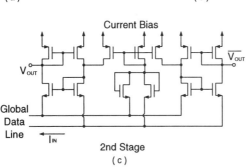

Fig. 5.18 Stabilized feedback current sense amp (SFCA). (Adapted from Seno et al. [10].)

the number of cells connected to the word line increases—the load is large. Therefore, the word line delay increases when the SRAM size increases. When the word line is high, all cells connected to the word line become active—power dissipation increases. Fig. 5.19 shows a divided word-line (DWL) structure in an SRAM[11]. As shown in the figure, using the block select signal ($BS_i, i = 1, 2, ...n_B$, where n_B is the block number) ANDed with the global word line generated by the global row decoder, the ith block of the memory cell array can be selected. Using this approach, the load of the word line is reduced to $\frac{1}{n_B}$ of the previous value. In addition, parasitic resistance and capacitance also become smaller due to the decrease in the length of the local word line. As a result, the propagation delay is reduced. During the read/write access, only the word line of a selected block of memory cells is active—power dissipation can also be reduced. As shown in the figure, the arrangement of the x address bits stays unchanged. From the y address bits, $log_2 n_B$ bits need to be taken out for block select—they are also called z address bits. Thus, pre-decode of the z address bits can be carried out with the x address decode.

For mega-bit SRAMs, the divided word line (DWL) architecture is not sufficient for use—at an increased size, the propagation delay of the 2-stage decoder and the increased word line load is still too large. Fig. 5.20 shows the hierarchical word-line

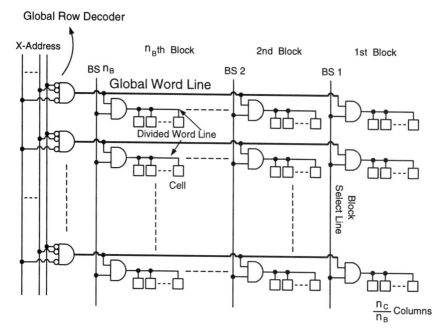

Fig. 5.19 Divided word-line (DWL) structure in a SRAM. (Adapted from Yoshimoto et al. [11].)

decoding (HWD) architecture for high-density SRAMs[12]. The HWD technique is used to further extend the DWL technique. The DWL architecture is a 2-layer hierarchy. The HWD architecture is a 3-layer hierarchy, which includes the global word-lines, sub-global word-lines, and local word-lines. The signal produced by the global word decoder is ANDed with the block group select signal to generate the sub-global word-line signal. The sub-global word-line signal is ANDed with the block select to produce the actual local word-line signal. Since the global word-line needs to drive only sub-global row decoders, which are located in the center of memory arrays, the length of the global word-line is reduced to one-half of the previous DWL one. In addition, the sub-global word-line only drives one-half of the local word-line drivers. By this arrangement, the local word-line driver drives a much smaller number of memory cells. In addition, the load of the global and the sub-global decoders can be effectively reduced via the hierarchical approach. Thus the speed of the word-line decoding can be enhanced.

5.1.6 Critical Path Analysis

For a SRAM, the read address access time determines the speed performance. The delay associated with the critical path determines the address access time. Fig. 5.21 shows the floor plan of a 4-Mb SRAM[13]. Predecoders and global row decoders are placed in the center column. Column decoders, sense amps, write drivers, I/O

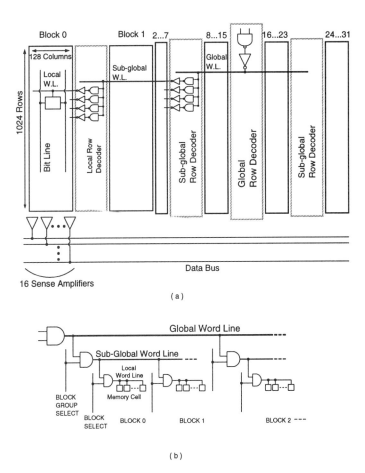

Fig. 5.20 Hierarchical word decoding (HWD) architecture. (Adapted from Hirose et al. [12].)

buffers, and I/O pads are placed in the center horizontal area. By this arrangement, the memory cell array is divided into four 512×1024 (0.256cm×1.024cm) areas using 5µm × 10µm memory cells. In order to facilitate decoding, in the horizontal direction, the memory cell array is divided into 32 blocks. Each block has 1024×128 cells. Among blocks, there 16 local row decoders and two sub-global row decoders. Fig. 5.22 shows the hierarchical architecture of the row decoding structure. In the hierarchical architecture of the row decoding, the row address is 10-bits wide (the column address is 12-bits wide). The pre-decoder processes the first four bits. The global, the sub-global, and the local row decoders handle 2 bits each. The block group select (BGS) and block select (BS) signals provided by the column decoder are also used to reduce the active region such that power consumption can be shrunk. By this arrangement, the load of every stage can be uniformly distributed such that delay and power consumption can be optimized. Fig. 5.23 shows the hierarchical architecture

252 CMOS MEMORY

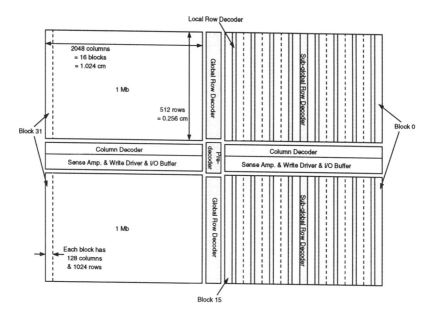

Fig. 5.21 Floor plan of a 4Mbit SRAM.

of the data-line. In each block, for every eight columns, there is a first-stage sense amp, which is responsible for sensing the difference between the bit-lines. In each block, all 16 first-stage sense amps are connected to the second-stage sense amps. The outputs of the second-stage sense amps are connected to the global data-line and the output buffer via the pre-driver. By an appropriate arrangement, the size and the power of the sense amps can be reasonably allocated. Also shown in the figure is the write-in path. Fig. 5.24 shows the critical path from the x-address input to the memory cell. In the decoders of all levels, the logic AND circuits have been widely used. During the pull-up transient, the propagation delay of the address buffer is:

$$\tau_1 = \frac{(C_{M1} + C_{G1}) \times \frac{V_{DD}}{2}}{\frac{I_{D1}}{2}} = \frac{(1.02 \times 10^{-12} + 0.06 \times 10^{-12}) \times \frac{5}{2}}{\frac{3.5 \times 10^{-3}}{2}} = 1.54ns,$$

(5.9)

where I_{D1} is the driving current of the address buffer, which has an aspect ratio of $20\mu m/0.8\mu m$ for the PMOS device ($I_{D1} = \frac{1}{2}\mu_n C_{ox} \frac{W}{L}(V_{GS} - V_T)^2 \cong 3.5mA$, with a gate oxide of 19nm.), C_{M1} is the parasitic capacitance associated with the level-2 metal line from the output of the address buffer to the input of the pre-decoder. With a width of $2.5\mu m$ and a length of 1.024cm across the 16 blocks (2048 columns), the parasitic capacitance of the metal line C_{M1} is 1.02pF. C_{G1} is the equivalent input capacitance to the pre-decoder. The pre-decoder delay can be approximated by considering the parasitic capacitance of the metal line from the output of the pre-decoder to the input of the global row decoder (C_{M2}) and the equivalent input

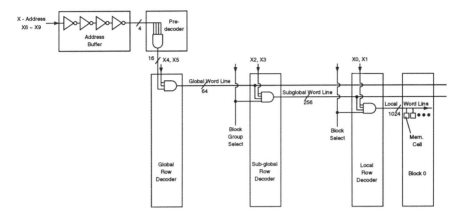

Fig. 5.22 Hierarchical architecture of the row decoding structure.

capacitance to the global row decoder (C_{G2}):

$$\tau_2 = \frac{(C_{M2} + C_{G2}) \times 2 \times \frac{V_{DD}}{2}}{\frac{I_{D2}}{2}} = \frac{(0.447 + 0.01) \times 2 \times 10^{-12} \times \frac{5}{2}}{\frac{3.5 \times 10^{-3}}{2}} = 1.3ns, \quad (5.10)$$

where C_{M2} is the parasitic capacitance of the level-1 metal line with a width of $1.2\mu m$ from the output of the pre-decoder to the input of the global row decoder. Accounting for crossing 512 rows (0.256cm), $C_{M2} = 0.447pF$. C_{G2} is the equivalent input capacitance to the global row decoder. I_{D2} is the driving current of the PMOS device in the pre-decoder. Note that a factor of 2 is used to account for the fact the pre-decoder needs to drive the global row decoders at the top and the bottom.

The delay of the global row decoder is

$$\tau_3 = \frac{2(C_{M3} + C_{G3}) \times \frac{V_{DD}}{2}}{\frac{I_{D3}}{2}} = \frac{(0.5 + 0.03) \times 2 \times 10^{-12} \times \frac{5}{2}}{\frac{3.5 \times 10^{-3}}{2}} = 1.51ns, \quad (5.11)$$

where C_{M3} is the parasitic capacitance of the level-2 metal line from the output of the global row decoder to the input of the sub-global row decoder, C_{G3} is the equivalent input capacitance to the sub-global row decoder, and I_{D3} is the driving current of the global row decoder during the pull-up transient. Note that a factor of 2 is used to account for that the global row decoder needs to drive the sub-global row decoders at the left and the right. The delay of the sub-global row decoder is:

$$\tau_4 = \frac{2(C_{M4} + C_{G4}) \times \frac{V_{DD}}{2}}{\frac{I_{D4}}{2}} = \frac{2 \times (0.23 + 0.06) \times 10^{-12} \times 2.5}{3.5 \times 10^{-3} \times 0.5} = 0.8ns, \quad (5.12)$$

where C_{M4} is used to account for the parasitic capacitance of the level-2 metal line from the output of the sub-global row decoder to the input of the local row decoder, C_{G4} is the equivalent input capacitance to the local row decoder, and I_{D4} is the

254 CMOS MEMORY

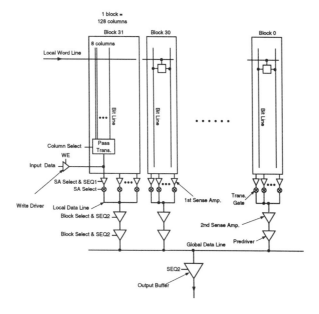

Fig. 5.23 Hierarchical architecture of the data line.

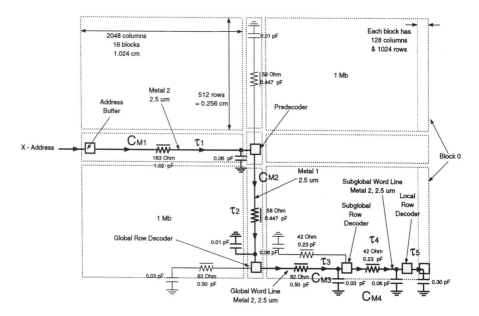

Fig. 5.24 Critical path from the x-address input to the memory cell.

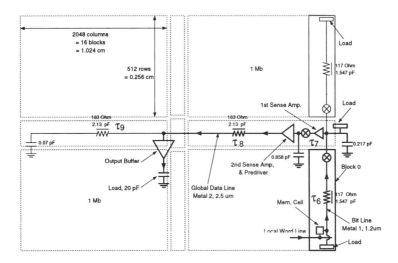

Fig. 5.25 Critical path from the memory cell to the sense amp and the output I/O pad.

driving current of the sub-global row decoder during the pull-up transient. The delay of the local row decoder is:

$$\tau_5 = \frac{C_{G5} \times \frac{V_{DD}}{2}}{\frac{I_{D5}}{2}} = \frac{0.3 \times 10^{-12} \times 2.5}{1 \times 10^{-3} \times 0.5} = 1.5ns. \quad (5.13)$$

The propagation delay from the x-address to the memory cell is $\tau_1+\tau_2+\tau_3+\tau_4+\tau_5 = 6.65ns$. Fig. 5.25 shows the critical path from the memory cell to the sense amp and the I/O related circuits using the circuits as shown in Fig. 5.26. At the data line, in addition to the propagation delay due to the metal-line parasitics, the parasitic capacitances associated with the related pass transistors and drivers are large. The delay associated with cell bit-line is:

$$\tau_6 = \frac{C_{bl} \times \Delta V_{bl}}{I_{D6}} = \frac{1.547 \times 10^{-12} \times 0.04}{4.2 \times 10^{-5}} = 1.47ns, \quad (5.14)$$

where C_{bl} is the equivalent load capacitance of the level-1 bit-line from the memory cell to the sense amp, which accounts for the metal line parasitics and the parasitics of related pass transistors and drivers. ΔV_{bl} is the bit-line swing (40mV) when the first-stage sense amp is active. I_{D6} is the driving current of the NMOS device in the memory cell ($2\mu m/0.8\mu m$). The delay of the 2-stage sense amp is mainly due to the delay in the first stage. When the output of the first-stage sense amp reaches a swing of 0.5V, the output of the second-stage sense amp has already changed completely. Therefore, the propagation delay of first-stage sense amp is approximated by considering the time when its output changes by $\Delta V_{sl} = 0.5V$. The output load of the first-stage sense amp is estimated to be $C_{sl} = 0.858pF$. If half of the biasing current in the first-stage sense amp is used to charge its output load

256 CMOS MEMORY

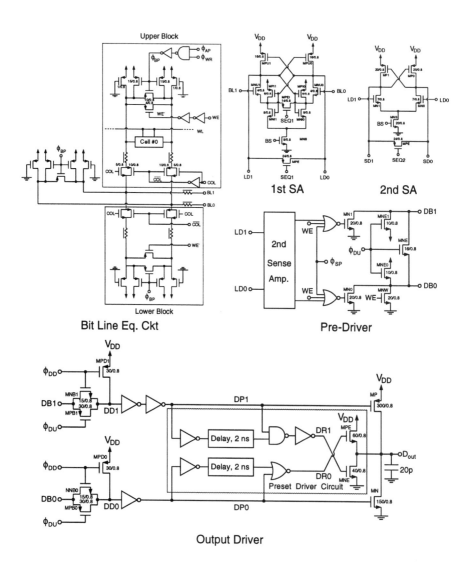

Fig. 5.26 Sense amp and the I/O related circuits.

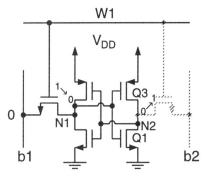

Fig. 5.27 Structure of a conventional single-bit-line memory cell.

($\frac{I_b}{2} = 0.55mA$), the delay of the sense amp is:

$$\tau_7 = \frac{C_{sl} \times \Delta l V_{sl}}{\frac{I_b}{2}} = \frac{0.858 \times 10^{-12} \times 0.5}{0.55 \times 10^{-3}} = 0.78ns. \quad (5.15)$$

After the second-stage sense amp, the pre-driver needs to drive the global data line, which is 1.024cm long—$C_{dl} = 2.13$pF. Considering the pull-down driving current of the pre-driver, which has an aspect ratio of 20μm/0.8μm for the NMOS device, the delay from the output of the predriver to the input of the output buffer can be computed using the formula for the t_{PHL} described in Chapter 3:

$$\tau_8 = \frac{C_{dl}}{\beta_n(V_{DD} - V_t)} \left[\frac{V_t}{V_{DD} - V_t} + \frac{1}{2} ln \left(\frac{3V_{DD} - 4V_t}{V_{DD}} \right) \right] = 1.497ns. \quad (5.16)$$

The output buffer needs to drive a 20pF load. Considering the aspect ratio of 150μm/0.8μm for the NMOS device in the output buffer, the delay of the output buffer is $\tau_9 = 1.137ns$. The delay from the memory cell via the data line, and the sense amp, to the output buffer is $\tau_6 + \tau_7 + \tau_8 + \tau_9 = 4.9ns$. The total delay from the address input to the output buffer is 11.5ns.

5.1.7 Application-Specific SRAM

For large-size and low-power SRAMs, due to the complementary bit lines of the memory cell, layout density cannot be increased effectively. Using narrow bit lines, the wiring resistance and thus the delay increase substantially. In addition, the electromigration problems become serious—reliability problems arise. Instead of two bit lines, single-bit line SRAMs have been proposed to decrease layout area and power dissipation and to enhance the bit-line reliability. Fig. 5.27 shows the structure of a conventional single-bit-line memory cell. Assume that logic 0 is to be written to one end of the memory cell, which stores logic 1 originally. The other end of the memory cell needs to rise from logic 0 to logic 1, which will be accomplished via the pull-up of the polysilicon TFT PMOS load. Due to the small on current of the

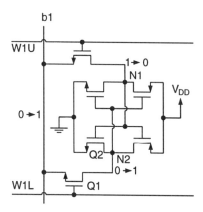

Fig. 5.28 Single-bit-line cell with two word lines. (Adapted from Sasaki et al. [14].)

TFT PMOS load, the write time is long—the major drawback of the single-bit-line cells. Fig. 5.28 shows the single-bit-line cell with two word lines[14]. (In a conventional SRAM memory cell, the single word line structure with two bit lines is used.) As shown in the figure, in the two word-line structure, assume that logic 0 is to be written to one end of the memory cell, which stored logic 1 previously. After logic 0 has been written, the data on the bit line are changed to logic 1. Then logic 1 is written to the other end. For example, if N1 is written with logic 0 and N2 is written with logic 1. W1U needs to be high to facilitate the write-in of logic 0 to N1. At this time, the word line W1L needs to maintain low to prevent logic 0 from being written to N2. After the write logic-0 process for N1 has been done, the bit line switches to high. During the high state of W1L, N2 is written with logic 1. By this arrangement, the write time can be shortened. Fig. 5.29 shows the write/read circuits for the single-bit-line cell array[14]. During the read operation, the bit line of a cell array and its dummy bit line are connected to the sense amp. At the dummy bit line there is a dummy cell, which is always on. During the read operation, the driving current of the dummy cell is one-half that of the memory cell. Thus the value stored in the memory cell can be obtained by comparing it with the dummy cell.

Applications of SRAM may be diversified in terms of throughput and output bandwidth. Here, an embedded mid-capacity, wide-word SRAM suitable for microprocessor processing unit (MPU) is described. Fig. 5.30 shows a SRAM structure using read/write shared sense amp and self-timed pulsed word lines[15], which is used to design a 2-way set associative cache with separate data and instruction for 64-bit wide bus. For such a specification, an I/O bus of 256 bits is required. Under this requirements, a bidirectional read/write shared sense amplifier (BSA) is designed. Conventionally, sense amp is used for read only. Here, BSA functions like a sense amp in DRAM—BSA is used for read and rewrite. As shown in the figure, this chip has four sections. Each section has 32 8K-memory blocks. Each 8K-bit memory block has 8 I/O pairs, which is divided into two 128 row × 32 column (4K) arrays. Among 8 bit lines, one is selected (via the 8-to-1 column selector) and connected to

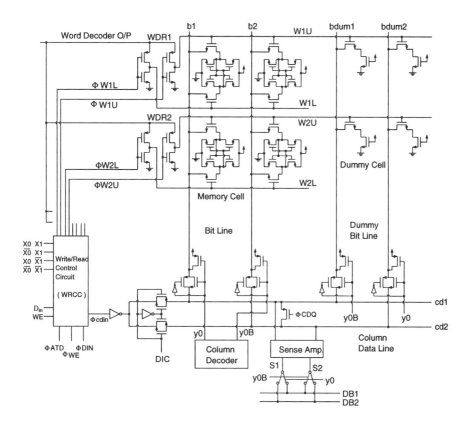

Fig. 5.29 Write/read circuits for the single-bit-line cell array (From Sasaki et al. [14], ©1993 IEEE.)

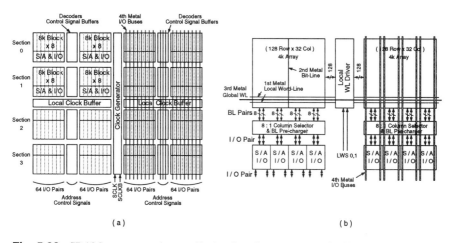

Fig. 5.30 SRAM structure using read/write shared sense amp and self-timed pulsed word lines. (From Kushiyama et al. [15], ©1995 IEEE.)

260 CMOS MEMORY

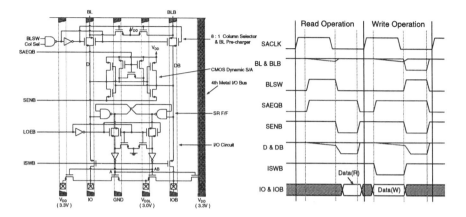

Fig. 5.31 The bidirectional read/write shared sense amp (BSA) circuit. (From Kushiyama et al. [15], ©1995 IEEE.)

BSA for the read/write shared sense amplification. Fig. 5.31 shows the bidirectional read/write shared sense amp (BSA) circuit[15]. BSA is used to serve as a sense amp for the read operation. During the write operation, BSA is used as the write circuit and the data input buffer. As shown in the figure, BSA contains an 8-to-1 column selector and a bit-line precharger, a CMOS dynamic sense amp, an SR flip-flop, and an I/O circuit. BLSW and the column select signal control the 8-to-1 column selector and the BL precharger. When both BLSW and the column select signal are high, the column switch turns on and the BL precharger stops precharging. Therefore, BL and BLB are connected to D and DB, respectively. If one or both is/are low, the operation is reversed. SAEQB and SENB are used to control the sense amp. When SAEQB is low, D and DB are equalized and precharged to V_{DD}. When both SAEQB and SENB are low, the operation of the dynamic sense amp is started. LOEB is used to control the SR flip-flop to latch the output data at IO and IOB. When LOEB is low, the transmission gate controlled by LOEB turns on. Hence, the SR flip-flop provides the output. The low level of the output is at $V_{DDL} = 3V$. The high level of the output is at $V_{DD} = 3.3V$. This is accomplished by setting one of A and AB to low such that one output is connected to V_{DDL} and the other output to V_{DD}. Therefore, the reduced voltage output swing leads to reduced power dissipation at the I/O bus. ISWB is used to control the write operation. When ISWB is low, IO and IOB are connected to D and DB for developing the signal, followed by amplification by the sense amp. As shown in the timing waveform, when SACLK is high, it is dedicated to the D and DB signal development. During the read cycle, the cell bit lines (BL and BLB) are connected to D and DB (when BLSW is high). During the write cycle, the outputs (IO and IOB) are connected to D and DB (ISWB is low). When SACLK is low, it is the sense/write period. During the read operation, the cell bit lines (BL and BLB) are separated from D and DB for precharging the bit line (BLSW is low). During the write cycle, since BLSW is high, BL and BLB are connected to D and DB and the data signal amplified by the sense amp is written to the cell.

5.1.8 Low-Voltage SRAM Techniques

Low voltage operation has been a general trend for high-density SRAMs. For a SRAM operating at a very low voltage, for example at 1V, due to the threshold voltage of the access transistor considering body effect, during the write cycle, it is difficult for logic 1 to be written into the memory cell. If the logic-1 voltage is too small during the write logic-1 operation, the logic-0 at the other end of the memory cell cannot be maintained since the PMOS device may be turned on. Therefore, for low-voltage operation, word line voltage during the write operation should have a boosted level. Under this situation, the boosted word-line voltage leads to a drop in the $\beta_{driver}/\beta_{access}$ ratio. An increase in β_{access} may cause a quick rise in the inverter output low-voltage V_L. As a result, during the read operation, the data in the memory cell may not be easily maintained. During the write operation, the cells connected to the same word line not accessed for write may be under quasi-read operation—similar problems exist. Therefore, a two-step word-voltage method (TSW) as shown in Fig. 5.32 should be used[16]. As shown in the figure, during the read cycle, the word line voltage is increased to V_{DD} first. At the same time, the pull-up transistor is turned off (ϕ_{LD} is high) such that the read cell can maintain a stabilized state during the read cycle since the $\beta_{driver}/\beta_{access}$ ratio is still maintained at a safe value. During the write cycle, the quasi-read cells can maintain a stabilized state too. Then, the word-line voltage is raised to a boosted level V_{ch}. As a result, during the read cycle, since one of the two: $N1$ and $N2$ is already discharged to 0 by the memory cell, logic 0 is not damaged by the boost in the $\beta_{driver}/\beta_{access}$ ratio. During the write cycle, due to the boosted word-line voltage, the memory cell can be written with logic-1.

In addition to the word-line boost technique, the read/write speed can be enhanced by changing the source-body biasing voltage (V_{SB}). Fig. 5.33 shows the driving source-line (DSL) memory cell architecture [17][18]. As shown in the figure, the source and the body of the NMOS driver transistors are separate. Body is still connected to ground. The source is connected to SL. During the read cycle, SL is negative, thus the source-body voltage (V_{SB}) of the NMOS driver transistor in the memory cell becomes positive. As a result, the threshold voltage is lowered. The increased drain current helps shrink the sense time. Note that before the sense cycle, the bit lines are precharge to $V_{DD}/2$ instead of V_{DD} as for conventional SRAMs. During the write operation, SL is floating. Thus the NMOS driver transistor is inactive. As long as there is a small voltage swing at the bit lines, the internal node potentials in the memory cell can be adjusted to a level near the bit-line voltage. At the end of the write cycle ($\overline{WE}$ goes back to high), SL is back to the ground level. The memory cell itself functions like a latch-type sense amp—the internal node voltage is restored to full-swing. Fig. 5.34 shows the comparison of the driving source-line (DSL) cell structure with the conventional structure in terms of read and write operations of the SRAM bit line. In a conventional SRAM, during the write cycle, the precharged-to-V_{DD} bit lines cause a large swing at the bit lines—power dissipation is large. By the arrangement described here, this problem can be avoided.

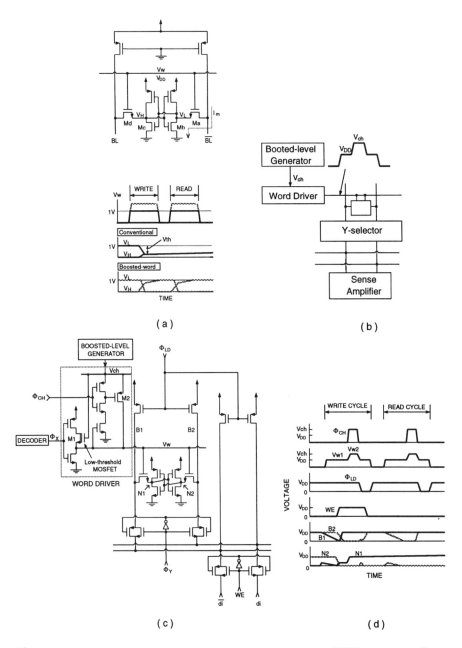

Fig. 5.32 1V TFT-load SRAM with a two-step word-voltage (TSW) method. (From Ishibashi et al. [16], ©1992 IEEE.)

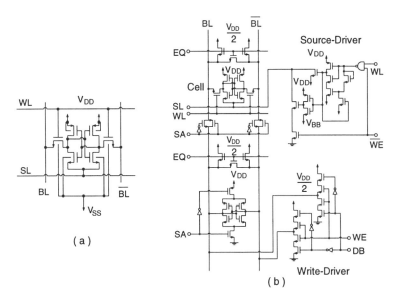

Fig. 5.33 Driving source-line (DSL) memory cell structure. (Adapted from Mizuno and Nagano [17][18].)

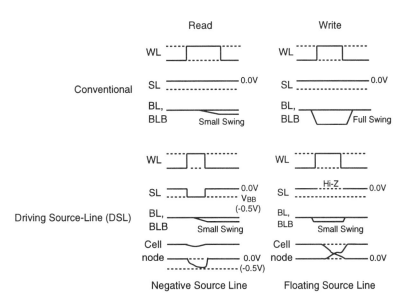

Fig. 5.34 Timing waveforms for the driving source-line (DSL) cell structure and the conventional cell structure. (From Mizuno and Nagano [17][18], ©1995, ©1996 IEEE.)

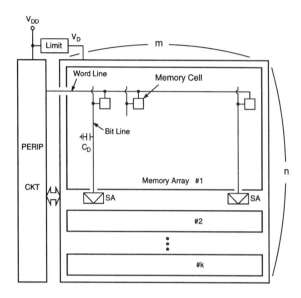

Fig. 5.35 General block diagram of a DRAM chip. (Adapted from Itoh [19].)

5.2 DRAM

Fig. 5.35 shows the general block diagram of a DRAM chip architecture[19]. As shown in the figure, it contains m bit lines and n word lines in k sub-arrays. Each sub-array contains m columns (m bit lines) and n/k rows (n/k word lines). Each bit line is divided into k sections—a smaller bit line capacitance C_B can be obtained. The internal supply voltage V_D is obtained from the external V_{DD} using the on-chip voltage limiter (converter). During the read operation, the word line is pulled high. Therefore, all m cells connected to the accessed word line enter the read operation. Via the sense/restore amplifier, the small signal swing on the bit line is then amplified to the full swing of V_{DD}. Then, from m bit lines, one of the bit line is selected—the data in the corresponding memory cell are transferred to the data output. Thus it concludes the read operation. At the same time, the amplified signals on the m bit lines are fed back to each individual memory cell for rewrite of the data. The write operation is similar to the read operation—during the write operation, data are transferred from the data input terminal to the memory cell via the selected bit line. Due to junction and subthreshold leakage currents of the access transistor, the data stored in the memory cell may be lost gradually. Therefore, a refresh operation is required to retain the data in the memory cell. During the refresh operation, the above read cycle without selecting a bit line is repeated. A complete refresh operation for the whole chip is done by executing the read operation without selecting a specific bit line for n word lines sequentially. The read/write operations can be carried out only when not in the refresh cycle. If the data in a memory cell need to be refreshed at least once within the maximum refresh time (t_{REFmax}), the maximum refresh cycle time becomes: t_{REFmax}/n. Ratio β is defined as the maximum refresh cycle time

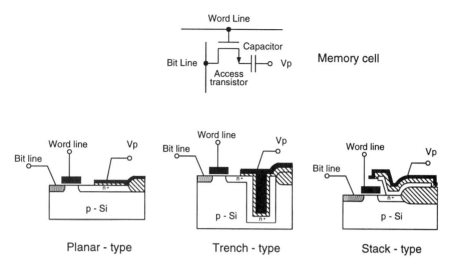

Fig. 5.36 Three types of the DRAM memory cell: planar, trench, and stack.

to the minimum normal read cycle time (t_{RCmin}):

$$\beta = (t_{REFmax}/n)/t_{RCmin}, \quad (5.17)$$

where β represents the maximum chip utilization under a refresh operation overhead—the larger the better. Thus, the data retention time (t_{REFmax}) should be lengthened or the access time (t_{RCmin}) should be shortened as much as possible.

5.2.1 Memory Cells

Memory cell is a key component in a DRAM. As shown in Fig. 5.36, the DRAM memory cell is composed of a storage capacitor and an access MOS transistor. The binary data are stored in the storage capacitor in terms of charge. Generally speaking, there are three basic types of DRAM memory cells—planar-type, trench-type, and stack-type. When the integration size is less than 1-Mbit, the memory cell is planar-type, where the storage capacitor is placed next to the access transistor. At 4-Mbit, in the stack-type memory cell, the storage capacitor is located above the access transistor to increase the device density. At 4-Mbit, the trench-type memory cell is also used. Greater than 4-Mbit, such as 64-Mbit and 256-Mbit, both stack-type and trench-type approaches can be used to implement the memory cell. At 1-Gbit, in addition to the trench and the stack structures, the capacitor is realized by using a special dielectric material with a high dielectric constant. To implement a $0.35\mu m^2$ DRAM memory cell for 1-Gbit DRAM, advanced photolithography and etching techniques are required. In order to facilitate the correct readout of the stored data, the charge stored in the capacitor ($C_s(V_{DD} - V_p)$) should be greater the sum of the noise ($C_B V_N$), the leakage charge ($I_L t_{REFmax}$), and the α-particle induced critical

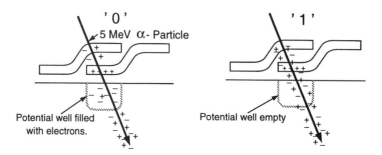

Fig. 5.37 Soft error created by alpha particles in a DRAM. (Adapted from May and Woods [20].)

charge (Q_C):

$$C_s(V_{DD} - V_p) \geq C_B V_N + I_L t_{REFmax} + Q_C, \quad (5.18)$$

where V_p is the capacitor plate voltage, C_B is the capacitance of the bit line, V_N is the noise voltage at the bit line, I_L is the leakage current, t_{REFmax} is the maximum refresh time, and Q_C is the critical charge induced by the α-particle. Fig. 5.37 shows the cross-section of the device with a soft error created by alpha particles in a DRAM[20]. Alpha particle is a doubly charged helium nuclei (1 proton and 2 neutrons). When an alpha particle is hitting silicon, about one million electron-hole pairs are generated along the path of the α-particle. The generated electrons and holes in the depletion layers surrounding diffusion or gate regions are separated by the electric field. For an n-channel memory device and the storage node, the electrons are swept into the storage well and holes are expelled into substrate. The electron-hole pairs generated outside the depletion region diffuse through bulk silicon. Therefore, at a dynamic node storing logic-0, its storage well is full of electrons. More electrons added do not cause an error at the storage node. In contrast, at a dynamic node storing logic-1, its storage well does not have any electrons. When the amount of the injected electrons exceeds the amount of the critical charge, stored logic-1 may become logic-0—a soft error occurs. Note that this does not cause permanent damage to the memory cell.

When the density of the memory cells becomes higher, the layout size of the memory cell is also scaled down. As described above, various techniques have been used to integrate memory cells—planar, stack, trench, and using the insulator material with a high permittivity. When the memory cell size is scaled down, the signal charge becomes smaller. In addition, down-scale of the supply voltage V_{DD} also further shrinks the signal charge in the storage capacitor. Therefore, reducing the three noise sources becomes important, especially for the Gbit DRAMs. In spite of the setback, the number of electrons, which are generated by the α-particle and collected by the storage well depletion layer, is linearly proportional to the diagonal length of the memory cell depletion layer. When the size of the memory cell is scaled down, the α-particle induced critical charge can be reduced accordingly. The leakage charge comes from the junction leakage and the subthreshold leakage of the access transistor.

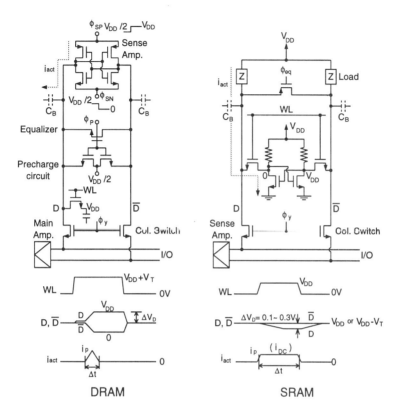

Fig. 5.38 Comparison of the read operation between DRAM and SRAM. (From Itoh et al. [21], ©1995 IEEE.)

When the junction temperature rises, the junction leakage increases exponentially, which can be reduced by lowering power dissipation and thermal resistance of the package. Compared to other noise sources, subthreshold leakage is much more difficult to reduce, which is the most bothering problem while scaling down the supply voltage. For a high-density DRAM, lowering the power supply voltage is a method to reduce the device reliability problems. Lowering the threshold voltage can also be used to increase the readout speed of the memory cell and the speed performance of other peripheral circuits. At a lower threshold voltage, subthreshold leakage is more serious due to the exponential relation of the subthreshold current with respect to the gate voltage. Therefore, the threshold voltage of the DRAM memory cell transistors should not be too small while the threshold voltage of the devices in the peripheral circuits can be lowered appropriately. As described before, using the multi-division of the bit lines, parasitic capacitance at the bit line (C_B) can be reduced. While the layout density increases, the distance between bit lines becomes smaller. Coupling of noise via bit lines becomes serious. Thus reduction of the bit line noise should be taken care of.

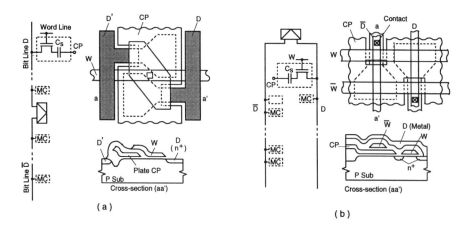

Fig. 5.39 Open and folded bit-line structures. (Adapted from Itoh [19].)

Fig. 5.38 shows comparison of the read operation between DRAM and SRAM[21]. For a DRAM, before the read/write access, the bit lines are precharged to $V_{DD}/2$ via the precharge and equalize circuit. When the selected word line rises to high, the charge in the capacitance C_s of the memory cell redistributes with the bit line parasitic capacitance C_B. The bit line at the other end of the sense amp is connected to a dummy cell such that the noise coupled from the word line to the bit line can be removed. Thus a voltage difference of $\frac{V_{DD}}{2} \frac{1}{1+\frac{C_B}{C_s}}$ exists between the bit lines. After the sense amp starts sensing, a fall from $\frac{V_{DD}}{2}$ to 0 in ϕ_{SN} triggers the operation of the sense amp, which is made of the NMOS cross-coupled pair. A rise from $\frac{V_{DD}}{2}$ to V_{DD} triggers the restore amp, which is made of the PMOS cross-coupled pair. The small voltage difference is amplified by the sense amp, followed by the reconstruction to the full swing by the restore amp such that it can be rewritten to the memory cell. Different from SRAM, in DRAM, only when the sense/restore amp is active, is there an active current. In contrast, due to the DC-path between the pull-up load and the memory cell, there is a fixed DC current I_{DC} in SRAM when the word line is high.

5.2.2 Bit-Line Architecture

Fig. 5.39 shows the open and the folded bit-line structures[19]. In the conventional open bit-line structure, the sense amp is located between two bit lines. As shown in the figure, the adjacent two bit lines D and D′ represent two different bit-line pairs (D and $\overline{D}$ are a bit-line pair). In the folded bit-line structure, the sense amp is located at the end of the two bit lines—two adjacent bit lines are the bit-line pair (D and $\overline{D}$). In the open bit-line, when W is high, both D and D′ may be sensed as high or low simultaneously. In the folded bit-line structure, when W is high, D and $\overline{D}$ are complementary. Its influence in the bit-line noise can be seen using the schematic memory array and the noise generation as shown in Fig. 5.40. When the bit lines are

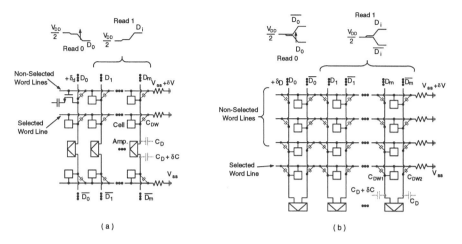

Fig. 5.40 Schematic memory array and typical noise generations in the (a) open; and (b) the folded bit-line structures. (Adapted from Itoh [19].)

used for large-swing amplification, due to influence of the bit line-word line coupling capacitance C_{DW}, noise is coupled from the word line of the unselected cell (δW) to the bit line. These word line noise may affect the D_0 read operation. As shown in the figure, for the open bit-line case (a), if the bit D_0 is logic-0 during the read operation while other bits D_1–D_m are logic-1. Therefore, the word line noise (δW) caused by C_{DW} is the most critical. The noise coupled via the unselected word line can be approximated as proportional to $C_{DW}^2 \delta C/C_D$, where δC is the capacitance difference between D_i and $\overline{D_i}$. For the folded bit-line structure, since D_i and $\overline{D_i}$ are adjacent, the noise coupled to the unselected word line (δW) is proportional to $C_{DW1} - C_{DW2}$. The bit-line noise coupled from the word line to D_0 bit line is approximated as proportional to $(C_{DW1} - C_{DW2})^2 \delta C/C_D$, where C_{DW1} and C_{DW2} represent the coupling capacitance from the D_i and $\overline{D_i}$ bit lines to the word line, respectively. Thus the folded bit-line structure has much less coupled noise as compared to the open bit line structure. The key point is that the outputs of the adjacent bit line pair are complementary, thus the noise coupling effects are cancelled out.

Fig. 5.41 shows the existence of the inter-bitline coupling noise in a folded bit-line scheme. There are two noise sources—(1) inter-pair coupling noise N1 between adjacent bit-line pairs, and (2) intra-pair coupling noise N2 within bit lines. (Only the coupling between adjacent bit lines is considered here.) Also shown in the figure is the cross-section of the bit-line pairs in the direction of the word line. The empty circles represent the selected bit-line. Solid circles represent the reference bit line connected to the dummy cell. As shown in the figure, during the precharge period, bit-line pairs are precharged to $\frac{V_{DD}}{2}$, followed by the read operation. As shown in Fig. 5.41(b), for the worst case, the selected bits have an identical value (in (b), all selected cells have logic-1). Thus, via inter-pair coupling noise N1, the voltage of the reference bit changes in the same direction (going up in (b)). Therefore, for the

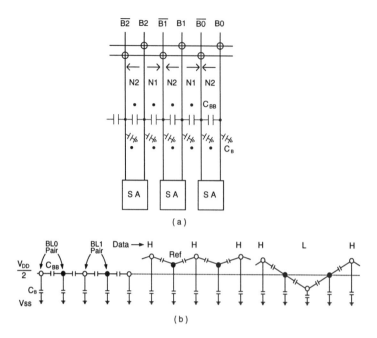

Fig. 5.41 Inter-bitline coupling noise in a folded bit-line scheme. N1 is the inter-pair coupling noise and N2 is the intra-pair coupling noise. (Adapted from Hidaka et al.[22].)

sense amp, the voltage difference is decreased relatively. In Fig. 5.41(b), for the optimum case, there is no influence. For the twisted bit-line (TBL) architecture as shown in Fig. 5.42, the influence of the inter-pair coupling noise N1 of the adjacent bit is uniform in the whole bit-line pair, instead of concentrated in one of the bit-line pair as in the folded bit-line architecture. Therefore, it can be regarded as the common-mode noise, which can be eliminated by the sense amp owing to the differential amplification properties. The bit lines are divided into four parts—A, B, C, D. At B and D, each pair of the bit lines cross each other. The adjacent pair of the bit lines cross each other at A and C. The memory cells are equally placed in the four parts—before A, between A and B, between B and C, and between C and D. The cross-couple at D is used to make the cross-coupled bit lines identical. Fig. 5.43 shows the principle of the twisted bit line operation[22]. Assume that BL_0, BL_1, and BL_2 are selected. The value of BL_0, BL_1, and BL_2 are all high. The N1 noise from BL_0 to BL_1 and $\overline{BL_1}$ is equivalent. Similarly, the N1 noise from BL_2 to BL_1 and $\overline{BL_1}$ is also equal. Thus, the level at BL_1 and $\overline{BL_1}$ rises simultaneously, while its voltage difference stays unchanged. For the sense amp, there is nothing different—the shift in the common mode only causes the speed difference in the n-portion pulldown and the p-portion pullup. As shown in Fig. 5.42(b), in addition to the cross-couple of the bit-line pair, between bit-line pairs, they can also be crossed. Thus, a further reduction in the influence of the intra-pair N2 noise can be reached. Using the twist bit-line (TBL) structure, the selection of the dummy cell is different

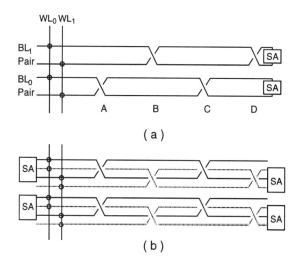

Fig. 5.42 Twisted bit line (TBL) configurations. (Adapted from Hidaka et al. [22].)

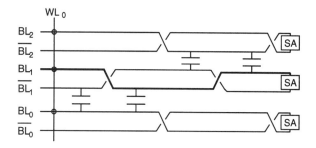

Fig. 5.43 Principle of the twisted bit line operation. (Adapted from Hidaka et al. [22].)

from that in the standard DRAM configuration. In the previous folded bit-line pair, for every bit line, one dummy cell is required. For the memory cell selected by the word line, at its complementary bit line, a selected dummy cell is used as a reference bit line. In the twisted bit-line configuration, each bit line needs two dummy cells. In addition, its decode logic is more complicated.

When memory size becomes big, it is necessary to divide the whole memory array into many subarrays. In addition, bit line is also divided into several pieces such that bit-line capacitance C_B can be reduced—during the read operation the voltage difference on the bit lines ($\frac{V_{DD}}{2} \frac{1}{1+C_B/C_S}$) can be increased. Fig. 5.44 shows the multi-divided data line structure: (a) conventional, (b) shared amp, (c) shared I/O lines parallel to sub-data lines, and (d) shared I/O lines across sub-data lines[19]. In a general bit-line structure, it contains a pair of bit lines (one bit line is connected to the selected cell and the other is connected to the dummy cell for reference) to achieve noise cancellation. For the conventional multi-divided bit-line structure as

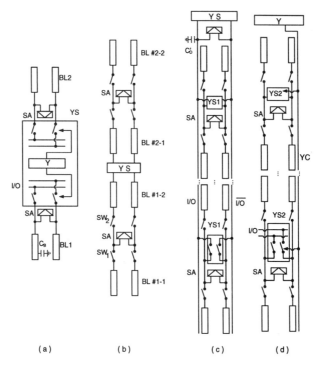

Fig. 5.44 Multi-divided data line structures: (a) conventional; (b) shared amp; (c) shared I/O lines parallel to subdata lines; and (d) shared I/O lines across subdata lines. (Adapted from Itoh [19].)

shown in (a), the Y switch signal (YS) controlled by Y decoder is used to connect the selected bit-line pair to the corresponding I/O line pair. In (a), a sense amp is used for each bit-line pair. For the shared sense-amp scheme used in the multi-divided bit-line structure as shown in (b), two bit-line pairs share a sense amp—thus the number of sense amps can be reduced. The multi-divided bit-line structure using the shared sense amp scheme is used to realize 256-Kb to 4-Mb DRAMs. The operation of the multi-divided bit-line structure using the shared sense amp scheme is described here. Assume that bit line # 1-1 is selected. The small read signal at the bit line is fed to the sense amp (SW_1 turns on and SW_2 turns off). Then, the amplified signal is transferred to bit line #1-2 and the I/O line (SW_2 turns on and the left bit line is selected by YS). Therefore, the unselected bit line #1-2 is unnecessarily charged/discharged. In addition, a more complicated timing sequence is required. Although the power consumption of the sense amp related circuits can be reduced, the bit-line related power consumption increases. As a whole, the power consumption still increases. The only advantage in this approach is its small layout area. As shown in (c) and (d), the shared I/O scheme and the shared sense amp scheme are taken advantage at the same time. As shown in (c), the I/O scheme and the I/O line are parallel to the bit line. Thus the drawback in (b) can be improved. In addition, the selected bit line is allowed

DRAM 273

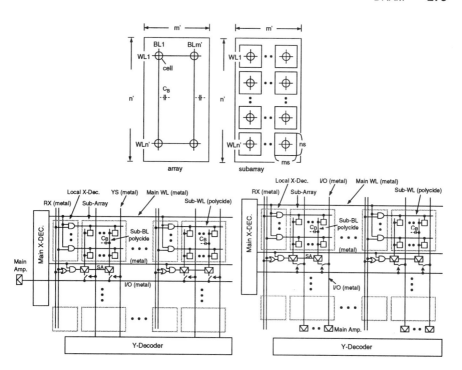

Fig. 5.45 Multi-divided arrays with bit line orthogonal to I/O line and bit line parallel to I/O line. (From Itoh et al. [23], ©1997 IEEE.)

to communicate with the common I/O, which is controlled by YS at the common I/O line. Using this approach, power dissipation and chip size can be reduced at the same time. However, all capacitances connected to the I/O lines are charged/discharged simultaneously. In addition, there exists an extra parasitic capacitance C'_D at the I/O line. The increased power dissipation can be reduced via the approach in (d), where signal YC is used to determine if the bit line to be connected to the I/O line or not. Thus the parasitic capacitances can be reduced. Using the multi-divided array architecture, the RC propagation delay can be reduced by reducing the length of the bit line and the number of the connected memory cells. Thus high speed can be obtained.

Fig. 5.45 shows the multidivided arrays with the bit line orthogonal to I/O line and the bit line parallel to I/O line[23]. The shared sense amp and I/O techniques have been used. As shown in the figure, the structure with data line orthogonal to I/O line is more often used. For the data line parallel to I/O line structure, the memory is divided into sub-arrays, which contain $m_s \times n_s$ memory cells, where n_s is 1K or 2K in the Giga bit era, and m_s is 256–512, to limit its word-line delay. From the figure, every sub-array can be used simultaneously for read/write operations. Since the data line parallel to I/O line structure has the capability of providing multiple bits at the output, thus it is suitable for use in the memory with wide data output. However, power dissipation is high.

274 CMOS MEMORY

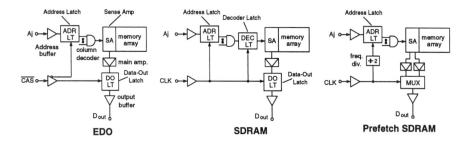

Fig. 5.46 High-speed column modes used in the DRAM with two-stage pipeline extended data-out (EDO), three-stage pipeline synchronous DRAM (SDRAM) and prefetch SDRAM. (Adapted from Itoh et al. [23].)

5.2.3 Advanced Architecture

For asynchronous DRAMs, since no clock is used, the ATD circuit for column strobe is needed. For synchronous DRAM (SDRAM) using latches, pipelined structure and parallelism can be used to enhance throughput of the write/read (W/R) process. Fig. 5.46 shows the high-speed column modes used in the DRAM with two-stage pipeline extended data-out (EDO), three-stage pipeline synchronous DRAM (SDRAM) and pre-fetch SDRAM[23]. As shown in the figure, the extended data-out (EDO) is based on adding latches at the address input and the data output to form a pipeline structure, which is controlled by column access strobe ($\overline{CAS}$). When $\overline{CAS}$ is low, the column address has arrived. At this time, the address input latch is transparent such that address is decoded by the column decoder and data are amplified and transferred to the output latch via the sense amp and the main amp. When $\overline{CAS}$ is high, the address latch holds the data and the output latch is transparent for output. Thus, EDO has a shorter CAS cycle time as compared to the conventional fast page (FP) mode. As shown in the figure, in the 3-stage synchronous DRAM (SDRAM), a 3-stage pipeline synchronously controlled by a system clock is used to provide a higher throughput. In the pre-fetch SDRAM, parallelism of the address fetch is used to enhance the throughput. In addition, parallel-to-serial throughput can be increased by using multiplexers. Therefore, pipeline and parallelism techniques for CPU designs have been used in the DRAM design to increase throughput.

Fig. 5.47 shows the timing in a synchronous DRAM (SDRAM). Also shown in the figure is the timing for a conventional fast page (FP) mode DRAM. As shown in the figure, in the conventional FP DRAM, continuous outputs are produced by controlling the $\overline{CAS}$ strobe. Column address comes in before $\overline{CAS}$ becomes low (row address arrives before ($\overline{RAS}$)). In the synchronous DRAM, there are no $\overline{RAS}$ and $\overline{CAS}$ asynchronous strobe signals. Instead, ACT (active) and RED (read) are used with CLK as the synchronous clock signal. In addition, when the ACT command is received, which represents the $\overline{RAS}$ falling edge in asynchronous mode, the row address comes in. RED command represents the $\overline{CAS}$ falling edge in asynchronous mode. After a 3-clock latency, the data throughput is one output per clock cycle (for

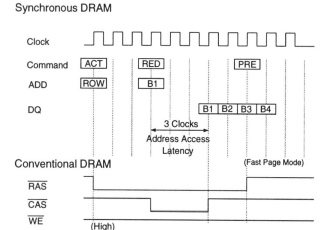

Fig. 5.47 Timing in a synchronous DRAM (SDRAM). (Adapted from Takai et al. [25].)

continuous column address outputs). With the pipeline structure, its access time is unchanged. Only at a cost of a 3-cycle latency, its throughput is three times higher. This is the effect of the pipeline technique used in designing memory. Fig. 5.48 shows a 3-stage-pipelined structure[25]. As shown in the figure, the first stage is the column decoder. The second stage is the read peripheral circuit (sense amp and main amp). The third stage is the data output buffer. While doing the 3-stage-pipelined design, the delay among three paths should be balanced such that the performance of the pipeline is not severely restricted by a specific critical path. As shown in the figure, the rising edge of each clock cycle represents the beginning of a pipelined stage. Before the start of the next rising edge of the next clock cycle, the operation of this stage should be accomplished. As shown in the figure, CLK1, CLK2, and CLK3 represent the clock signals to control latches 1, 2, and 3. With more pipeline stages, throughput is higher. However, the latch induced delay may limit the enhancement in throughput. In addition, latency will increase—the effective access delay becomes larger.

In order to further enhance the DRAM performance, implementing the logic circuit with DRAM to become an embedded DRAM is a good approach. Fig. 5.49 shows a standard DRAM and an embedded DRAM[23]. Generally speaking, for DRAM chips, a substantial amount of propagation delay and power consumption are at the I/O. In addition, when the I/O count is large, arrangement of pads is a problem. If the logic circuits and DRAM can be implemented in the same chip, wiring capacitance can be substantially reduced—a higher throughput can be obtained. Due to the rapid progress in the CMOS processing technology, realizing the embedded DRAM chips becomes less difficult. The bottleneck is on the adjustment of the difference between the memory and the logic devices. In addition, CMOS technology generalized for

276 CMOS MEMORY

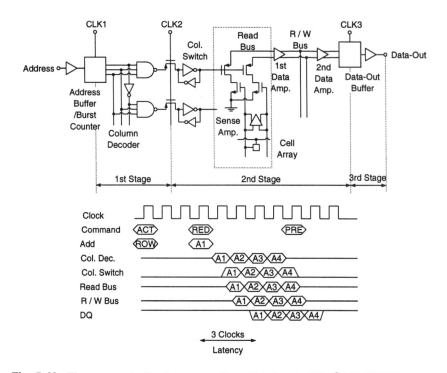

Fig. 5.48 Three-stage-pipelined structure. (From Takai et al. [25], ©1994 IEEE.)

realizing logic and memory together should be designed and the control of the multi-layer interconnects and the power dissipation is important.

5.2.4 Low-Power DRAM Techniques

Fig. 5.50 shows the switched-source-impedance CMOS circuits with shaded MOS devices in subthreshold conduction mode[26]. As shown in the figure, it is the concept of the multi-threshold (MT) CMOS active/standby mode logic (type II). The switch used in the switched-source-impedance CMOS circuits is referred to the power switch transistor. When it is off, the circuit enters the standby mode—the leakage current of the MOS device flows to the high-impedance load connected to the supply. As a result, the source voltage increases and the subthreshold current decays exponentially. During the standby mode, by connecting the device, which is certain to be turned off, to the virtual power line controlled by the power switch transistor, the subthreshold leakage current during the standby mode can be reduced. As shown in the figure, the transistors to be turned off use the virtual power line with one end connected to the power supply line. If the devices to be turned off cannot be known in advance (as in (e)), then both ends need to connect to the power supply lines. This arrangement is particularly useful for the memory design—when the memory enters

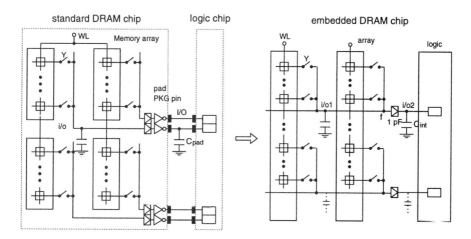

Fig. 5.49 Standard DRAM and embedded DRAM. (Adapted from Itoh et al. [23].)

the standby mode, its node voltages in general are at certain precharged values.

The switched source impedance method can be used to reduce the standby subthreshold current in the DRAM array. Fig. 5.51 shows the partial activation of multi-divided power line[21], which is realized by adding a power switch transistor in each memory array to reduce the subthreshold leakage for the memory array not accessed. For the memory array, which is accessed by turning on the power switch transistor, it can operate normally. Via 2-D array selection, the subthreshold power consumption can be reduced by mk times. (Note that mk is the array number.)

Fig. 5.52 shows the conventional hierarchical power line structure when entering active mode from standby mode using the multi-threshold (MT) active-standby logic. When entering the active mode from the standby mode, the off-transistors during the standby mode are connected to the power supply via the power switch transistor to reduce the leakage current. From the standby mode to the active mode, there is a slow recovery problem—the power switch transistor turns on to charge/discharge large capacitance of the virtual supply line. Fig. 5.53 shows the level controllable local power line (LCL)[27]. As shown in the figure, two reference voltages 0.15V and $V_{DD} - 0.15V$ are used as the power supplies during the standby mode operation. Thus the source-induced subthreshold reduction technique is still applicable for the turn-off transistor. At the same time, during the standby mode, the extent of the voltage drop/rise on the virtual supply lines can be controlled to within a limit. When the circuit enters the active mode, the power switch transistor turns on and the virtual power line is charged to full swing faster. As shown in the figure, V_{ref1} and V_{ref2} can be tuned to a precise designated value by fuse-blowing after the wafer fabrication process to avoid process variation effects.

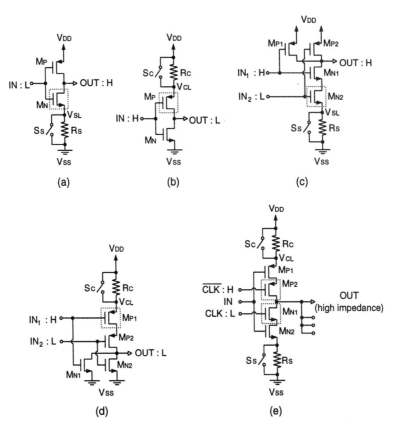

Fig. 5.50 Switched-source-impedance CMOS circuits with marked MOS devices in sub-threshold conduction mode. (a) inverter with low input; (b) inverter with high input; (c) NAND gate with low and high inputs; (d) NOR gate with low and high inputs; (e) clocked inverter. (Adapted from Horiguchi et al. [26].)

5.2.5 Low-Voltage DRAM Techniques

When memory size becomes larger, density and power dissipation of a DRAM becomes higher. Thus junction temperature in the devices used in the DRAMS rises, which results in an increased junction leakage current. As a result, the data-retention time is decreased. Therefore, lowering power dissipation becomes more and more important in the future DRAM designs. Fig. 5.54 shows an important concept—a universal-V_{DD} used in a 1.5–3.6-V 64-Mb DRAM with a two-way power supply unit[28]. As shown in the figure, the power supply of the memory cell array is set at a fixed value—usually the smallest permissible value under the V_{DD} specification. At such a supply voltage, devices and circuits are optimized. Regardless of the external supply voltage, the designed DRAMs have the best reliable performance. By adding an on-chip voltage-down converter (limiter), when V_{DD} is greater than 1.5V, the internal voltage is $V_{int} = 1.5V$. When V_{DD} is smaller than 1.5V, the internal voltage

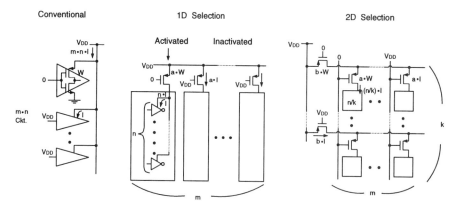

Fig. 5.51 Partial activation of the multi-divided power line. (Adapted from Itoh et al. [21].)

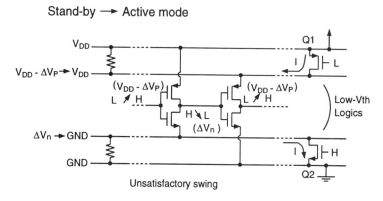

Fig. 5.52 Conventional hierarchical power line structure when entering active mode from stand-by mode.

is $V_{int} = V_{DD}$. No matter how much change for the V_{DD}, the performance of the memory core is not substantially affected. Note that compared to the DRAM core, the peripheral circuit is susceptible to a larger stress. In addition to the voltage-down converter, Fig. 5.55 shows another on-chip supply voltage conversion scheme to lower the supply voltage for large-size DRAMs. As shown in the figure, the DRAM circuits are stacked, which is similar to the charge-recycling bus (CRB) architecture used in the low-power circuit described in Chapter 3. Therefore, the voltage drop between the gates becomes $V_{int} = \frac{1}{2}V_{ext}$. Using this approach, the power loss in the voltage converter can be reduced—efficient power utilization can be reached. As shown in the figure, in the driver circuit, the signal voltage is reduced from V_{DD} to $\frac{1}{2}V_{DD}$. In the receiver circuit, signal voltage of the receiver circuit is reconstructed from $\frac{1}{2}V_{DD}$ to V_{DD}. In the driver circuit, standard inverters are used. In the receiver circuit, DCVS logic circuits are used. The output level of the stacked logic in the

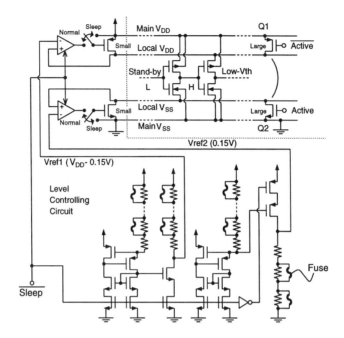

Fig. 5.53 Level controllable local power line (LCL) (From Yamagata et al. [27], ©1995 IEEE.)

driver circuit is $0 \to \frac{1}{2}V_{DD}$ or $\frac{1}{2}V_{DD}$ to V_{DD}.

In this paragraph, low-voltage circuit techniques for the word line driver and the sense amp are described. Fig. 5.56 shows the negative-voltage word-line driver[27]. As shown in the figure, when the word line is not selected, its voltage is $V_{bb} < 0$. Thus the leakage current of the access transistor can be reduced. When the word line is selected, it is at a boosted level as for a conventional word line driver. The circuit structure and the operation principle of the negative voltage word line driver are similar to the low-power bus architecture described in Chapter 3. Fig. 5.57 shows a conventional DRAM with the $\frac{V_{DD}}{2}$ bit-line precharge method. As shown in the figure, it contains a memory cell, an n-channel sense amp, a p-channel restore amp, and an equalizer circuit. When EQ is high, both the memory bit line and the sense bit line (BL and $\overline{BL}$) are precharged to $\frac{V_{DD}}{2}$. When WL rises from low to high to surpass V_{DD} at V_{pp} ($V_{pp} = V_{DD} + V_{TN}(V_{DD}) + \alpha$) such that the memory storage capacitor C_s has the top amount of charge "$C_s(V_{DD} - V_p)$", where $V_p = \frac{V_{DD}}{2}$, the selected memory cell shares its stored charge with the bit line. As a result, the voltage at the bit line is $\frac{V_{DD}}{2} \frac{C_s}{C_s + C_B} = \frac{V_{DD}}{2(1 + C_B/C_s)}$, where C_B is the bit-line capacitance, and C_s is the storage capacitance of the memory cell. The reference bit line is connected to the dummy cell. Using the dummy cell, the effect of the word line-to-bit line coupling on the bit line when the word line is boosted to V_{pp} can be eliminated. When the voltage difference at the bit lines reaches a value, while the sense amp can correctly operate,

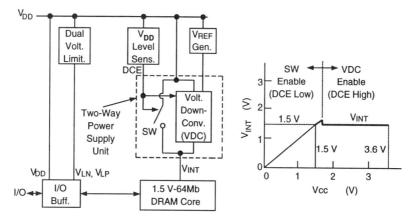

Fig. 5.54 Concept of universal-V_{CC} (1.5–3.6-V) 64-Mb DRAM with a two-way power supply unit (Adapt from Nakagome et al. [28], ©1991 IEEE.)

the common source node of the n-channel sense amp SN is pulled to ground. In addition, the common source node of the p-channel restore amp SP is pulled to V_{DD}. Thus, the cross-coupled latch-type sense amp formed by the n-channel sense amp and the p-channel restore amp amplifies the difference between BL and $\overline{BL}$ to full-swing. In addition, the full-swing data are rewritten into the memory cell, where the stored data have been damaged due to charge sharing. The use of the $\frac{V_{DD}}{2}$ precharge and sense scheme has several advantages: (1) Minimized array noise: This approach is suitable for the folded-bit line architecture since substrate noise and capacitively coupling of the word line pulse can be compensated. (2) The full-swing V_{CC} restores operation: The CMOS latch-type sense amp can pull the bit lines to full-swing. (3) Low power dissipation and small leakage current: At a half V_{CC} swing, a smaller power dissipation is expected. It also has several drawbacks: (1) Large sensing delay: At the beginning of the sense period, the source node of both n-channel and p-channel is at $\frac{V_{DD}}{2}$, therefore its threshold voltage increases due to body effect. When the difference between SN of the n-ch sense amp and the bit line voltage exceeds the threshold voltage, the NMOS device in the n-channel sense amp turns on. Thus the increased threshold voltage delays turn-on of the n-channel and the p-channel devices, thus lengthens the sensing delay, which is especially serious at a low supply voltage. (2) Increased leakage: During the final portion of the active period, BL and $\overline{BL}$ are pulled to full-swing. Therefore, the threshold voltage of the n-channel sense amp and the p-channel restore amp becomes the smallest—increased leakage current. Fig. 5.58 shows the well-synchronized sensing and equalizing circuit[31]. As shown in the figure, when equalizing, SN is connected to its n-channel well to have a smaller threshold voltage such that the sensing speed can be enhanced. As shown in the figure, after sensing is completed, the bias of the n-channel well becomes negative to increase the threshold voltage to reduce leakage. As shown in the figure, VW_p is the body potential for the n-channel sense amp and the equalizer. VW_n is the body potential for the p-channel restore amp. When S0 is low, the source node of

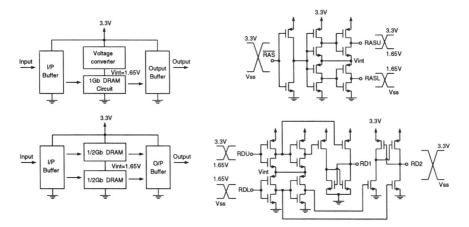

Fig. 5.55 Block diagram of conventional and on-chip supply voltage conversion schemes in a 3.3-V 1-Gb DRAM. (From Takashima et al. [29], ©1993 IEEE.)

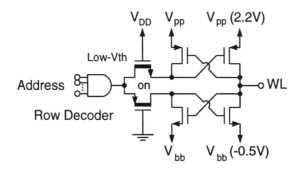

Fig. 5.56 Negative-voltage word line driver. (Adapted from Yamagata et al. [27].)

VW_p, VW_n, the n-channel sense amp, and the p-channel restore amp is equalized to $\frac{V_{DD}}{2}$ via the equalizer. When S0 is 1, SN and VW_p are discharged to ground; SP and VW_n are charged to V_{DD}. When sensing and restoring are done, VW_p and SN are separated; VW_n and SP are separated. In the meantime, VW_p is discharged to $V_{BB} < 0$ and VW_n is charged to $V_{pp} > V_{DD}$. Thus a higher threshold voltage and a smaller leakage can be obtained. Fig. 5.59 shows another low-voltage sense amp[37]. As shown in the figure, the common source voltage (SP) of the Pch-RA is boosted to above V_{DD} when the sense amp starts operation. When the gate voltage difference is greater than $|V_{TP}|$, the PMOS turns on. Thus a larger current drive capability and an earlier turn-on of the PMOS device lead to a higher sensing speed.

Fig. 5.60 shows the sense circuits using the conventional voltage sensing scheme and the complementary current sensing scheme[32]. As shown in the figure, in the conventional voltage sensing scheme, when CD is high, the voltage signal on the sense line is transferred to the I/O line via a pass transistor. Thus the voltage on the

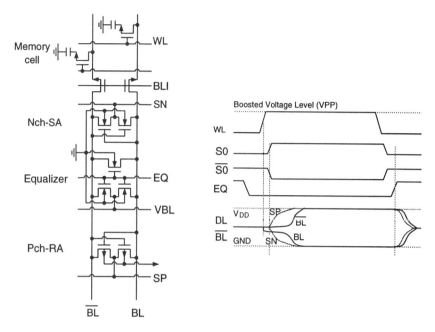

Fig. 5.57 Conventional DRAM with the $\frac{V_{DD}}{2}$ bitline precharge method. (From Ooishi et al.[31], ©1994 IEEE.)

sense line is influenced by the I/O line. The amplification operation of the main amp will not be started until the voltage difference on the sense line is not substantially affected by the I/O line. As shown in the figure, in the complementary current-sensing scheme, the sense amp input signal is changed to current with its magnitude determined by the voltage on the sense line. The sense line is connected to the gate instead of the drain. Therefore, the sense line is not affected by the I/O line. The signal can be transferred to RO line and the current sense amp earlier. Fig. 5.61 shows the current-sensing circuits. As shown in (a), the bipolar sense amp based on the cascode structure has a high output resistance and gain. As shown in (b) and (c), the MOS current sense amp is based on the comparator, which is used to compare the voltage difference between V_R and RO line to produce the output. In (c) the PMOS device $M2$ is used instead of the NMOS device $M1$ in (b). Therefore, the voltage swing at the RO line is larger—complementary sensing scheme. In (c), the complementary current-sensing circuit is composed of the NMOS device used as the readout gate, the complementary PMOS device used for driving the transistor, and the feedback. A large swing range at the RO line makes it suitable for low-voltage operation. Fig. 5.62 shows the word line driver with the feedback charge-pump circuit[32]. As shown in the figure, $CP1$ and $CP3$ comprise the complementary boost generator. $P1$ and $P1B$ are the complementary signals. Node 7 and node 8, which are also complementary, have a voltage value between V_{DD} and $2V_{DD}$. If only the single boost circuit is used, its maximum voltage is $V_{DD} - V_T$. Using the complementary-type boost

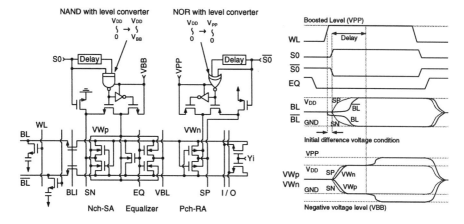

Fig. 5.58 Well-synchronized sensing and equalizing circuit. (From Ooishi et al. [31], ©1994 IEEE.)

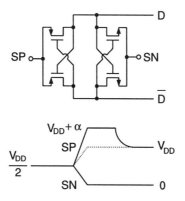

Fig. 5.59 Overdrive low-voltage sense amp. (Adapted from Kawahara et al. [37].)

circuit, node 7 can have the boosted voltage exceeding V_{DD} since when $P1$ is low and $P1B$ is high, node 8 can be charged to V_{DD} instead of $V_{DD} - V_T$. Consequently, node 8 can be boosted to $2V_{DD}$ when P1 rises to high. Similarly, node 5 is between V_{DD} and $2V_{DD}$. Node 4 is between $2V_{DD}$ and $3V_{DD}$. Since V_{CH} is $2V_{DD}$, similar to $CP1$, the complementary boost circuit is formed by $P2$ and $P2B$. The purpose of $P2B$ is to charge node 4 to $2V_{DD}$ when $P2$ is low. When P2 is high, node 4 is boosted to $3V_{DD}$. At this time, node 5 is at V_{DD}. Thus node 6 is boosted from V_{DD} to $3V_{DD}$ to turn on $M19$. Thus node 3 is boosted to $2V_{DD}$ to drive the word line.

Fig. 5.63 shows the charge transfer presensing scheme (CTPS) with $\frac{1}{2}V_{DD}$ bit-line precharge[33]. As shown in the figure, when the voltage difference at the bit lines is constructed, the sense-line voltage is preset to VSAP(= $V_{DD}(1 + \gamma) > V_{DD}$). When the control voltage VTG to the transfer gate TG rises to $\beta + V_{thn}$, charge sharing between the sense line with BL and $\overline{BL}$ occurs. At this time, the sense

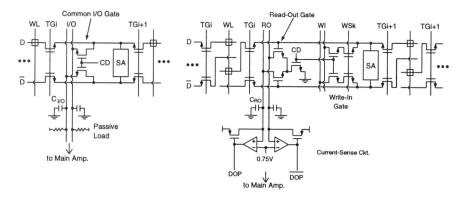

Fig. 5.60 Sense circuits using the conventional voltage-sensing scheme and the complementary current-sensing scheme. (Adapted from Nakagome et al. [32].)

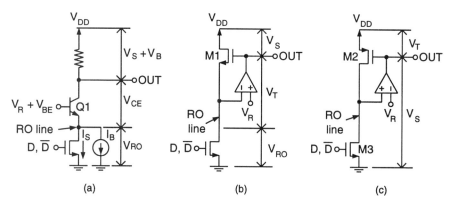

Fig. 5.61 Current-sensing circuits with (a) the common-base sensing scheme; (b) the NMOS-driven current-sensing scheme; and (c) PMOS-driven current-sensing scheme. (Adapted from Nakagome et al. [32].)

amplifier is not active yet. Since the sense-line capacitance is smaller than the bit-line capacitance, the small voltage difference at the bit line is pre-amplified at the sense line due to the transferred charge difference to the sense line. After amplification, the voltage difference is expressed as $\alpha = \alpha_0 \frac{C_B + C_s}{C_{SA}}$, where $\alpha_0 = \frac{V_{DD}}{2} \frac{1}{1 + \frac{C_B}{C_s}}$, and $C_B = C_{BL} + C_{SA}$. C_{BL}, C_{SA}, and C_s are the capacitances of the bit line, sense line, and the storage node, respectively. The previous voltage difference at the bit line (α_0) is amplified by $\frac{C_B + C_s}{C_{SA}} = 1 + \frac{C_{BL} + C_s}{C_{SA}}$ times, followed by the amplification by the sense amp. This is especially useful for low-voltage large-size DRAM at the cost of the extra delay due to presensing.

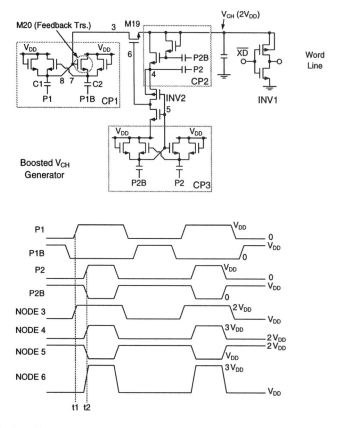

Fig. 5.62 Word line driver with the feedback charge-pump circuit. (From Nakagome et al. [32], ©1991 IEEE.)

5.3 BiCMOS MEMORIES

Since 1980, capacity of SRAMs has been expanding and access time of SRAMs has been scaled down continuously. Fig. 5.64 shows the trends on the SRAM access time in the past[34]. Using CMOS technology, TTL I/O SRAMs with an access time 10–20ns have been reported. In the past, their sizes have been increased four times every three years. High-speed SRAMs with an access time smaller than 10ns were first implemented by the bipolar technology to realize the 64-Kb ECL I/O SRAM. The drawback of the bipolar SRAM is its large power dissipation, which prohibits a further increase in the integration size. For a size larger than 256Kb, $0.8\mu m$ BiCMOS technology has been used to realize 256Kb–1Mb SRAM's with ECL I/O. $0.5\mu m$ BiCMOS technology has been used to implement 4-Mb TTL I/O and ECL I/O SRAMs. Generally speaking, the speed of the BiCMOS SRAM is about two times faster as compared to the CMOS SRAM at the same size. Fig. 5.65 shows the evolution of SRAM in terms of chip size and memory cell size versus SRAM

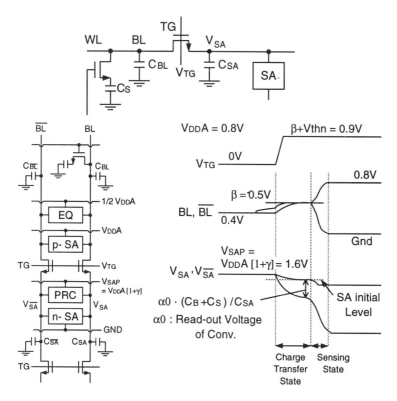

Fig. 5.63 Charge transfer presensing scheme (CTPS) with $\frac{1}{2}V_{DD}$ bit-line precharge. (From Tsukude et al. [33], ©1997 IEEE.)

capacity[34]. As shown in the figure, at each generation, the memory cell size is reduced by 1/3 and the chip size is increased by 1.5 times. BiCMOS SRAMs follow the same trend as for the CMOS SRAMs.

5.3.1 BiCMOS SRAM

Fig. 5.66 shows the characteristics of submicron BiCMOS SRAMs[34]. As shown in the figure, when the capacity of a BiCMOS SRAM is increased, its memory cell is scaled down. Due to an identical memory cell structure as for the CMOS SRAM, when the capacity is greater than 4-Mb, poly-TFT load has also been used in the memory cell for the BiCMOS SRAM. The first-generation BiCMOS SRAM is used to replace bipolar ECL I/O SRAM. The BiCMOS SRAM combines the advantages of low power and high density of the CMOS memory cell and high current-driving capability and high speed of the bipolar device. The first-generation ECL I/O BiCMOS SRAMs, which are used in the ECL I/O mainframe systems, use a $-5.2/-4.5$V ECL supply voltage, a high-resistive load (HRL) 4-T memory cell, and totem-pole BiCMOS gates. The second-generation BiCMOS SRAMs, which are designed based

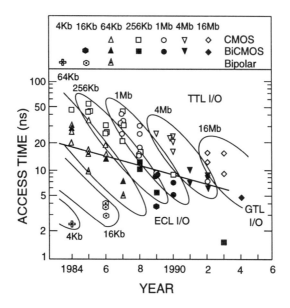

Fig. 5.64 Trends on the SRAM access time. (From Takada et al. [34], ©1995 IEEE.)

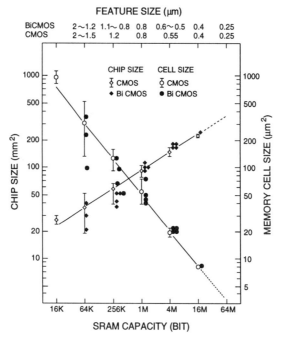

Fig. 5.65 Chip size and memory cell size versus SRAM capacity. (From Takada et al. [34], ©1995 IEEE.)

Technology	0.8 µm	0.55 µm	0.4 µm
Size	1M	4M	16M
V_{DD}	5V	5V / 3.3V	3.3V
Memory Cell Load	HRL	HRL / TFT	TFT
BJT	Single poly-Si	Single/Double poly-Si	Single/Double poly-Si
Shared Layer Emitter / Cell	HRL	HRL/GND Pad	TFT/GND Pad
Well	Twin	Twin/Triple	Twin/Triple
I / O	ECL	5V/3.3V-TTL	3.3V-TTL

Fig. 5.66 Submicron BiCMOS SRAM characteristics. (Adapted from Takada et al. [34].)

on 0.5-0.6µm BiCMOS technology, are used to improve the speed of the CMOS TTL I/O SRAMs. The second-generation BiCMOS SRAMs use a 5V/3.3V supply voltage, a high-resistive or poly-TFT PMOS load memory cell, and BiNMOS gates. The third-generation BiCMOS SRAMs, which are based on 0.4µm BiCMOS technology and a 3.3V supply voltage, poly-TFT PMOS load memory cell, triple-well structure and BiNMOS gates, are used to transfer data to and from the high-speed RISC microprocessors synchronously with a more than 100MHz clock rate. In addition, they have been used in the second cache system in high-performance engineering workstations. Fig. 5.67 shows the 4.5V ECL and 3.3V TTL interface circuits[34]. As shown in the figure, open-emitter ECL can output GHz frequency signals. The speed of push-pull CMOS 3.3V-TTL can reach 100MHz.

Fig. 5.68 shows the cross-sections of the bipolar transistors and memory cells of three generations of BiCMOS technology[34]. When the bipolar devices are included in the BiCMOS technology, extra processing steps should be reduced. To realize the bipolar devices, the existing CMOS processing steps should be utilized as much as possible. As shown in the figure, in the first-generation BiCMOS technology, the high-resistive polysilicon load layer in the CMOS memory cell can also be used to implement the polyemitter in the bipolar device. The emitter is formed by the dopant diffusion from the doped polysilicon layer into the silicon p-type base region. In the second-generation BiCMOS technology, an additional polycide layer is used to form the self-aligned bit-line contact landing pad in the memory cell and the global interconnect ground line to the cell. Based on these two poly layers, the poly-base and the poly-emitter structures can be obtained. The first polysilicon layer forms the extrinsic base contact, which can also be the MOS gates. The second polysilicon layer forms the polysilicon emitter. For the BiCMOS SRAM using the third-generation BiCMOS technology, the polysilicon TFT PMOS load has been used to integrate the memory cell such that low-voltage operation of the BiCMOS SRAM with a good soft error immunity can be obtained. Using the polysilicon gate and polysilicon

	ECL	TTL (CMOS)
Output Driver	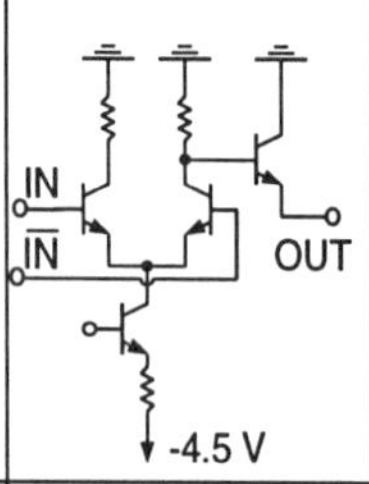 (circuit, -4.5 V)	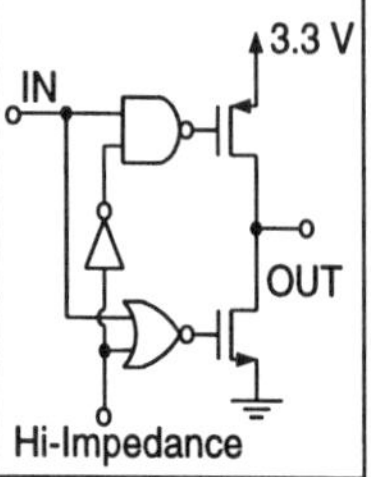 (circuit, 3.3 V, Hi-Impedance)
Type	Open-Emitter	Push-Pull
Swing	0.8 V	3.3V
Termination	-2.0 V (50 Ω)	—
Speed	1 GHz	100 MHz

Fig. 5.67 4.5V ECL and 3.3V TTL interface circuits. (Adapted from Takada et al. [34].)

TFT techniques, the stacked emitter polysilicon (STEP) structure has been used to implement the emitter electrode of the bipolar device such that the perimeter and plug effects for the narrow emitter window can be improved—at a small emitter width, such as $0.4\mu m$, the current gain is not degraded too much.

Fig. 5.69 shows the critical path in an ECL I/O asynchronous BiCMOS SRAM[34]. As shown in the figure, at the first-stage address buffer served as a level converter, the ECL level input signals (-0.9V/-1.7V) are transformed into full-swing CMOS signals. Via the BiCMOS decoder, the word line driver can drive the word line at a faster speed. Via the bit lines, the data in the selected memory cell are passed to the ECL-level cascoded bipolar differential sense amp for amplification of the signal difference at the bit lines. Finally, the output buffer is used to transform the output signal of the sense amp into the ECL-level output. Since the signal swing at the bit lines, sense lines, and I/O is much smaller as compared to the CMOS SRAM, higher speed can be expected. Fig. 5.70 shows the critical path in a TTL I/O asynchronous BiCMOS SRAM[34]. Due to the full-swing CMOS level at the I/O, no level converter is needed. In the circuit, only the address decoder and the sense amp are BiCMOS. As for the word line driver, it is implemented by CMOS such that at 3.3V a full-swing voltage at the word line can be obtained—optimized read/write access of the memory cell. In the TTL I/O asynchronous BiCMOS SRAM, the output of the cascoded bipolar sense amp is not CMOS TTL level. An ECL-to-CMOS level converter and an output driver are used to transform the signal swing to CMOS TTL level. At a supply voltage of 3.3V, the operation region of the bipolar sense amp is small. Thus, when designing the cascoded bipolar amplifier, the operation region should be carefully designed.

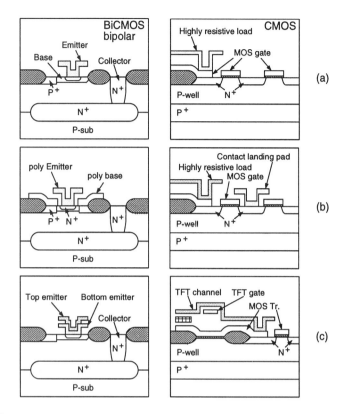

Fig. 5.68 Cross-sections of the bipolar transistors and the memory cells of three generations of BiCMOS technology. (From Takada et al. [34], ©1995 IEEE.)

In the conventional BiCMOS ECL I/O SRAM described in Fig. 5.69, after the input ECL-level signal is converted into the full-swing CMOS level in the bipolar address buffer, the BiCMOS address decoder is used to drive the word lines. Fig. 5.71 shows an improved address buffer and the decoding circuit in the BiCMOS ECL I/O SRAM[35]. As shown in the figure, in the improved BiCMOS ECL I/O SRAM, the bipolar wired-OR pre-decoder is added. With the inclusion of the partial decoder function in the level converter, the number of the effective decoder gates is decreased—the delay time of the x-decoder can be shrunk. The address buffer outputs are connected in the wired OR emitter follower configurations. Thus the wired-OR pre-decoder can be used to carry out the pre-decode function. In the partial decoder level converter, when A, B, and C are low, the current in the current mirror is at its maximum. Thus N is low, the inverter output is high. As long as one of A, B, and C is high, the current in the current mirror is shut off. Thus, node N is pulled up to high by the PMOS device $Q1$—the inverter output is low. Its output function is $\overline{A + B + C}$. Combining the address buffer with the wired-OR function of the pre-decoder, the output of the partial decoder level converter becomes:

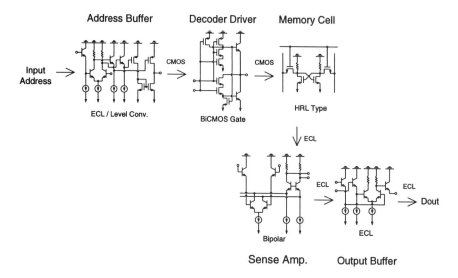

Fig. 5.69 Critical path in an ECL I/O asynchronous BiCMOS SRAM. (Adapted from Takada et al. [34].)

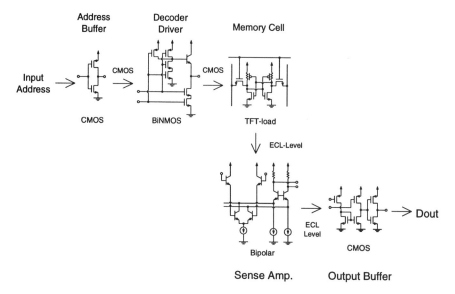

Fig. 5.70 Critical path in a TTL I/O asynchronous BiCMOS SRAM. (From Takada et al. [34], ©1995 IEEE.)

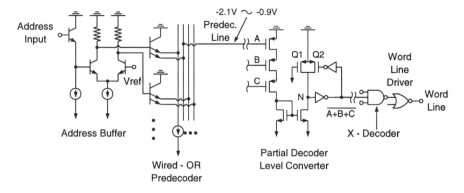

Fig. 5.71 X-address decoding circuits. (Adapted from Takada et al. [35].)

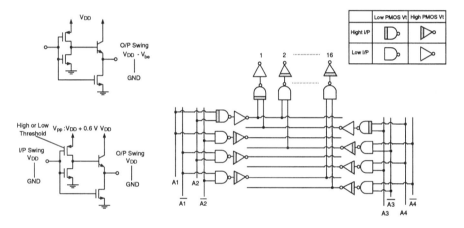

Fig. 5.72 Conventional BiNMOS inverter gate and the boosted-BiNMOS inverter gate with high and low $|V_{tp}|$ PMOS devices. (Adapted from Toyoshima et al. [36].)

$\overline{A_1} + \overline{A_2} + \ldots + \overline{A_N} = A_1 \cdot A_2 \ldots \cdot A_N$ — the decode function is accomplished.

Fig. 5.72 shows the conventional BiNMOS inverter gate and the boosted-BiNMOS inverter gate with high and low V_{tp} PMOS devices[36] used in the decoder driver for driving a large word line load. As shown in the figure, due to the excess minority carriers in the base of the npn bipolar device, its base voltage can reach V_{DD} only. Considering the V_{be} drop of the npn bipolar device, the output swing of the BiNMOS gate is from ground to $V_{DD} - V_{be}$. During the pull-up transient, due to the base-emitter capacitance (C_{be}) there is a slight voltage overshoot at the base—its output can be a little bit higher than $V_{DD} - V_{be}$. However, it is still not full-swing. At a supply voltage of 1.5V, this problem becomes a major drawback. As shown in the figure, in the boosted-BiNMOS inverter gate, the supply voltage to the PMOS device, which drives the npn bipolar device, is $V_{pp} = V_{DD} + 0.6V$ such that the output high voltage can be close to V_{DD} for full-swing operation. In order to speed

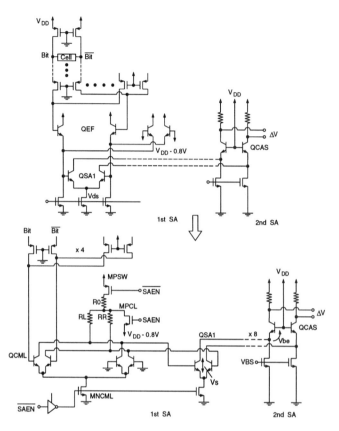

Fig. 5.73 Conventional cascoded sensing circuit and the stepped-down current-mode logic (CML) bipolar cascoded sensing circuit. (Adapted from Toyoshima et al. [36].)

up the charge buildup in the base of the npn bipolar device, the threshold voltage of the PMOS device can be reduced. Thus the switching speed of the BiNMOS gate can be enhanced. Due to the smaller threshold voltage, the subthreshold leakage current of the PMOS device increases. The compromised approach is the use of two thresholds for the PMOS devices. As shown in the 4-input decoder design, when A1–A4 are high, for the 2-NAND gates in the first layer, only 1/4 of them have a high value for both inputs— all PMOS devices are off. In the rest 3/4 2-NAND gates, at least one PMOS device is turned on to charge the bipolar device. Thus the low-threshold PMOS devices can be used in the 2-NAND gates to enhance the switching speed while not causing too much leakage. In the inverter following the 2-NAND gate, 3/4 of the gates are with high inputs (NMOS are off) and 1/4 are with low inputs (PMOS are on). In order to reduce leakage, high-threshold PMOS devices should be used here. Similarly, the NAND gates in the second layer have a higher chance in having low inputs—low-threshold PMOS device can be used.

Fig. 5.73 shows the conventional cascoded sensing circuit and the stepped-down current-mode logic (CML) bipolar cascoded sensing circuit[36]. As shown in the figure, the conventional cascoded sensing circuit is not suitable for low-voltage BiCMOS SRAMs. In the conventional sensing circuit, the bit line is connected to the emitter follower. The level-shifted signal is amplified by the cascoded amplifier formed by the two sense amps. Thus, the minimum operating voltage is: $V_{DD}(min) = 2V_{be} + V_{DS}$, where V_{DS} is the drain-source voltage of the NMOS current source. For V_{be} of 0.8V and V_{DS} of 0.2V, the minimum operating voltage is 1.8V, which is not appropriate for 1.5V operation. Thus the emitter follower stage needs to be improved. As shown in the figure, the improved circuit—using the stepped-down current-mode logic (CML) bipolar cascoded sensing circuit is suitable for low-voltage operation. In this circuit, the current mode-logic (CML) gate is made of the resistive load (RL, RR), the differential pair (QCML), and the current source (MNCML). The values of RL and RR are designed to have a sense output signal swing of 40mV, which is similar to the bit line signal swing. The differential signal is further amplified by the cascoded amplifier consisting of QSA1 and QCAS. The PMOS devices MPSW and MPCL are used to shut down the sense amp QSA1. When SAEN is low, the collector of the CML gate is connected to V_{DD}-0.8V— the sense amp is not selected. Thus QSA1 does not function. Only when SAEN is high does QSA1 function—the cascoded amplifier (QSA1 and QCAS) operates. V_{ce} of QSA1 is set to $V_{CE} = V_s$. If V_s is set to 0.3V, QSA1's $V_{CE} = 0.3$, which is sufficient to keep the bipolar device from entering saturation. The V_{CE} of QCML is $V_{CE} = V_{be} - V_s$ =0.5V. Thus the minimum value of V_{DD} is: $V_{DD}(min) = V_s + V_{be} + V_{DS}$ (path 1); $V_{DD}(min) = V_{be} + V_{CE} + V_{DS}$ (path 2). Neglecting the differential signal swing at RL and RR, since at QSA1 $V_{CE} = V_s$, the above two equations are identical—$V_{cc}(min)$=1.3V<1.5V, which can be used in a system using a supply voltage of 1.5V.

5.3.2 BiCMOS DRAM

Fig. 5.74 shows the block diagram of an ECL I/O 0.3μm BiCMOS DRAM[38]. As shown in the figure, bipolar devices are used in the input buffer, the main amplifier, and the output buffer. In other portions, CMOS devices are used. Due to the reliability consideration for the miniaturized CMOS devices, the voltage swing in the CMOS decoder is 2V. In the word line driver, it is 2.5V to boost the word line voltage. In the memory cell, a voltage swing of 1.5V is adopted. Fig. 5.75 shows the sensing circuit with the overdrive rewrite amplifier in the 4Mb DRAM[38]. As shown in the figure, 256 cells are connected to each bit line. In order to enhance the speed performance, bipolar cascoded main amp and output buffer have been used. In order to speed up the sense and restore operations at a voltage swing of 1.5V, the overdrive rewrite amplifier scheme is used. During the sense/restore mode, the common source node of the PMOS device is connected to V_{CH}, which has a voltage swing greater than 1.5V. Thus, the restore operation, which is similar to the techniques described in Fig. 5.59, can be at high speed. V_{CH} is provided by the supply of the word line driver.

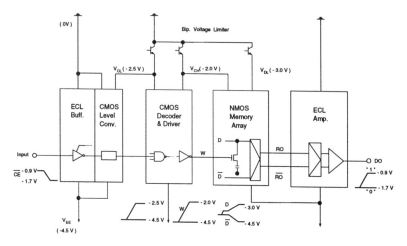

Fig. 5.74 Block diagram of an ECL I/O 0.3μm BiCMOS DRAM. (From Kawahara et al. [38], ©1991 IEEE.)

For BiCMOS SRAMs, reference voltages are frequently used. Fig. 5.76 shows the comparisons of the voltage limiter circuits[40]. As shown in the figure, the PMOS voltage limiter circuit has a drawback in the current-driving capability. In the bipolar voltage limiter circuit, the base current may still be large, therefore a Darlington pair is needed. In the Darlington voltage limiter, the output voltage swing is limited by the influence from $2V_{be}$ voltage drop in the Darlington pair. The BiPMOS voltage limiter, which combines the PMOS device without the gate-driving current with the bipolar device having a strong current-driving capability, is suitable for use in the ECL-level BiCMOS DRAMs. Fig. 5.77 shows the voltage-limiting circuit, which includes the bandgap-reference circuit[39] and the voltage up-converter. The V_L generator is based on the bandgap reference[39]. As shown in the figure, the V_L generator circuit is designed to have a temperature coefficient close to zero. In the circuit, $V_{BG} = V_{BE1} + R_1(V_{BE3} - V_{BE2})/R_2 = V_{BE1} + lnr \cdot \frac{kT}{q} \cdot \frac{R_1}{R_2}$, where r is the emitter area ratio of Q_2 to Q_3, and V_{BEi} is the base-emitter voltage of Q_i. In order to make the temperature coefficient of V_{BG} close to zero, one needs that $\frac{dV_{BG}}{dT} = \frac{dV_{BE1}}{dT} + lnr \cdot \frac{k}{q} \cdot \frac{R_1}{R_2} = 0$. Thus $\frac{R_1}{R_2} = \frac{-q}{klnr} \frac{dV_{BE1}}{dT}$. Based on $\frac{dV_{BE1}}{dT}$, by selecting an appropriate R_1/R_2, V_{BG} can be made to be close to temperature independent. As shown in the figure, a burn-in test circuit is used to to provide a stabilized V_{ref}, which is limited by R_6 and Q_5, when V_{EE} is very negative. Using the op amp with the output circuit, V_L can be adjusted to be equal to V_{ref}, where $V_{ref} = V_{BG}(1 + \frac{R_4}{R_5})$. By using various $\frac{R_4}{R_5}$ ratios, various limiting voltages can be made. In the figure, C_B is used for phase compensation of the voltage limiter frequency response. For the three different limiting voltages, three different capacitances are needed for individual compensation.

In a BiCMOS SRAM, with the merged bipolar and CMOS devices in the substrate, for latchup consideration substrate current can't be overlooked. Fig. 5.78 shows the

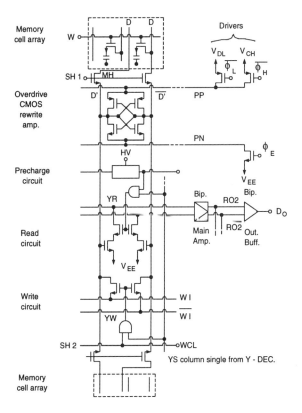

Fig. 5.75 Sensing circuit with the overdrive rewrite amplifier. (From Kawahara et al. [38], ©1991 IEEE.)

substrate current-generation mechanism in the BiCMOS driver[41]. As shown in the figure, the substrate current in the BiCMOS driver is due to the hot carrier effect of the MOS devices and I_{BB} from the effects of the BJT collector resistance (r_{cs}). As shown in the figure, due to the collector resistance, during the final portion of the pull up process when V_{out} is close to V_{DD}, the voltage at node CO is not V_{DD}. Instead, it is close to V_{out}. Due to the base-emitter capacitance, internal voltage overshoot occurs at the base of $Q1$. Therefore, the parasitic pnp structure (base-collector-substrate) may have an extra saturation current with the hot carrier related current of the MOS device—the substrate current I_{BB}. During pull-down, a similar situation exists. Node CO is pulled down earlier than the output node. As a result, the bipolar device enters saturation earlier. The increased saturation current due to the forward biased of B2-CO also raises I_{BB}. In order to avoid saturation of the bipolar device, the voltage at the internal node CO should not be much smaller than the base. Fig. 5.79 shows the substrate current-reduction technique[41]. As shown in the figure, using a two-collector structure, the substrate current problem can be reduced. When the base voltage of $Q1$ is higher than CO, M1 and r'_{cs} function as a negative feedback, the voltage difference between the base voltage and the CO node

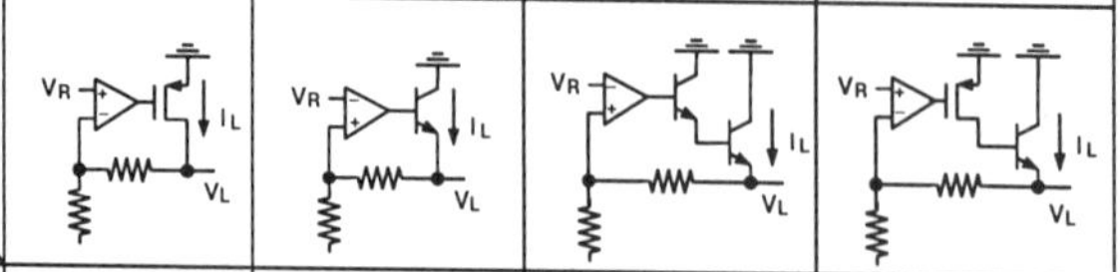

298 CMOS MEMORY

Output Trs. Features	PMOS	Bipolar	Bipolar Darlington	PMOS+Bipolar
Circuit	(circuit)	(circuit)	(circuit)	(circuit)
Margin for the maximum voltage applicable to PMOS	$V_B - |V_L|$	—	—	$V_B - |V_L + V_{BE}|$
V_L : V_{CH}	1.0 V			1.8 V
V_L : V_{CL}	0.5 V	—	—	1.3 V
V_L : V_{DL}	0			0.8 V
Margin for bipolar active operation	—	$|V_L| - V_{BE}$	$|V_L| - 2 V_{BE}$	$|V_L| - V_{BE}$
V_L : V_{CH}		1.2 V	0.4 V	1.2 V
V_L : V_{CL}	—	1.7 V	1.1 V	1.7 V
V_L : V_{DL}		2.2 V	1.4 V	2.2 V
Output current of op-amp	0	I_L / h_{FE}	I_L / h_{FE}^2	0
PMOS size (ratio)	1	—	—	$1 / h_{FE}$

V_B : maximum voltage application ($= 3.0$ V) V_{BE} : 0.8 V

Fig. 5.76 Comparisons of the voltage limiter circuits (From Kawahara et al.[40], ©1992 IEEE.)

does not become too large. Using a similar method, the substrate current during the pull-down transient in the BiNMOS device (Q_2, M_3) in Fig. 5.79 can be lowered.

Fig. 5.80 shows the conventional common I/O sensing and the direct sensing techniques used in BiCMOS DRAMs[42]. As shown in (a), in the common I/O sensing, when YS is high, I/O($\overline{I/O}$) and D($\overline{D}$) are connected, and the influence I/O ($\overline{I/O}$) in D($\overline{D}$) cannot be avoided. After D and $\overline{D}$ become large enough, the influence from $I/O(\overline{I/O})$ does not cause any bad effect, when YS becomes high. As shown in Fig. 5.80(b), in the direct sensing, $D/\overline{D}$ is separated from $RO/\overline{RO}$ by using a bipolar differential amplifier and clocked inverters. Thus the signal at YR becomes high earlier—without considering the effect on $D/\overline{D}$. However, the clocked inverter is needed to separate $D/\overline{D}$ from the bipolar device. As shown in the figure, the improved direct sensing structure uses a V-I converter controlled by YR. When YR is high, a current signal proportional to the difference between D and $\overline{D}$ is produced. This important current-sensing principle is used in the low-voltage CMOS DRAMs mentioned before. Thus YR can come in earlier—it does not need to worry about the damage on $D/\overline{D}$. In addition, the circuit is concise.

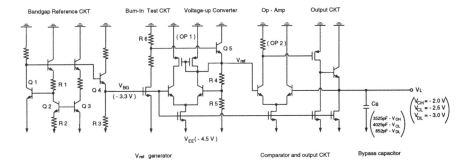

Fig. 5.77 Voltage-limiting circuit. (Adapted from Kawahara et al. [40].)

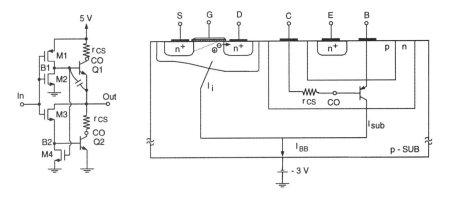

Fig. 5.78 Substrate current-generation mechanism in the BiCMOS driver. (Adapted from Kawahara et al. [41].)

5.4 SOI MEMORY

SOI technology has been used to integrate memory circuits. Large-size SOI CMOS DRAM and SRAM have been developed. In this section, circuit techniques for SOI SRAM and DRAM are described.

5.4.1 SOI SRAM

In order to increase packing density, devices in the memory cell of SOI SRAMs do not have body contact—body is floating. The parasitic BJT effect due to the floating body structure in the memory cell can degrade the performance of an SOI SRAM. Due to the floating body of the SOI CMOS device, accumulation of holes in the neutral region of the silicon thin-film makes the body potential rise. As a result, the threshold voltage of the device drops. In addition, the leakage current of the parasitic bipolar device increases. For the memory cell and the column structure as shown in Fig. 5.81, consider the effect of the floating body potential on the memory cell[43]. When the potential of the floating body rises, the threshold voltage of the

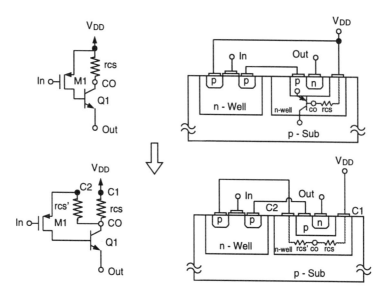

Fig. 5.79 Substrate current-reduction technique. (Adapted from Kawahara et al. [41].)

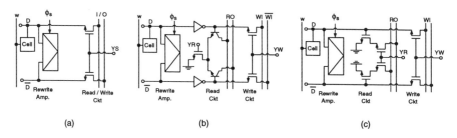

Fig. 5.80 Conventional common I/O sensing and direct sensing techniques used in BiCMOS DRAMs. (Adapted from Kitsukawa et al. [42].)

access transistor in the memory cell drops and the leakage current of the parasitic bipolar device increases. Consider the write 0 transient for the memory cell 0. Write 0 implies that node A will be discharged to low. In the figure, write delay 'max' is referred to data 1 stored in memory cell 1-511 such that node A in the memory cell is high. Write delay 'min' is referred to data 0 stored in memory cell 1-511 such that node A in the memory cell 1-511 is low. When the 'write 0' procedure is initiated, word line LWL becomes high. At this time, via the read/write switch, the input data (DIN=0) forces the bit line (BL) to pull low. The access transistors (T1, T2) of cell 0 are turned on. Therefore, node A is pulled low. Thus the 'write 0' function is accomplished. Fig. 5.82 shows the transients for the 'max' case of the unselected cell disturb during a write cycle at V_{DD}=1.95V and 105°C[43]. Comparing the transient waveforms referred to 'min' and 'max', for the max case 'write 0' takes a longer time. This is due to the fact that for the 'max' case, the drain node (A) of the pass transistor (T1) in cell 1-511 maintains high. When the source end (at BL) of T1 changes from

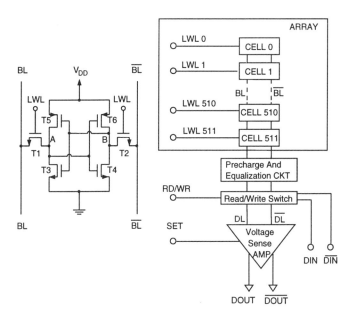

Fig. 5.81 Block diagram of an SOI SRAM column and its 'write 0' transient for the memory cell 0. (Adapted from Kuang et al. [43].)

high to low, the substantial leakage current due to the parasitic BJT results in a slow increase in node A. Consequently, the pull-down of the internal node A becomes slower. Therefore, 'write 0' is slower. For the 'min' case, no such situation occurs, since for the 'min' case, node A in cell 1-511 maintains low. No leakage as in the 'max' case exists. Therefore, speed is faster. When designing the memory cell, the aspect ratio of the PMOS devices should be arranged carefully such that the leakage current from the parasitic BJT does not change the potentials at the storage nodes.

Three-dimensional IC technology has been used to realized SRAM. Fig. 5.83 shows the diagram of a 3D SRAM[44]. As shown in the figure, at the bottom layer bulk MOS devices are used to integrate memory cells. At the top layer, the SOI devices realized by the selective laser recrystallization technique are used for address decoders, sense amps, I/O buffers, and chip controllers. Thus a higher packing density can be obtained.

In some SRAM designs, SOI devices and bulk devices are simultaneously used. Fig. 5.84 shows the memory cell of an SRAM using laser-recrystallized SOI PMOS load[45]. As shown in the figure, in the memory cell, bulk NMOS device and SOI PMOS load are used. Different from standard poly-TFT PMOS load, the SOI PMOS load is based on the top-gate structure. Thus, the self-aligned source/drain can be obtained and the performance of it is similar to that of the bulk PMOS device. The SOI PMOS load is realized by the laser recrystallization of the CVD-deposited polysilicon layer. In this memory cell, the P^+ poly-gates of the PMOS load devices

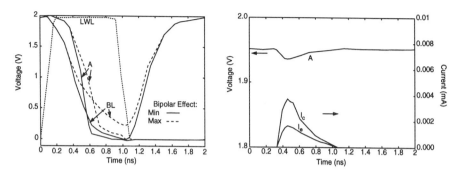

Fig. 5.82 Transients of the unselected cell disturb during a write cycle at $V_{DD}=1.95$V and 105°C. (From Kuang et al. [43], ©1997 IEEE.)

are connected to the drain of the NMOS device, which has an N^+ poly-gate. As a result, effective diodes exist in the memory cell. Since the doping density of the pn junctions are high, their effect on the swing at the storage node is small. Compared to the conventional poly-TFT PMOS load, the SOI PMOS load offers a higher on/off ratio and a smaller leakage current.

5.4.2 SOI DRAM

Due to the properties in high-density, low-parasitic capacitances, low-leakage current, and good radiation hardness, SOI technology has been regarded as a good candidate for realizing low-voltage low-power large-size high-speed DRAMs. Due to smaller junction capacitances (as compared to bulk) and smaller junction areas, the load associated with the bit line of an SOI memory (C_B) is much smaller. When the read operation begins, the voltage difference at the bit lines is $\frac{V_{DD}}{2} \frac{1}{1+C_B/C_s}$. Owing to smaller C_B (parasitic capacitance at the bit line) and C_s (capacitance at the storage cell), a higher speed at lower power consumption can be reached. The subthreshold slope of the SOI MOS device is steeper (closer to the ideal 60mV/dec) as compared to the bulk device. Therefore, the threshold voltage can be made smaller for the SOI NMOS device. As a result, SOI CMOS is suitable for low-voltage operation. However, the floating body effect is serious for partially-depleted SOI MOS devices.

Fig. 5.85 shows the schematic diagram representing the soft error immunity in an SOI DRAM[46]. As shown in the figure, when the SOI DRAMs are illuminated by the alpha particles, owing to the buried oxide, which provides an isolation from the substrate, a large amount of electron-hole pairs produced by the alpha particles do not affect the storage nodes of the memory cell. Hence, a good soft error immunity can be obtained. In contrast, for the bulk DRAM, the electron-hole pairs produced by the alpha particles may be absorbed by the source/drain of the devices. Therefore, the potential of the storage nodes may change. Hence, the soft-error immunity of the bulk DRAM is inferior.

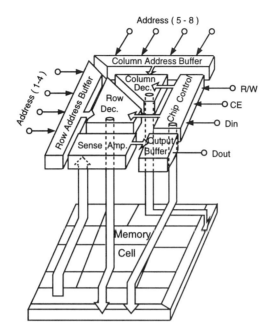

Fig. 5.83 Diagram of a 3D SRAM. (From Nishimura et al. [44], ©1985 IEEE.)

Due to the influence of the floating body, the performance of an SOI DRAM may be affected. The floating body effect can be reduced by adding body contacts. Fig. 5.86 shows another way to add body contacts for SOI DRAM circuit applications[47]. As shown in the figure, only in the driver, the buffer, and the sense amplifier are body contacts used. The performance of these circuits are susceptible to the floating body effect. On the other hand, in the memory cell and the logic gate, sometimes the floating body configuration is not used to increase the packing density.

In order to increase the cell density, the body contacts mentioned before should not be used in an SOI DRAM cell. Without the body contact, the SOI device is susceptible to the floating body effect. Due to junction leakage and thermal generation, the body potential may rise since holes are accumulated in the neutral region of the floating body. Thus the threshold voltage is further reduced—subthreshold leakage current may increase. The leakage problems in an SOI DRAM memory cell can be resolved by the body refresh operation as shown in Fig. 5.87[48]. When the body potential of the pass transistor in an SOI DRAM memory cell does not reach a designated value, the bit line is pulled low intentionally such that the piled holes are expelled from the body. As a result, the body potential is maintained at a low value as shown in the figure. Therefore, the parasitic BJT cannot be turned on. By using this technique, the body voltage is maintained at a low value. Therefore, during the latency period, the leakage current is smaller—a better data-retention time can be obtained. Fig. 5.88 shows the timing diagram of the hidden body refresh approach in an SOI DRAM[48].

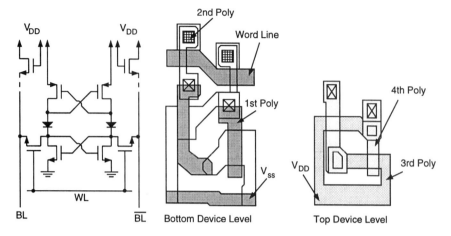

Fig. 5.84 Memory cell of an SRAM using laser-recrystallized SOI PMOS load. (From Takao et al.[45], ©1992 IEEE.)

As shown in the figure, for the cell without the read operation during the refresh cycle, its bit line drops from $\frac{V_{DD}}{2}$ to 0V to pull down the body potential. If the word line $WL0$ is high during the refresh cycle, read operation is carried on in the cells on the word line $WL0$ of block B0. All cells of blocks B1, B3, B5, and B7 are inactive. Thus write 0 is carried out for the bit lines of the bit line pair BL0. Since no cells are active, no false data are written to the cell by accident. The potential of the floating body can be adjusted. During the next refresh cycle, $WL1$ of block B0 is high. Hidden body refresh can be carried out for bit line pair BL1 of blocks B1, B3, B5, and B7.

In the previous section, 3D techniques for SRAMs were described. Similarly, the 3D techniques have been used to design DRAM. As shown in Fig. 5.89, in a stacked SOI DRAM memory cell, there are three device layers[49]. The bulk devices are at the bottom layer. The SOI devices are at the other two layers. Between device layers, doped polysilicon layers have been used to function as shielding such that interference can be reduced. The planar capacitors are with the pass transistors at the same layer or two adjacent layers to increase density and to reduce parasitic capacitances.

In the SOI CMOS device, its threshold voltage is a function of the body voltage. Fig. 5.90 shows the V_{bs}-controlled threshold voltage of the SOI NMOS devices[50]. As shown in the figure, when the body-source is reverse biased, its threshold voltage increases proportional to the magnitude of the reverse bias. On the other hand, when the body-source is forward biased, its threshold voltage decreases with the forward bias. Based on the body bias characteristics, the SOI MOS devices are especially suitable for implementation of the active/standby mode logic gates as described in Chapter 3. Due to the isolation structure, the body contact of each SOI MOS device can be individually controlled. During the standby mode, the SOI device is operating

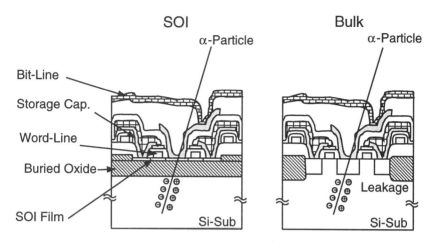

Fig. 5.85 Schematic diagram representing the soft error immunity in an SOI DRAM. (Adapted from Yamaguchi and Inoue[46].)

at a reverse bias of V_{bs} to reduce subthreshold leakage and standby power. Thus, using the SOI MOS devices, no multi-threshold (MT) devices as for the bulk CMOS technology are required. Fig. 5.91 shows the body-synchronous sensing scheme and the super body-synchronous sensing scheme used in the SOI DRAM[50]. As shown in Fig. 5.91(a), it is similar to the well-synchronized sensing scheme in bulk CMOS DRAM. In the SOI circuit, due to a smaller body capacitance, SAN and SAP can be pulled down/up more easily. Thus a better performance can be expected for the SOI circuit. In the super body-synchronous scheme in Fig. 5.91(b), the body-source junction is forward biased before the completion of the sense operation. (Before the sense operation, since SAN and SAP are preset to 1V, the forward bias is the smallest.) After the sense operation, during the restore operation, the p-body voltage of the NMOS device changes from 1.5V to 0.5V and the n-body voltage of the PMOS device changes from 0.5V to 1.5V. Thus the magnitude of the threshold voltage increases to reduce leakage current. Fig. 5.92 shows the body controlling technique in the active/standby logic circuits for SOI DRAMs[50]. As shown in the figure, during the standby mode, the off devices are with their body connected to V_{nb} and V_{pb} with a reverse biased value. Due to the increase in the magnitude of the threshold voltage, the leakage current can be reduced. Thus, no multi-threshold devices are required. In addition, no power switch transistor and virtual power line are required either. Using the SOI CMOS technology, active/standby mode logic is easier to implement.

In addition to suppressing floating body effects, body contacts can be utilized for low-voltage operation. Since the body of an SOI MOS device can be controlled by adding the body contact, the DRAM circuit technique based on adjusting the threshold voltage by controlling V_{SB} described before can be implemented more conveniently in SOI DRAMs. Fig. 5.93 shows the body-pulsed sense amplifier (BPS) circuit and its waveforms compared with super body synchronous sensing (SBSS)[51]. As shown

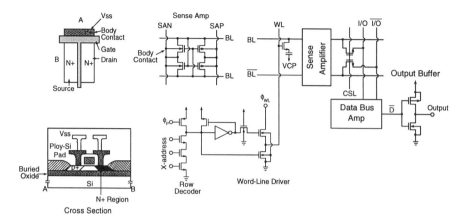

Fig. 5.86 Data path of read operation in an SOI DRAM with body contacts for their devices using 0.5μm SOI CMOS technology. (Adapted from Suma et al. [47].)

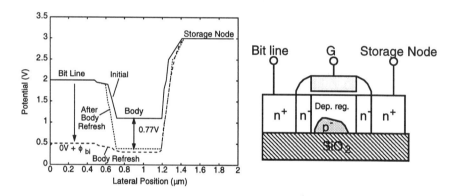

Fig. 5.87 Potential distribution in an SOI DRAM memory cell using body refresh techniques. (From Tomishima et al. [48], ©1996 IEEE.)

in the figure, it is based on adjusting the body potential of some circuits dynamically such that the conducting capability of these devices can be enhanced for some appropriate time. For other period of time, their original states are maintained. By this arrangement, the power consumption does not increase substantially. As shown in the figure, in period (1), word line (WL) becomes high and the bit line (BL) starts to change. In period (2), S0N becomes high and the sense amp begins to become active. At this time, the body voltage (SBN) of $M1$ and $M2$ is raised and the body voltage (SBP) of $M3$ and $M4$ is lowered temporarily. As a result, the magnitude of the threshold voltage of $M_1–M_4$ becomes smaller. Hence, the speed performance of the sense amp becomes faster. In period (3)—the restoring period—the body voltage (SWP and SWN) of $M5$ and $M6$ is changed accordingly such that the conducting capability of $M5$ and $M6$ can be upgraded. This results in the enhanced restoring process. Fig. 5.94 shows the body-driven equalizer (BDEQ) circuit and its related

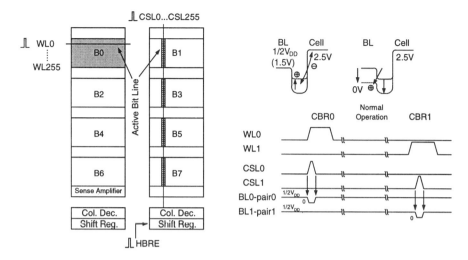

Fig. 5.88 Timing diagram of the hidden body refresh approach in an SOI DRAM. (From Tomishima et al.[48], ©1996 IEEE.)

transient[47]. When the equalizer becomes active (BLEQ is high), the body voltage of the NMOS device is raised simultaneously. Hence, the threshold voltage of the NMOS device is reduced, consequently the conducting capability is enhanced. Thus the equalizing time is reduced.

Fig. 5.95 shows the body current clamper (BCC) circuit. The objective of this circuit is to adjust the body voltage of the sense amp. If the body voltage is too high for the NMOS or too low for the PMOS, the body-source diode turns on, which may result in a large amount of current flowing through V_{DDA} and G_{ndA}. Therefore, some voltage bounces exist on V_{DDA} and G_{ndA}, which may affect the operation of the sense amp. This drawback can be improved by inserting diodes between body-bias line and V_{DDB}/G_{ndB}. By this arrangement, the body voltage can be clamped to within an appropriate value to suppress voltage bounces on V_{DDA} and G_{ndA}.

5.5 NONVOLATILE MEMORY

DRAM and SRAM (described before) are volatile memories. When the power supply is turned off, the data in the volatile memory are lost. When the power supply is turned off, the data in a nonvolatile memory can still be kept. Nonvolatile memories include ROM (read only memory), PROM (programmable ROM), EPROM (erasable PROM), EEPROM (electrically EPROM), flash memory, and FRAM (ferroelectric RAM).

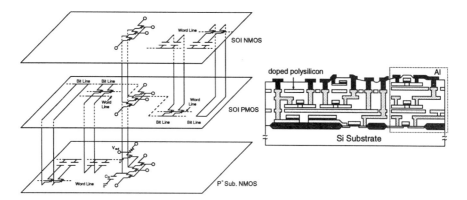

Fig. 5.89 Structure of a stacked SOI DRAM memory cell. (From Ohtake et al.[49], ©1986 IEEE.)

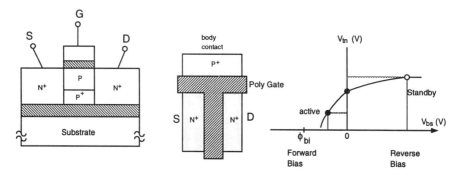

Fig. 5.90 V_{bs}-controlled threshold voltage of SOI NMOS devices. (Adapted from Kuge et al.[50].)

5.5.1 ROM

ROM (read only memory) is used in a computer system as program memory. The memory cell structure in a ROM is simple. Therefore, among all memory circuits, ROM has the lowest cost per bit value, which is the greatest advantage. However, the speed of a VLSI ROM may be slow. Fig. 5.96 shows the core circuit in a ROM based on the NOR array structure. The NOR array in the ROM is composed of a matrix made by the intersections of the lateral word lines and the vertical output bit lines. At the early stage, the lateral word lines are made of polysilicon. Due to its large sheet resistance, when the polysilicon word line is long for a large integration size, the associated delay cannot be neglected. By using the polycide word lines, the parasitic resistance and hence the delay have been reduced. The vertical bit lines are made of metal interconnects. The cross region is the diffusion region. Each intersection of the cross diffusion region and the horizontal polysilicon region is a prearranged transistor. In order to minimize the area of a ROM cell, two ROM cells share a diffusion contact to the ground line. The add-on of this transistor to the ROM

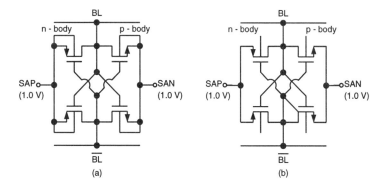

Fig. 5.91 (a) Body-synchronous; and (b) super body-synchronous sensing schemes used in the SOI DRAM. (Adapted from Kuge et al. [50].)

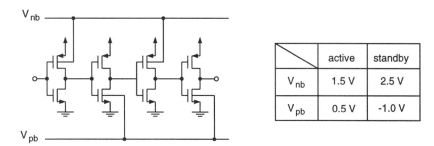

	active	standby
V_{nb}	1.5 V	2.5 V
V_{pb}	0.5 V	-1.0 V

Fig. 5.92 Body-controlling technique in the active/standby logic circuits for SOI DRAMs. (Adapted from Kuge et al. [50].)

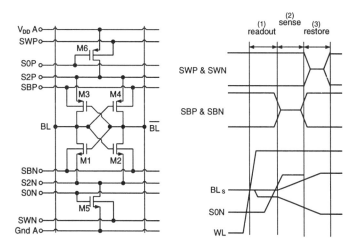

Fig. 5.93 Body-pulsed sense (BPS) amplifier circuit and its waveforms compared with super body-synchronous sensing (SBSS). (Adapted from Shimomura et al. [51].)

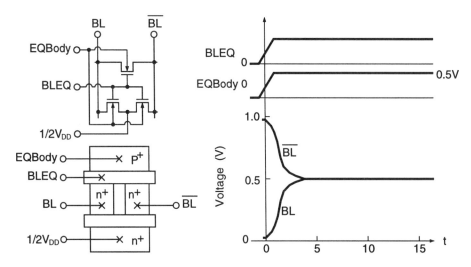

Fig. 5.94 Body-driven equalizer (BDEQ) circuit and its related transients. (Adapted from Shimomura et al.[51].)

circuit depends on the logic function. ROM can be programmed via the contact mask—contact mask programmable. Without the contact, the prearranged transistor is not used.

The read access time of ROM is:

$$t_{aa} = t_w + t_c, \qquad (5.19)$$

where the word line delay (t_w) can be computed by estimating the length of the word line. Based on a 0.8μm CMOS technology to design a 4M-b ROM, if the area of a ROM cell is 3.2μm × 4.8μm, the length of a polycide word line in a 4Mbit ROM is $2048 \times 3.2 = 6554\mu$m. Considering the sheet resistance of the polycide word line of $0.2\Omega/\square$ and the width of the word line of 0.8μm, the parasitic resistance of the word line is 1.05KΩ. If the unit-area parasitic capacitance of the word line is 0.6nF/cm^2, the parasitic capacitance of the whole word line is 0.31pF. Therefore, the RC delay of the worst-case word line is $t_w = 0.33$ns. The delay of the output column line can be estimated by the parasitic capacitance at the output:

$$t_c = \frac{C_L \Delta V}{I_{av}}, \qquad (5.20)$$

where ΔV is half of the voltage swing. If V_{DD}=5V, ΔV=2.5V. The parasitic capacitance of the vertical output line can be estimated as follows: The unit-area junction capacitance of the drain is C_j=3.5 × 10^{4}F/m^2. The unit-length sidewall junction capacitance is C_{jsw}=3×10^{-10}F/m. The drain area of each bit on the vertical line is 2.4μm × 2μm. The length of the perimeter of the sidewall surrounding the drain area is 2 × 4.4μm = 8.8μm. In addition, at the vertical output, there is gate-drain overlap capacitance of 2 × 10^{-4} × 1.1pF/cm=0.22fF. The parasitic capacitance

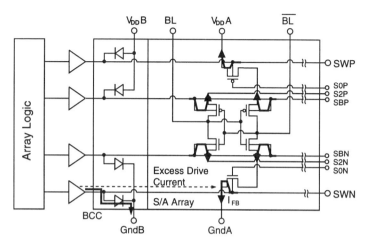

Fig. 5.95 Body current clamper (BCC) circuit. (Adapted from Shimomura et al. [51].)

per bit at the output line is $(2.4 \times 10^{-6}) \times (2 \times 10^{-6}) \times (3.5 \times 10^{-4}) + (8.8 \times 10^{-6}) \times (3 \times 10^{-10}) + 0.22\text{fF} = 4.54\text{fF}$. The parasitic capacitance of the whole output line is $4.54\text{fF} \times 2048 = 9.3\text{pF}$. The average current is $I_{av} = \frac{1}{2} \times 60(\mu A/V^2) \times 2 \times 0.8 \times (5 - 0.8)^2 = 1.32\text{mA}$. Therefore, the delay of the output line is $t_c = 5.3 \times 10^{-12} \times 2.5/1.32 \times 10^{-3} = 17.6\text{ns}$. Therefore, the address access time from the word line to the bit line is 17.94ns.

5.5.2 EPROM

EPROM is erasable programmable ROM. Fig. 5.97 shows the cross-section of an EPROM cell and its threshold voltage characteristics of the erased and programmed states[52][53]. As shown in the figure, it is composed of a double-gate MOS transistor. The top gate is the select gate used for the access of the transistor. The threshold voltage of the device is changed when charge is stored in the floating gate. Programming an EPROM memory cell is by applying a high voltage at the top select gate and the drain. Due to the positive voltage applied at the top gate, the large vertical electric field affects the electron transport. If a large drain voltage is applied, due to the high electric field in the lateral channel, impact ionization exists—electrons and holes are generated. Due to the large vertical electric field, electrons are attracted to move toward the vertical double-gate direction. Since the oxide under the floating gate is very thin, the electrons tunnel through the thin oxide to be trapped at the floating gate. This is the channel hot electron (CHE) injection effect. Since the floating gate is surrounded by the isolating oxide, the injected electrons cannot move anywhere else but stay at the floating gate. As a result, the threshold voltage is shifted toward the positive direction as shown—logic 0. The change in the threshold voltage is determined by the amount of the electrons stored at the floating gate. By applying a gate voltage with a magnitude between V_{T1} and V_{T0}, the stored logic value of the

312 CMOS MEMORY

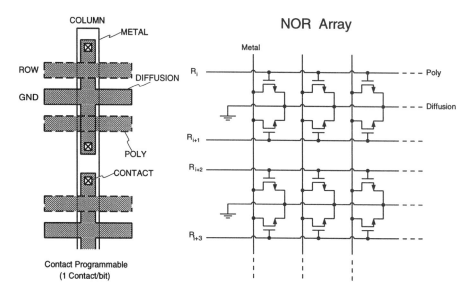

Fig. 5.96 ROM cell.

device can be determined by detecting the magnitude of the drain current. Due to the injected electrons at the floating gate, at such a gate bias, the device will turn off. For the device without the electron injection, its threshold voltage will not change. Thus the device turns on with a substantial drain current. Therefore, the stored data in a double-gate memory cell can be identified. In a typical EPROM, the electrons at the floating gate can be retained for several years. The program time of a memory cell in an EPROM is in the order of 1ms. In order to reduce the programming time, the gate injection current should be increased. It can be done by increasing the electric field along the channel in the gate dielectric material, which depends on the channel length and the floating gate voltage. By making the dielectric layer between polysilicon layers thinner, the vertical electric field can be increased. At a higher vertical electric field, the programming time can be shrunk. However, the breakdown problem may exist at a high electric field. Using ONO oxide-nitride-oxide dielectric layer (a high dielectric constant) to reduce the internal electric field, its breakdown voltage can be increased. By exposure to the ultraviolet light, the electrons at the floating gate of memory cells in the EPROM can be erased. The erase time is on the order of 30 minutes. Partial erasure of the memory cells in an EPROM is not possible.

Fig. 5.98 shows the sense amp in the EPROM[53]. By detecting the drain current, the stored data can be determined. As shown in the figure, the read operation can be done by applying a voltage between the programmed threshold voltage and the original threshold voltage at the top select gate. During the read access of the selected memory cell storing logic "1" (small threshold), the drain voltage of the memory cell should be small. If the applied drain voltage is not small, a small amount of impact ionization current may cause soft (unintentional) writing—the threshold voltage will

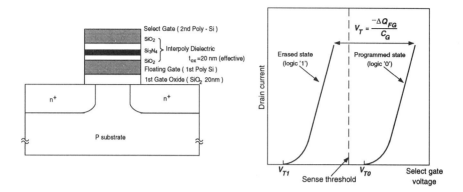

Fig. 5.97 Cross-section of an EPROM memory cell and its threshold voltage characteristics. (Adapted from Ohtsuka et al. [53].)

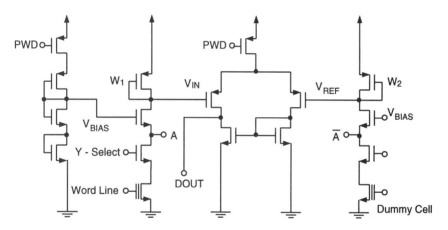

Fig. 5.98 Sense amp in the EPROM. (Adapted from Ohtsuka et al. [53].)

gradually rise. In the EPROM, dummy cells are still required. In the EPROM dummy cell, logic 1 (small threshold voltage) is stored. During the read access, if the memory cell also stores logic 1 (small threshold), its bit-line voltage is smaller than the dummy bit-line voltage. Thus the difference between V_{IN} and V_{REF} is detected by the sense amp. Since A and $\overline{A}$ can only rise to $V_{BIAS} - V_T$, the maximum bit-line voltage is limited by V_{BIAS}. The bit-line swing can be reduced by adjusting the aspect ratio of the PMOS device (W_1, W_2). Similarly, by adjusting the aspect ratio of the PMOS device W_2, V_{REF} is designed to be at the half-swing of the sense amp input (V_{IN}) such that a maximum voltage difference can be obtained for comparison. V_{BIAS} is obtained by the diode connected circuit—$V_{BIAS} \cong 2V_T$. Thus the maximum bit-line voltage is V_T.

Fig. 5.99 shows another sense amp using complementary current mirror in an EPROM[54]. As shown in the figure, V_{BIAS}=3V. By arranging that V_{BL}<1.5V,

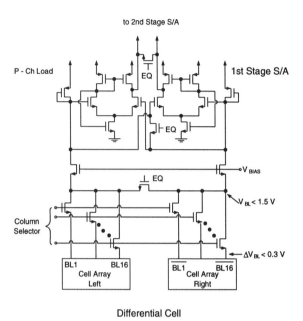

Fig. 5.99 Sense amp using complementary current-mirror in an EPROM. (Adapted from Kuriyama et al. [54].)

unintentional writing can be avoided. In addition, by limiting the bit line swing to within 0.3V, the charging/discharging delay of the bit line can be shortened. The column selector is used to connect one of the 16 bit-line pairs to the sense amp. In addition, an equalize transistor (EQ) is added. Fig. 5.100 shows the bit-line bias circuit[54]. As shown in the figure, $V_{BIAS} = 2V_T$ is generated. When the supply voltage is changed, the biasing current varies—the standard V_{BIAS} generator is sensitive to the change in the supply voltage. Thus the precision of the sense amp is affected. Using a depletion NMOS device with its source and gate connected, its drain current is fixed—$I_D = \frac{1}{2}C_{ox}\frac{W}{L}V_{TD}^2$, which is not influenced by the bounce in V_{DD}—its V_{BIAS} is more stable and its power supply rejection ratio (PSRR) is higher.

5.5.3 EEPROM

For the EPROM described before, erasure of the internally stored data can only be done by exposure to the ultraviolet light. When exposed, all data in the chip are eliminated. Here, an EEPROM (E^2PROM), which has the capability to perform a single-bit erasure electrically, is introduced. Fig. 5.101 shows the cross-section of an EEPROM memory cell[55]. It is composed of (1) the access transistor, (2) the memory transistor, and (3) the erase area. As shown in the figure, the floating gate potential, which determines the maximum threshold voltage shift (V_{TN}) and the gate

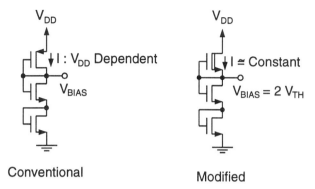

Fig. 5.100 Bit-line bias circuit. (Adapted from Kuriyama et al. [54].)

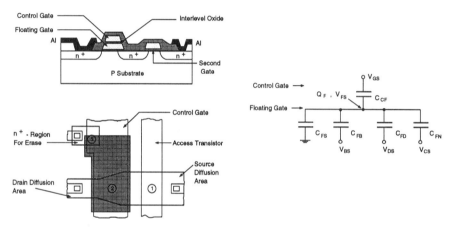

Fig. 5.101 Cross-section of an EEPROM memory cell. (From Scheibe and Krauss, [55], ©1980 IEEE.)

current, is expressed as:

$$V_{FS} = \frac{C_{CF}}{C_T}V_{GS} + \frac{C_{FD}}{C_T}V_{DS} + \frac{C_{FB}}{C_T}V_{BS} + \frac{C_{FN}}{C_T}V_{CS} + \frac{Q_F}{C_T},$$
$$C_T = C_{CF} + C_{FD} + C_{FB} + C_{FN}, \qquad (5.21)$$

where V_{CS} is an additional potential that can be applied during programming to the n^+-region formed outside the channel. C_{FB}, C_{FD}, and C_{FN} are the coupling capacitances between the floating gate and the substrate contact V_{BS}, the drain contact V_{DS}, and the n^+-region formed outside the channel, respectively. The programming process is identical as for EPROM—by injection of hot electrons from the channel to the floating gate to raise the threshold voltage of the memory transistor (logic 0). The erase process is done by applying an erase voltage (high voltage) at V_{CS} and grounding the top control gate. Via Fowler-Nordheim tunneling[56], the electrons in the floating gate can tunnel to the n^+ region labeled 3—its threshold voltage can

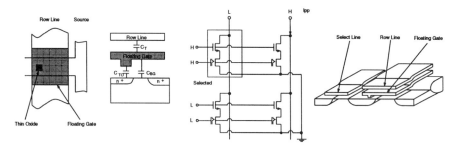

Fig. 5.102 EEPROM memory transistor structure. (From Yaron et al. [57], ©1982 IEEE.)

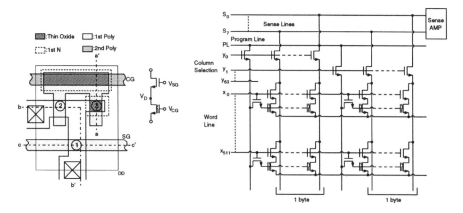

Fig. 5.103 Cell arrays of a 5V-only EEPROM. (From Miyamoto et al. [58], ©1986 IEEE.)

be lowered. Different from the optical erase as for EPROM, the erase process for EEPROM may have the over-erase problem. If the erase time is too long, in the floating gate, holes will be accumulated—the transistor becomes depletion mode. Since the access transistor is still enhancement-mode, the access of the memory cell is dominated by the access transistor.

Fig. 5.102 shows another EEPROM memory transistor structure[57]. As shown in the figure, in the overlap region between the drain and the floating gate, there is an area with a thinner gate oxide. The erase to logic-1 (low-threshold) process can be accomplished via Fowler-Nordheim tunneling directly from the floating gate to the drain area. By grounding the top control gate and applying a high voltage to the drain, the electrons in the floating gate will tunnel to the drain terminal. Programming to logic-0 (high threshold) is done by imposing a high voltage on the top control gate and grounding the drain. The electrons are tunneling from the drain through the bottom gate oxide to the floating gate. Thus the threshold voltage is raised. Different from the standard EEPROM cell described before, the programming process here does not use the channel hot-electron injection method.

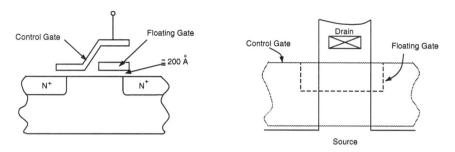

Fig. 5.104 Cross-section of a flash EEPROM cell. (Adapted from Samachisa et al. [59].)

Fig. 5.103 shows the cell arrays of a 5V-only EEPROM[58]. The overlap region between the control top gate and the floating gate is not the active region of the memory device. Instead, it is the memory transistor region labeled region 2 controlled by the floating gate. As shown in the figure, region 1 is the access transistor. Region 3 is for the erase operation. As shown in the figure, it has an 8-bit (1-byte) output. y0–y63 represents the 64-to-1 selection. One of the 64 bit-line pairs is connected to the sense amp for amplification and for write of the data signal.

5.5.4 Flash Memory

The property of the flash memory is between EPROM and EEPROM. For flash memory, its cells can be written and read electrically. In addition, flash memory has the same advantages in density as EPROM. Fig. 5.104 shows the cross-section of a flash EEPROM cell [59]. As shown in the figure, different from the EEPROM cell, no select transistor (access transistor) is required in the flash memory cell. The flash memory still has the capability of being erased electrically. During the electrical erase process, the entire memory array or the entire memory block is erased at the same time—it does not have the capability of individually erasing only a single bit as for EEPROM. This is why it is called flash. The erase process is done via FN tunneling as described before. As shown in the figure, the floating gate occupies a portion of the channel region near the drain. From the cross-section, it is compatible to an enhancement-mode NMOS device (formed by the control gate) connecting with a floating gate EEPROM cell in series. The programming process of the flash EEPROM is accomplished via the channel hot-electron (CHE) injection as described before. The erase process is done via FN tunneling by imposing a high voltage on the drain—the electrons in the floating gate tunnel through the oxide to the drain. There is also the over-erase problem in flash memory as in EEPROM. (In contrast, UV-erase EPROM does not have the over-erase problem.) When over-erased, the floating gate becomes positively charged. As a result, it becomes a depletion-mode device. Due to the dominance of the enhancement-mode control gate, its threshold voltage is still determined by the control gate. Thus the layout area of the flash EEPROM is slightly larger than that of the EPROM and smaller than that of the EEPROM. Therefore,

318   CMOS MEMORY

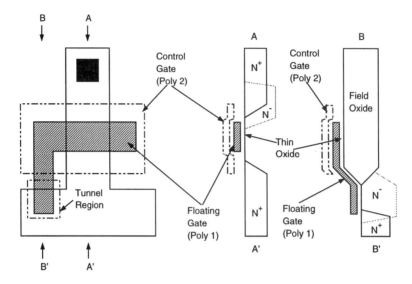

Fig. 5.105 Source-coupled split-gate (SCSG) flash EEPROM cell. (From Kuo et al. [60], ©1992 IEEE.)

flash memory has the high density and electrical programmable/erasable properties.

Fig. 5.105 shows the source-coupled split-gate (SCSG) flash EEPROM[60]. As shown in the figure, it is different from the flash EEPROM cell described before, where the floating gate tunnel region for the erase operation is at the drain side. In the SCSG flash EEPROM cell, it is at the source—source-coupled split-gate cell. Due to the separation of the erase region from the drain region, the program region (drain) and the erase region (source) can be individually optimized. In the tunneling region under the bird's beak marked as BB' in the figure, an n^- region is used to avoid direct band-to-band tunneling and impact ionization of the gated diode breakdown under the thin oxide. Thus a large amount of erase current can be avoided. The SCSG flash EEPROM cell is also based on the split-gate structure, which has an enhancement-mode transistor in series with a floating gate transistor. By this arrangement, the over-erase problem can be eliminated.

There is another flash EEPROM memory cell, which is similar to the memory cell described in Fig. 5.97 except that the thin oxide between the floating gate and the substrate is even thinner, to accomplish the erase operation by FN tunneling. However, this kind of the memory transistor may have over-erase problems. Thus a circuit to resolve the over-erase problem should be implemented. Fig. 5.106 shows the block diagram of an in-system programmable $32K \times 8$ flash EEPROM[61]. During the erase process, in order to avoid over-erase, all bytes are programmed to the high threshold voltage (00H) via two write sequences. In the first cycle, the erase code is written into the command register. In the second cycle, the erase confirm code is written. Thus the command decoder triggers the high-voltage flash erase switch. A 12V supply is

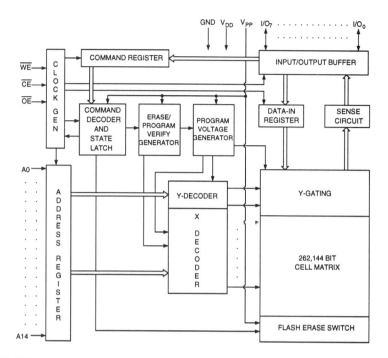

Fig. 5.106 Block diagram of an in-system programmable 32K×8 flash EEPROM. (From Kynett et al. [61], ©1988 IEEE.)

connected to the source of all array cells and the gate is grounded. Erase is done via FN tunneling between the floating gate and the source—the threshold voltage of all array cells drops simultaneously. At the same time, the internal erase verify voltage is set up by the erase/program verify generator. Thus the microprocessor can initiate the read operation for the memory cells referred to the address bytes. As long as an address byte cannot match FFH (all low-threshold voltage), the erase process will be repeated once to achieve the goal in tight threshold control. The programming process is similar to the erase process except that the programming operation is for the addressed byte.

Fig. 5.107 shows the erase/program verify voltage generator circuit for flash E^2PROM[61]. As shown in the figure, when $Enable$ is high, the source follower $M5$ outputs an output verify signal: $V_{verify} = V_{pp} \frac{R_2}{R_1 + R_2 + R_3}$, which was derived from $I_D(M5) = I_D(M6)$ without considering the body effect of $M5$. Fig. 5.108 shows the high-voltage flash erase switch and the program load circuit[61]. As shown in Fig. 5.108(a), during the erase process, $ERASE$ is high and the erase switch circuit outputs V_{pp} to the source terminal of the array cell. The zero threshold device $M5$ pulls the array source to $V_{pp} = 12V$. $M7$ is used to fix the array source to 0V during the programming and the read processes. Fig. 5.108(b) shows the program load circuit. When the gate of the column select transistor M7 (YSEL) is connected

320 CMOS MEMORY

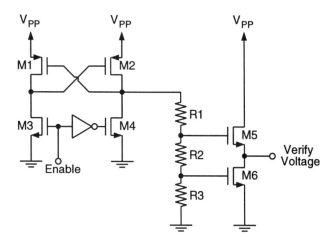

Fig. 5.107 Erase/program verify voltage generator for flash EEPROM. (Adapted from Kynett et al.[61].)

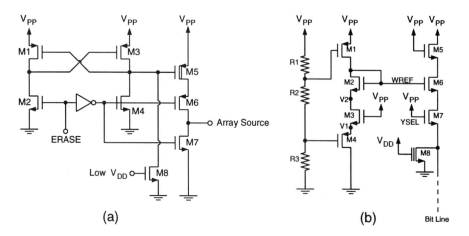

Fig. 5.108 High-voltage flash (a) erase switch; and (b) program load circuit. (Adapted from Kynett et al.[61].)

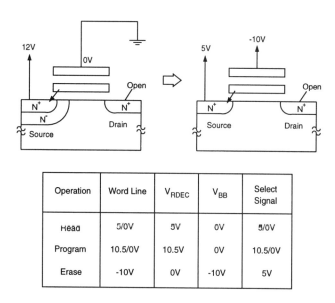

Fig. 5.109 Negative-gate-biased source erase scheme. (Adapted from Umezawa et al. [62].)

to V_{pp}, if $M2$ and $M6$ are matched and $M3$ and $M7$ are matched, the bit line voltage can be designed to be identical to V1. Thus the variation on the bit line voltage during the programming operation can be reduced to lower the program disturb problem—a high bit line voltage may result in tunneling of electrons to the floating gate of the unselected cell to cause a reduction in the cell threshold voltage.

Fig. 5.109 shows the negative-gate-biased source erase scheme[62]. As shown in the figure, in the conventional erase scheme, the source is connected to a large voltage—12V—and the gate is grounded. The electrons in the floating gate tunnel to the source via the FN tunneling mechanism. Due to the large electric field at the source end, an n^- double diffusion structure is needed to increase the breakdown voltage, which may not be advantageous for further device miniaturization. As shown in the figure, in the second approach for the source erase scheme, the source end is connected to 5V and the control gate is biased at −10V. Via capacitively coupling of the oxide from the control gate and the source to the floating gate, the electric field between the floating gate and the source junction still can trigger electron tunneling. Owing to a small voltage at the source, the n^- double diffusion is not needed. Thus, during the programming process, the gate needs a large positive voltage. During the erase process, it needs a negative voltage. Fig. 5.110 shows a row decoder, which provides an output high level up to V_{RDEC} and an output low level down to V_{BB}. As shown in Fig. 5.110(a), the frequently used row decoder contains two PMOS devices—a diode-connected PMOS and another PMOS used to isolate the negative voltage from the positive during the erase process. During the read and the programming process, in order to bias the non-selected word line to 0V, the bias voltage needs to be negative. On the other hand, the word line voltage of the selected word line

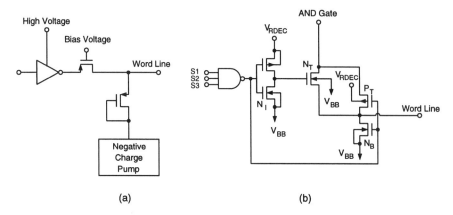

Fig. 5.110 Row decoder circuit in flash EEPROM. (Adapted from Umezawa et al.[62].)

needs to be high (10.5V). Therefore, the stress of the gate oxide in the PMOS device is large. As shown in Fig. 5.110(b), using an improved row decoder, this problem can be improved. During the program process, the gate voltage of P_T is 0V. Thus its maximum gate oxide stress is at 10.5V. During the erase operation, V_{RDEC} is set to ground and the AND gate is also set to ground. Thus the gate oxide stress in the erase P_T is at 10V—no thicker oxide is required.

Fig. 5.111 shows the negative charge pump circuit[62]. As shown in the figure, when F3 switches from high to low, V_1 is boosted to a negative voltage. At this time, node A stores the ground level. When F2 becomes low, node A is also boosted to a negative voltage. As a result, the PMOS device is turned on and V_1 is set to the ground level, which serves as the high level when F3 becomes low. When F3 becomes low, V_1 is boosted to negative. Then, node B is boosted to more negative when F4 becomes low. V_1 is passed to V_2. When F1 becomes low, V_2 is boosted to a voltage more negative than V_1. At the same time, node B is pulled to V_1. Thus, via each stage, the voltage can be boosted to more negative. By this arrangement, via cascading identical negative boost circuits, a designated negative bias can be obtained. Fig. 5.112 shows the programming circuit and the positive charge pump circuit[62]. As shown in the figure, a positive charge pump is used to realize the 10.5V supply voltage to the drive the control gate of the memory cell, the column selector, and the Din buffer. By this arrangement, the voltage for the drain can be 5V. The positive charge pump circuit has an identical operation principle except that phases F1–F4 are complementary to those in the negative charge pump. The PMOS devices in the negative charge pump are replaced by the NMOS devices. Ground is replaced by V_{DD}. Thus V_{pcp} can be pumped to a designated positive voltage by cascading the pump circuits. Fig. 5.113 shows the read circuit used in the flash EEPROM, which is similar to the read circuit used in the EPROM.

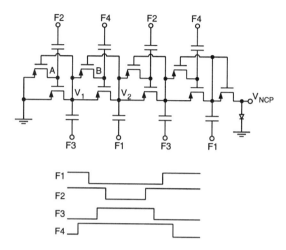

Fig. 5.111 Negative charge pump circuit in flash EEPROM. (Adapted from Umezawa et al.[62].)

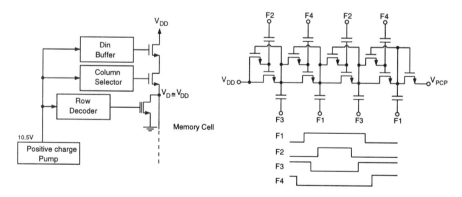

Fig. 5.112 Programming circuit and positive charge pump circuit for flash memory. (Adapted from Umezawa et al. [62].)

324   CMOS MEMORY

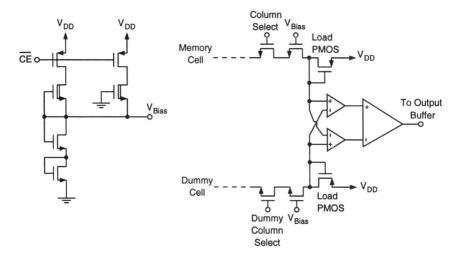

Fig. 5.113 Read circuit in flash EEPROM. (Adapted from Umezawa et al.[62].)

Fig. 5.114 shows another flash EEPROM cell with the sector erase mode[63]. In the previous flash EEPROM cell, the erase process is done by imposing a positive voltage at the source and a negative voltage at the control gate. Based on this structure, band-to-band tunneling current may exist—the erase current is large. As shown in the figure, the substrate is biased at a positive voltage, instead of the source voltage. By biasing the gate at −13V and leaving the drain and the source open, the electrons in the floating gate tunnel through the gate oxide to the substrate via FN tunneling. At the drain end, a diffusion self-aligned (DSA) structure using a large tilt-angle boron implant is used to increase the injection efficiency of CHE such that the programming time can be reduced. During the erase operation, p-well is biased at +5V. Thus, an additional deep n-well is needed to provide the isolation—the triple-well structure.

Fig. 5.115 shows the cross-section of the divided bit line NOR (DINOR) flash EEPROM cell[64]. As shown in the figure, eight memory cells are connected via the OR structure. Via the select gate, it is connected to the main bit line. The structure is similar to the EEPROM. In EEPROM, the select gate controls a cell. In DINOR, it controls eight cells. Thus the bit line is divided—erasure of the memory cells in the specific region can be reached. Fig. 5.116 shows the voltage conditions of the DINOR cell for the erase and the programming operations[64]. During the erase operation, the selected cell shifts to the high-threshold voltage since the electrons tunnel from the inverted channel to the floating gate via FN tunneling. Note that the definition of erase is opposite to that defined before. The p-well is biased at −8V. For the selected sector to erase, a 10V bias is applied at the gates of the internal cell. For the unselected sectors, the gates of the internal cells are biased to 0V and their sources are biased at −4V to avoid tunneling. During the program process, the drain voltage of the cell is set to 5V/0V and the control gate voltage to −8V, and the source is open. The p-well is biased at 0V. Via tunneling from the floating gate to drain

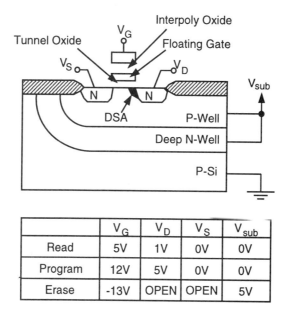

Fig. 5.114 A flash EEPROM cell with sector erase mode. (Adapted from Jinbo et al. [63].)

during the program process, their threshold voltage of the selected cells falls back. For the unselected cells, their control gate voltage is 0V. As shown in the figure, by setting the voltage of the bit line to 0V or 5V, the logic value to program in a selected cell can be done. Fig. 5.117 shows the memory array organization of the DINOR cell[64]. As shown in the figure, 8-bit sub-BLs are connected to the main BLs (MBL0–MBL511). Via the SL driver, a whole block of 1Kbyte (8K bits) can be erased.

Fig. 5.118 shows the cell arrays in flash memory[65]. As shown in the figure, NOR structure is the most common one. Metal bit lines are connected to the drains of the memory cells—it needs a large contact area. The advantage of the NOR structure

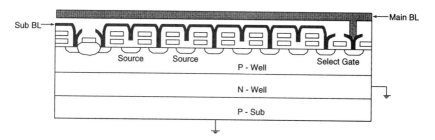

Fig. 5.115 Cross-section of the divided bit-line NOR (DINOR) flash EEPROM cell. (Adapted from Kobayashi et al. [64].)

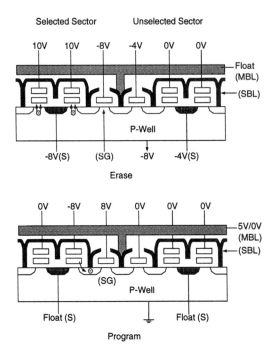

Fig. 5.116 Voltage conditions of the DINOR cell during the erase and the programming operations. (Adapted from Kobayashi et al. [64].)

are its small parasitics and high speed. The second structure is the virtual-ground structure, where diffusion area has been used as the connecting layer among cells before connecting to the bit line and to ground. Its strength is based on its density. However, parasitic effects may be more serious. The third and fourth approaches are the AND/NAND structures. Both structures are alike. In the AND structure, cells are connected via OR as the DINOR structure just described. In the NAND structure, they are connected via AND. The NAND structure has its strength in its high density. In addition, page program and page erase via BDS, BSS, and control gate can be obtained for the NAND structure. During the erase process, in the NAND structure, the memory cells are over-erased to become depletion-mode ($V_{TH} < 0$). During the program process, they are programmed to $V_{TH} > 0$. During the read access, the unselected gate is biased at $V_G > V_{TH}$ (if V_{TH}=3V, V_G=4.5V) and the select gate is biased at 0V. Thus all unselected gates turn on, regardless of V_{TH} (3V or <0V). The current in the serial path of the NAND path is determined by the select transistor. If V_{TH} is positive, there is no current. When V_{TH} is negative, it is with current. The drawback of the NAND structure is the long serial path, which may restrict the high-speed operation. Therefore, the bit-line parasitics and/or the voltage swing should be minimized.

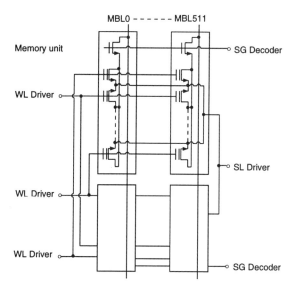

Fig. 5.117 Memory array organization of 3V 4Mb flash EEPROM. (Adapted from Kobayashi et al. [64].)

Fig. 5.119 shows the dynamic bandgap generator with pumping power supply and triple-well bipolar transistor[66]. As shown in the figure, the output of the charge pump circuit (V_p) is the non-regulated supply voltage. When $S1$ is high, $N1$ falls from V_p to 0V—the PMOS power switch transistor is turned on. Therefore, the bandgap reference voltage generator functions normally. The bandgap reference voltage generator is a simplified Brokaw's circuit[67] or Widlar's circuit[68]. $Switch2$ and CR act as a sample-and-hold circuit. When $S2$ is high, $Switch2$ turns on. Therefore, V_{RF} samples the bandgap reference output. When $S2$ becomes low, $Switch2$ turns off. Thus V_{RF} is held by CR. Fig. 5.120 shows the voltage doubler with the charge pump circuit for high-voltage generation at V_{DD}[66]. As shown in the figure, the charge pump cell is made of two diode-connected NMOS devices and two capacitors controlled by two non-overlapping phases F1 and F2. When F1 is low, node A is charged to $V_{DD} - V_T$. When F1 rises to V_{DD}, node A is boosted to $2V_{DD} - V_T$ and node B becomes $2V_{DD} - 2V_T$ (F2 is low). When F2 becomes high, node B is boosted to $3V_{DD} - V_T$, and node D becomes $3V_{DD} - 3V_T$. By cascading the charge pump circuits described above, the designated V_p can be obtained. In the voltage doubler, when S3 is high, node N2 is grounded, and node N0 is V_p. Thus C1 has the $C1 \cdot V_p$ amount of charge. At the same time, S1 is low. When S1 is high, via the bootstrap circuit, node N2 rises to V_p and node N0 is boosted to $2V_p$. Via the diode-connected NMOS device, V_H is equal to $2V_p - V_T$. The operation of S2 and S4 is identical to that of S1 and S3. Only the boosted nodes are replaced by N1 instead of N2. A negative voltage can be generated using this circuit. Replacing NMOS devices with the PMOS devices, and V_{DD} with ground, the negative charge pump circuit can be obtained.

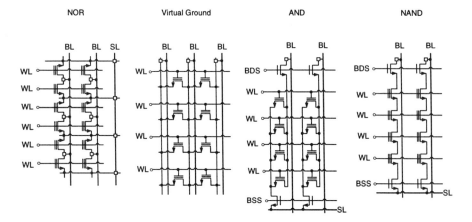

Fig. 5.118 Cell arrays in flash memory. (Adapted from Kume [65].)

	FRAM	Flash
Endurance	> 10^{12}	10^6
Data Transfer Speed	5MB/s	100kB/s
Write Time	1	10
Erase Time	—	10ms/blk
Write Voltage	2V	12V

Table 5.1 Comparison between FRAM and flash memory. (Adapted from Sumi [69].)

5.6 FERROELECTRIC RAM (FRAM)

FRAM (or FeRAM) uses ferroelectric material to realize the dielectric for the memory cell capacitor as for DRAM. Therefore, the structure of FRAM is similar to that of one-transistor-one-capacitor (1T1C) DRAM. Different from DRAM, FRAM is non-volatile. Thus, when power is off, it still retains data. FRAM has advantages in high density, high speed, and low-voltage operation. Compared to the conventional non-volatile memories such as flash memory, no internally produced high voltages are required for the programming and the erase operations of the memory cell in FRAM. In addition, no consecutive cycles as for flash memory are needed for the program and erase operations of FRAM. Therefore, FRAM is a promising non-volatile memory device for future low-voltage VLSI non-volatile memory applications.

Table 5.1 lists comparisons between FRAM and flash memory. As listed in the table, the endurance and the data-transfer speed of the FRAM are better. In addition, FRAM does not need an additional erase cycle and possesses the random access

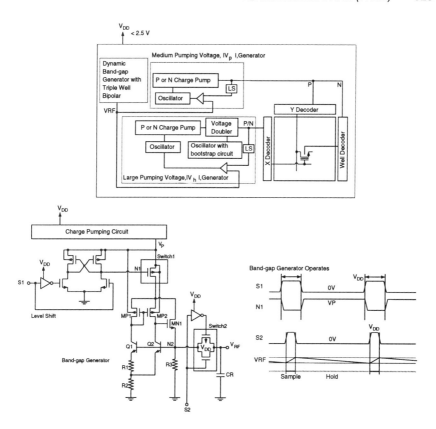

Fig. 5.119 Dynamic bandgap generator with pumping power supply and triple-well bipolar transistor. (From Kawahara et al. [66], ©1996 IEEE.)

property. The write voltage of FRAM is much lower than that of the flash memory, which is especially suitable for low-voltage operation. Therefore, FRAM is suitable for high-capacity, high-speed/high-performance, low-voltage non-volatile memory applications. Fig. 5.121 shows the lattice structure of the ferroelectric material used in the FRAM capacitor[70]. As shown in the figure, the generic molecular structure is ABO_3, for example $(Ba, Sr)TiO_3$. Using the ferroelectric material as the nonlinear dielectric, the capacitor is called ferroelectric capacitor, which has the property of permanent charge retention. In a ferroelectric capacitor, when an external electric field is imposed, its internal B atom in the lattice structure may move upward or downward to become an electrical dipole. The net ionic displacement causes the permanent charge. When the external electric field is removed, the majority of the domains remain poled in the direction of the applied electric field. At the same time, compensating charge is needed to remain on the plates of the capacitor. This compensating charge causes a hysteresis in the charge versus voltage plot of the ferroelectric capacitor as shown in Fig. 5.122[70]. If the applied voltage to the ferroelectric capacitor is reversed, the remanent domains change and require compensating charge

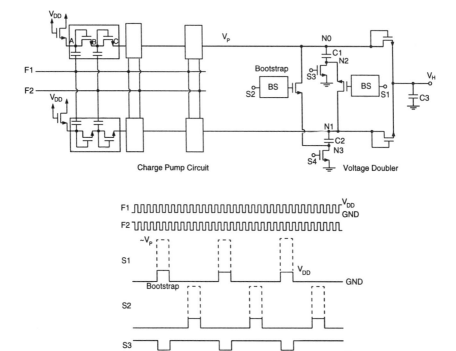

Fig. 5.120 Voltage doubler with the charge pump circuit for high-voltage generation at V_{DD}. (From Kawahara et al. [66], ©1996 IEEE.)

flowing to the capacitor plates. If the voltage is in the same direction, no change will be caused. At the same time, a smaller amount of charge flows to the capacitor. As shown in the figure, when a positive voltage ($> V_{sat}$) is imposed on the ferroelectric capacitor at the operating point C, the maximum change in the capacitor charge is $\Delta Q = Q_{sat} - Q_0$, where Q_0 is the charge at point C. If a negative voltage ($< -V_{sat}$) is imposed on the ferroelectric capacitor at point C, its polarity will reverse. At this time, $\Delta Q = Q_{sat} + Q_0$, which is larger than the positive charge difference. When the externally imposed voltage is zero, it will go to point D, instead of C. In order to return to C, an externally imposed positive voltage greater than $> V_{sat}$ is required. At point C, when the imposed voltage is larger than $> V_{sat}$, the charge remains at Q_{sat}. In addition, its state will not be changed. Therefore, for a ferroelectric memory cell, by imposing a positive voltage ($0 < V < V_{sat}$) on the capacitor, depending on the initial operating point, the charge needed to change its operating voltage to V ($V < V_{sat}$) varies. This implies the capacitance of the ferroelectric capacitor changes according to its state, which can be perceived by $C = \frac{dQ}{dV}$ in Fig. 5.122. Note that this externally imposed charge is used to change its polarity, which will not disappear when an externally imposed electric field disappears.

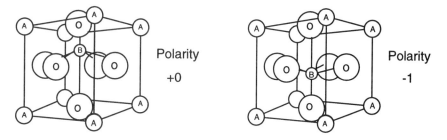

Fig. 5.121 Lattice structure of the ferroelectric material ABO_3. (Adapted from Evans and Womack [70].)

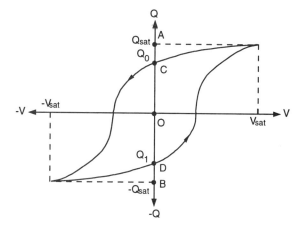

Fig. 5.122 Hysteresis of the ferroelectric capacitor. (Adapted from Evans and Womack [70].)

The storage of data in a FRAM cell is based on the difference in the capacitance, which is different from the difference in charge in a DRAM. When the externally imposed voltage is positive, the capacitance at C is smaller than that at D. Therefore, for two ferroelectric capacitors with different stored capacitances, their related charge/discharge currents vary. By sensing the voltage difference between the two ferroelectric capacitors with different capacitances, the stored logic values can be detected. Initially, when the external bias is not imposed, the ferroelectric capacitor, which does not have any polarity, is at point O. When a $V > V_{sat}$ external bias is imposed, it has a polarity. After the imposed bias of $V > V_{sat}$ is removed ($V = 0$), it is at point C (write 0). Similarly, after an imposed bias of $V < -V_{sat}$ is removed, at a bias of $V = 0$, it is at point D (write 1). When an imposed bias of $-V_{sat} < V < V_{sat}$ is removed ($V = 0$), it is at point O. In DRAM, the stored charge in the memory cell capacitor is used to identify logic-0 or logic-1. When the stored charge is lost, data are lost—refresh is required. In the ferroelectric memory, the stored logic value is identified by detecting its capacitance value. Therefore, even the top and the bottom plates of the ferroelectric capacitor are shorted, its remanent charge is not affected (still at C

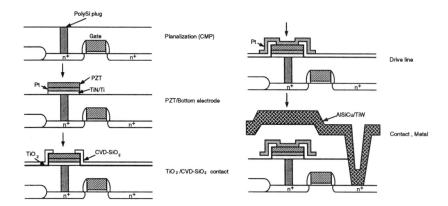

Fig. 5.123 FRAM 1T/1C stacked capacitor fabrication process. (From Onishi et al. [72], ©1994 IEEE.)

or D). Hence, its stored value will not be lost even after power off. From the hysteresis curve, for a good ferroelectric capacitor, the width of the hysteresis should be as narrow as possible—low-voltage operation can be accommodated. In addition, the height of the hysteresis should be as large as possible—a large capacitance difference.

Fig. 5.123 shows the FRAM 1T/1C stacked capacitor fabrication process[72]. After the bottom MOS access transistor is done, contact holes are filled with a layer of n^+-doped polysilicon, followed by planarization of the surface by CMP (chemical mechanical polishing). Then, a layer of Pt/TiN/Ti ($0.1/0.2/0.3 \mu$m) is deposited by DC magnetron sputtering, followed by the PZT ($Pb(Zr_{1-x}Ti_x)O_3$) film coated by the solgel method. After patterning the PZT/Pt/TiN/Ti layers, TiO_2 and SiO_2 films are deposited, where TiO_2 is used as the diffusion barrier for the inter-diffusion between SiO_2 and TiO_2 during anneal. After the contact-hole opening step, a layer of $Ti(0.02 \mu m)/Pt(0.1 \mu m)$ is deposited for the capacitor plate electrode. Bit line contact opening and formation of the $AlSiCu/TiW$ interconnects finalize the fabrication process. From the cross-section, FRAM capacitor looks like a DRAM stacked capacitor cell except that ferroelectric material is used as the dielectric material.

Fig. 5.124 shows the cell-plate-driven read scheme used in FRAM[69]. As shown in the figure, it is composed of the memory cells and the reference cells. During the read operation, the bit lines are pre-discharged to ground. Then, the cell plate voltage is raised. The access transistors are turned on by WL0 (WL1) and RWL0 (RWL1). Due to the difference in the capacitances, the rise in the voltage when charging the bit lines is different. By sensing the voltage difference, the logic value of the stored data can be detected. Due to the large cell plate capacitance, the delay in the pull-up of the cell plates connected to the bit lines is the bottleneck of the speed performance. Fig. 5.125 shows the non-cell-plate-driven read scheme in FRAM[74]. As shown in the figure, during the read process, the cell plates are set to $\frac{V_{DD}}{2}$ and the bit lines are predischarged to ground. When the word line of the access transistor

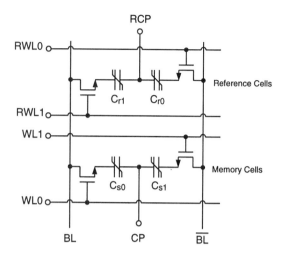

Fig. 5.124 Cell-plate-driven read scheme in FRAM. (Adapted from Hirano et al. [73].)

is high, bit lines will be charged up. Depending on the voltage difference during charge-up, the capacitance of the ferroelectric capacitor can be detected. After read, the bit lines are equalized to $\frac{V_{DD}}{2}$ such that there is no voltage drop at the capacitor. The disadvantage of this approach is that due to the junction leakage, during the read operation, the capacitor plate is not at $\frac{V_{DD}}{2}$. Therefore, the accuracy of the read operation can be affected. In addition, since the cell plate voltage is at $\frac{V_{DD}}{2}$, when the stored data need to be changed, a large voltage is needed—not suitable for low-voltage operation. Fig. 5.126 shows the FRAM memory cell and the peripheral circuit with bit line-driven read scheme[73]. As shown in the figure, T1–T3 are the read cycles. At T1, the bit lines are precharged to high ($\overline{DBLP}$ is low). At T2, the word line is high and the cell plates are at 0V. Thus the selected capacitor and the reference capacitors are connected to the bit lines—the voltage at the bit lines may change since charge sharing between the bit lines and the memory capacitor occurs. The voltage difference between the bit lines is sensed and amplified by the sense amp at T3 (SAE is high). If the selected capacitor stores logic 1, the voltage drop at the capacitor is at its peak. If the selected capacitor stores logic 0, there is no voltage drop at the capacitor. T4–T6 are the rewrite cycle. At T4, CP is high, therefore, for the capacitor storing logic 1 (bit line is high) there is has no voltage drop at the capacitor. For the capacitor storing logic 0, a negative voltage drop exists—it is rewritten to the most negative state. At T5, CP is low. The capacitor storing logic 1 has a positive voltage drop—it is rewritten to the most positive state. At this time, the reference cell should be rewritten to its reference state at the origin—after T4, RWLn is zero. At the same time, RRC is high. Thus one plate of the reference capacitance is set to 0. Applying a voltage of V_{RCP} at the other plate, the reference capacitor is biased at $-V_{RCP}$, which is used for the reference capacitor to return its state to origin when it is biased at 0V. At T6, the read and the rewrite processes are completed.

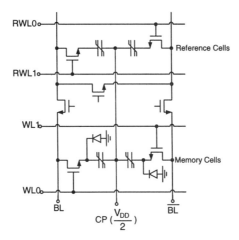

Fig. 5.125 Non-cell-plate-driven read scheme in FRAM. (Adapted from Hirano et al. [73].)

The write process is easier. T7 is for write logic 0 and T8 is for write logic 1. At T7, CP is high. If the input bit line is high, no voltage drop exists since there is no voltage change. When the input bit line is low, the cell capacitor is biased at a negative voltage. Therefore, it is written to the most negative voltage drop. At T8, CP is low. If the input bit line is high, the cell capacitance is biased at the most positive state. Thus it is rewritten to the most positive state. During the write cycle, the reference cell remains idle.

The flash memory NAND structure mentioned in the previous section can be used in the FRAM design. Fig. 5.127 shows the concept of the chain FRAM (CFRAM)[75]. As shown in the figure, the chain FRAM structure is based on 1T1C cell, which is different from the DRAM cell. In the FRAM cell, the function of the transistor is to short the ferroelectric capacitor for holding data. Since the bit line is connected to the block select transistor only, layout density can be high and the bit line load can be decreased. The read procedure of FRAM is similar to the flash memory with the NAND structure. When the FRAM circuit enters precharge/equalize period, the bit line is pre-discharged to ground and the word lines are high. When all access transistors turn on, the ferroelectric capacitor is shorted with the two electrodes equalized to $\frac{V_{DD}}{2}$. At this time, the block select transistor is off. When accessing the cell, the word line of the accessed cell turns from high to low—the access transistor turns off. Therefore, the charge in the accessed ferroelectric capacitor redistributes with the bit line capacitance. Due to the applied negative bias, the accessed ferroelectric capacitor storing logic-1 state releases a large amount of charge—a large capacitance. The accessed ferroelectric capacitor storing logic-0 state releases a smaller amount of charge—a smaller capacitance. As a result, the voltage on the bit line after charge redistribution between the ferroelectric capacitance and the bit line parasitic capacitance changes. The difference between the access bit line voltage and the reference bit line voltage is amplified by the sense/restore amplifier and restored to

FERROELECTRIC RAM (FRAM)

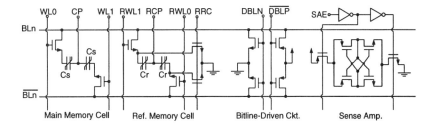

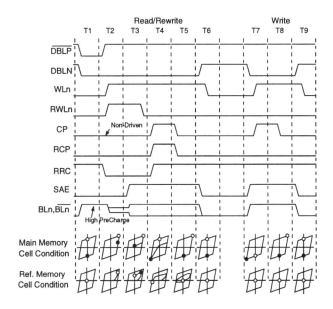

Fig. 5.126 FRAM memory cell and the peripheral circuit with bit line-driven read scheme. (From Hirano et al. [73], ©1997 IEEE.)

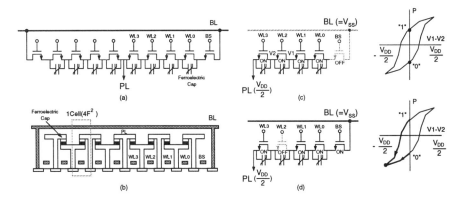

Fig. 5.127 Concept of the chain FRAM (CFRAM). (Adapted from Takashima et al. [75].)

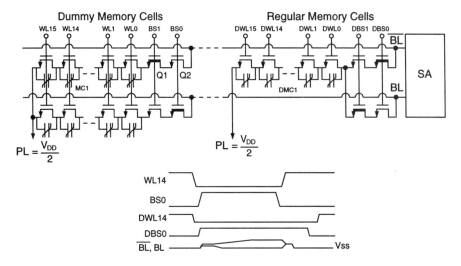

Fig. 5.128 Memory cell architecture of chain FRAM. (From Takashima[75], ©1998 IEEE.)

full-swing for writing back. When the read operation of the circuit is accomplished, the word line switches back to high (precharge/equalize period). At this time, the two electrodes of the ferroelectric capacitor are shorted. Thus the operating point of the ferroelectric capacitor goes back to the logic-1/logic-0 state as shown in the figure for holding the data. The advantage of this scheme is that charge/discharge of the cell plate voltage is not necessary. In addition, data hold is easily achieved. Density is high. Circuit structure and operation are simple. Fig. 5.128 shows the memory cell architecture of chain FRAM[75]. As shown in the figure, dummy cell is always stored with the logic-0 state. The capacitance of the dummy cell is much bigger than that of the regular memory cell. Therefore, after redistribution with the bit line, its voltage is between the high voltage of the regular memory cell and the low voltage. For each access, 16 memory cells are connected to the bit line via the block select transistor. For various memory cell accesses, due to various parasitic capacitances at the connected bit lines (largest for the $WL15$ access; smallest for the $WL0$ access), structure and operation should be identical for both the dummy and the regular memory cells such that a common-mode for the identical change can be obtained. The write logic-0 operation for the dummy cell is carried out when the memory cell access is completed and BL and $\overline{BL}$ are discharged to ground. Note that at this time the regular memory cell enters the equalize period. In the circuit, depletion mode devices have been added for cancellation of coupling noise from the bit lines due to the BS and DBS signals.

5.7 SUMMARY

In this chapter, volatile and non-volatile memory circuits are described. Starting from the memory cell, decoder, bit line related architecture, and sense amp for SRAM have been described. Then the critical path analysis and the application specific SRAM have been presented. Low-voltage circuit techniques for realizing DRAM and SRAM have been analyzed. In addition, BiCMOS SRAM and DRAM have been analyzed. Recently, SOI CMOS technology has been used to integrate SRAM and DRAM. SOI DRAM and SRAM, with strengths and weaknesses, have been described. The floating body problem and its circuit solutions for integrating SRAM and DRAM have been explained. In the next portion of this chapter, non-volatile memories have been described. Starting from ROM, EPROM and EEPROM have been introduced. Then, flash memory, which combines the advantages of the EEPROM and EPROM, has been explained. In the final portion of the this chapter, a recently emerged non-volatile ferroelectric RAM (FRAM) has been introduced. The data storage mechanism in the FRAM cell is based on the difference in the capacitance instead of the charge as for the DRAM. The operation of FRAM has also been described in this chapter.

REFERENCES

1. C.-W. Chen, J.-P. Peng, M.-Y. S. Shyu, M. Amundson, and J. C. Yu, "A Fast 32K × 8 CMOS Static RAM with Address Tansition Detection," *IEEE J. Sol. St. Ckts.*, **22**(4), 533–537 (1987).

2. S. Hayakawa, M. Kakumu, H. Takeuchi, K. Sato, T. Ohtani, T. Yoshida, T. Nakayama, S. Morita, M. Kinugawa, K. Maeguchi, K. Ochii, J. Matsunaga, A. Aono, K. Noguchi, and T. Asami, "A 1μA Retention 4Mb SRAM with a Thin-Film-Transistor Load Cell," *ISSCC Dig.*, 128–129 (1990).

3. K. Sasaki, K. Ishibashi, K. Shimohigashi, T. Yamanaka, N. Moriwaki, S. Honjo, S. Ikeda, A. Koike, S. Meguro, and O. Minato, "A 23-ns 4-Mb CMOS SRAM with 0.2-μA Standby Current," *IEEE J. Sol. St. Ckts.*, **25**(5), 1075–1081 (1990).

4. E. Seevinck, F. J. List, and J. Lohstroh, "Static-Noise Margin Analysis of MOS SRAM Cells," *IEEE J. Sol. St. Ckts.*, **22**(5), 748–754 (1987).

5. J. Lohstroh, E. Seevinck, and J. De Groot, "Worst-Case Static Noise Margin Criteria for Logic Circuits and Their Mathematical Equivalence," *IEEE J. Sol. St. Ckts.*, **18**(6), 803–807 (1983).

6. M. Kubo, I. Masuda, K. Miyata, and K. Ogiue, "Perspective on BiCMOS VLSI's," *IEEE J. Sol. St. Ckts.*, **23**(1), 5–11 (1988).

7. M. Matsumiya, S. Kawashima, M. Sakata, M. Ookura, T. Miyabo, T. Koga, K. Itabashi, K. Mizutani, H. Shimada, and N. Suzuki, "A 15-ns 16-Mb CMOS

SRAM with Interdigitated Bit-Line Architecture," *IEEE J. Sol. St. Ckts.*, **27**(11), 1497–1503 (1992).

8. T. Seki, E. Itoh, C. Furukawa, I. Maeno, T. Ozawa, H. Sano, and N. Suzuki, "A 6-ns 1-Mb CMOS SRAM with Latched Sense Amplifier," *IEEE J. Sol. St. Ckts.*, **28**(4), 478–483 (1993).

9. K. Sasaki, S. Hanamura, K. Ueda, T. Oono, O. Minato, Y. Sakai, S. Meguro, M. Tsunematsu, T. Masuhara, M. Kubotera, and H. Toyoshima, "A 15-ns 1-Mbit CMOS SRAM," *IEEE J. Sol. St. Ckts.*, **23**(5), 1067–1072 (1988).

10. K. Seno, K. Knorpp, L.-L. Shu, N. Teshima, H. Kihara, H. Sato, F. Miyaji, M. Takeda, M. Sasaki, Y. Tomo, P. T. Chuang, and K. Kobayashi, "A 9-ns 16-Mb CMOS SRAM with Offset-Compensated Current Sense Amplifier," *IEEE J. Sol. St. Ckts.*, **28**(11), 1119–1124 (1993).

11. M. Yoshimoto, K. Anami, H. Shinohara, T. Yoshihara, H. Takagi, S. Nagao, S. Kayano, and T. Nakano, "A Divided Word-Line Structure in the Static RAM and Its Application to a 64K Full CMOS RAM," *IEEE J. Sol. St. Ckts.*, **18**(5), 479–485 (1983).

12. T. Hirose, H. Kuriyama, S. Murakami, K. Yuzuriha, T. Mukai, K. Tsutsumi, Y. Nishimura, Y. Kohno, and K. Anami, "A 20-ns 4-Mb CMOS SRAM with Hierarchical Word Decoding Architecture," *IEEE J. Sol. St. Ckts.*, **25**(5), 1068–1074 (1990).

13. J. H. Lou and K. W. Su, "Design of an SRAM," *Final project report*, CMOS digital IC engineering course, NTUEE (1996).

14. K. Sasaki, K. Ueda, K. Takasugi, H. Toyoshima, K. Ishibashi, T. Yamanaka, N. Hashimoto, and N. Ohki, "A 16-Mb CMOS SRAM with a 2.3-μm^2 Single-Bit-Line Memory Cell," *IEEE J. Sol. St. Ckts.*, **28**(11), 1125–1130 (1993).

15. N. Kushiyama, C. Tan, R. Clark, J. Lin, F. Perner, L. Martin, M. Leonard, G. Coussens, and K. Cham, "An Experimental 295MHz CMOS 4K× 256 SRAM Using Bidirectional Read/Write Shared Sense Amps and Self-Timed Pulsed Word-Line Drivers," *IEEE J. Sol. St. Ckts.*, **30**(11), 1286–1290 (1995).

16. K. Ishibashi, K. Takasugi, T. Yamanaka, T. Hashimoto, and K. Sasaki, "A 1-V TFT-Load SRAM Using a Two-Step Word-Voltage Method," *IEEE J. Sol. St. Ckts.*, **27**(11), 1519–1524 (1992).

17. H. Mizuno and T. Nagano, "Driving Source-Line (DSL) Cell Architecture for Sub-1-V High-Speed Low-Power Applications," *Symp. VLSI Ckts Dig.*, 25–26 (1995).

18. H. Mizuno and T. Nagano, "Driving Source-Line Cell Architecture for Sub-1-V High-Speed Low-Power Applications," *IEEE J. Sol. St. Ckts.*, **31**(4), 552–557 (1996).

19. K. Itoh, "Trends in Megabit DRAM Circuit Design," *IEEE J. Sol. St. Ckts.*, **25**(3), 778–789 (1990).

20. T. C. May and M. H. Woods, "Alpha-Particle-Induced Soft Errors in Dynamic Memories," *IEEE Trans. Elec. Dev*, **26**(1), 2–9 (1979).

21. K. Itoh, K. Sasaki, and Y. Nakagome, "Trends in Low-Power RAM Circuit Technologies," *IEEE Proc.*, **83**(4), 524–543 (1995).

22. H. Hidaka, K. Fujishima, Y. Matsuda, M. Asakura, and T. Yoshihara, "Twisted Bit-Line Architectures for Multi-Megabit DRAM's," *IEEE J. Sol. St. Ckts.*, **24**(1), 21–27 (1989).

23. K. Itoh, Y. Nakagome, S. Kimura, and T. Watanabe, "Limitations and Challenges of Multigigabit DRAM Chip Design," *IEEE J. Sol. St. Ckts.*, **32**(5), 624–634 (1997).

24. T. Watanabe, R. Fujita, K. Yanagisawa, H. Tanaka, K. Ayukawa, M. Soga, Y. Tanaka, Y. Sugie, and Y. Nakagome, "A Modular Architecture for a 6.4-Gbyte/s, 8-Mb DRAM-Integrated Media Chip," *IEEE J. Sol. St. Ckts.*, **32**(5), 635–641 (1997).

25. Y. Takai, M. Nagase, M. Kitamura, Y. Koshikawa, N. Yoshida, Y. Kobayashi, T. Obara, Y. Fukuzo, and H. Watanabe, "250 Mbyte/s Synchronous DRAM Using a 3-Stage-Pipelined Architecture," *IEEE J. Sol. St. Ckts.*, **29**(4), 426–431 (1994).

26. M. Horiguchi, T. Sakata, and K. Itoh, "Switched-Source-Impedance CMOS Circuit for Low Standby Subthreshold Current Giga-Scale LSI's," *IEEE J. Sol. St. Ckts.*, **28**(11), 1131–1135 (1993).

27. T. Yamagata, S. Tomishima, M. Tsukude, T. Tsuruda, Y. Hashizume, and K. Arimoto, "Low Voltage Circuit Design Techniques for Battery-Operated and/or Giga-Scale DRAM's" *IEEE J. Sol. St. Ckts.*, **30**(11), 1183–1188 (1995).

28. Y. Nakagome, K. Itoh, K. Takeuchi, E. Kume, H. Tanaka, M. Isoda, T. Musha, T. Kaga, T. Kisu, T. Nishida, Y. Kawamoto, and M. Aoki, "Circuit Techniques for 1.5-3.6-V Battery-Operated 64-Mb DRAM," *IEEE J. Sol. St. Ckts.*, **26**(7), 1003–1010 (1991).

29. D. Takashima, S. Watanabe, T. Fuse, K. Sunouchi, and T. Hara, "Low-Power On-Chip Supply Voltage Conversion Scheme for Ultrahigh-Density DRAM's," *IEEE J. Sol. St. Ckts.*, **28**(4), 504–509 (1993).

30. T. Ooishi, M. Asakura, S. Tomishima, H. Hidaka, K. Arimoto, and K. Fujishima, "A Well-Synchronized Sensing/Equalizing Method for Sub-1.0V Operating Advanced DRAMs," *Symp. VLSI Ckts Dig.*, 81–82 (1993).

31. T. Ooishi, M. Asakura, S. Tomishima, H. Hidaka, K. Arimoto, and K. Fujishima, "A Well-Synchronized Sensing/Equalizing Method for Sub-1.0-V Operating Advanced DRAM's," *IEEE J. Sol. St. Ckts.*, **29**(4), 432–440 (1994).

32. Y. Nakagome, H. Tanaka, K. Takeuchi, E. Kume, Y. Watanabe, T. Kaga, Y. Kawamoto, F. Murai, R. Izawa, D. Hisamoto, T. Kisu, T. Nishida, E. Takeda, and K. Itoh, "An Experimental 1.5-V 64-Mb DRAM," *IEEE J. Sol. St. Ckts.*, **26**(4), 465–472 (1991).

33. M. Tsukude, S. Kuge, T. Fujino, and K. Arimoto, "A 1.2- to 3.3-V Wide Voltage-Range/Low-Power DRAM with a Charge-Transfer Presensing Scheme," *IEEE J. Sol. St. Ckts.*, **32**(11), 1721–1727 (1997).

34. M. Takada, K. Nakamura, and T. Yamazaki, "High Speed Submicron BiCMOS Memory," *IEEE Trans. Elec. Dev*, **42**(3), 497–505 (1995).

35. M. Takada, K. Nakamura, T. Takeshima, K. Furuta, T. Yamazaki, K. Imai, S. Ohi, Y. Sekine, Y. Minato, and H. Kimoto, "A 5-ns 1-Mb ECL BiCMOS SRAM," *IEEE J. Sol. St. Ckts.*, **25**(5), 1057–1062 (1990).

36. H. Toyoshima, S. Kuhara, K. Takeda, K. Nakamura, H. Okamura, M. Takada, H. Suzuki, H. Yoshida, and T. Yamazaki, "A 6-ns, 1.5-V, 4-Mb BiCMOS SRAM," *IEEE J. Sol. St. Ckts.*, **31**(11), 1610–1617 (1996).

37. T. Kawahara, Y. Kawajiri, G. Kitsukawa, T. Nakagome, K. Sagara, Y. Kawamoto, T. Akiba, S. Kato, Y. Kawase, and K. Itoh, "A Circuit Technology for Sub-10ns ECL 4Mb BiCMOS DRAMs," *Symp. VLSI Ckts. Dig.*, 131–132 (1991).

38. T. Kawahara, Y. Kawajiri, G. Kitsukawa, Y. Nakagome, K. Sagara, Y. Kawamoto, T. Akiba, S. Kato, Y. Kawase, and K. Itoh, "A Circuit Technology for Sub-10-ns ECL 4-Mb BiCMOS DRAM's," *IEEE J. Sol. St. Ckts.*, **26**(11), 1530–1537 (1991).

39. G. Kitsukawa, K. Itoh, R. Hori, Y. Kawajiri, T. Watanabe, T. Kawahara, T. Matsumoto, and Y. Kobayashi, "A 1-Mbit BiCMOS DRAM Using Temperature-Compensation Circuit Techniques," *IEEE J. Sol. St. Ckts.*, **24**(3), 597–602 (1989).

40. T. Kawahara, Y. Kawajiri, G. Kitsukawa, K. Sagara, Y. Kawamoto, T. Akiba, S. Kato, Y. Kawase, and K. Itoh, "Deep-Submicrometer BiCMOS Circuit Technology for Sub-10-ns ECL 4-Mb DRAM's," *IEEE J. Sol. St. Ckts.*, **27**(4), 589–596 (1992).

41. T. Kawahara, G. Kitsukawa, H. Higuchi, Y. Kawajiri, T. Watanabe, K. Itoh, R. Hori, Y. Kobayashi, and T. Matsumoto, "Substrate Current Reduction Techniques for BiCMOS DRAM," *IEEE J. Sol. St. Ckts.*, **24**(5), 1381–1389 (1989).

42. G. Kitsukawa, K. Yanagisawa, Y. Kobayashi, Y. Kinoshita, T. Ohta, T. Udagawa, H. Miwa, H. Miyazawa, Y. Kawajiri, Y. Ouchi, H. Tsukada, T. Matsumoto, and K. Itoh, "A 23-ns 1-Mb BiCMOS DRAM," *IEEE J. Sol. St. Ckts.*, **25**(5), 1102–1111 (1990).

43. J. B. Kuang, S. Ratanaphanyarat, M. J. Saccamango, L. L.-C. Hsu, R. C. Flaker, L. F. Wagner, S.-F. S. Chu, and G. G. Shahidi, "SRAM Bitline Circuits on PD SOI: Advantages and Concerns," *IEEE J. Sol. St. Ckts.*, **32**(6), 837–844 (1997).

44. T. Nishimura, Y. Inoue, K. Sugahara, M. Nakaya, Y. Horiba, and Y. Akasaka, "A Three Dimensional Static RAM," *Symp. VLSI Tech.*, 30–31 (1985).

45. Y. Takao, H. Shimada, N. Suzuki, Y. Matsukawa, and N. Sasaki, "Low-Power and High-Stability SRAM Technology Using a Laser-Recrystallized p-Channel SOI MOSFET," *IEEE Trans. Elec. Dev*, **39**(9), 2147–2152 (1992).

46. Y. Yamaguchi and Y. Inoue, "SOI DRAM: Its Features and Possibility," *SOI Conf. Dig.*, 122–124 (1995).

47. K. Suma, T. Tsuruda, H. Hidaka, T. Eimori, T. Oashi, Y. Yamaguchi, T. Iwamatsu, M. Hirose, F. Morishita, K. Arimoto, K. Fujishima, Y. Inoue, T. Nishimura, and T. Yoshihara, "An SOI-DRAM with Wide Operating Voltage Range by CMOS/SIMOX Technology," *IEEE J. Sol. St. Ckts.*, **29**(11), 1323–1329 (1994).

48. S. Tomishima, F. Morishita, M. Tsukude, T. Yamagata, and K. Arimoto, "A Long Data Retention SOI-DRAM with the Body Refresh Function," *Symp. VLSI Ckts. Dig.*, 198–199 (1996).

49. K. Ohtake, K. Shirakawa, M. Koba, K. Awane, Y. Ohta, D. Azuma, and S. Miyata, "Triple Layered SOI Dynamic Memory," *IEDM Dig.*, 148–151 (1986).

50. S. Kuge, F. Morishita, T. Tsuruda, S. Tomishima, M. Tsukude, T. Yamagata, and K. Arimoto, "SOI-DRAM Circuit Technologies for Low Power High Speed Multigiga Scale Memories," *IEEE J. Sol. St. Ckts.*, **31**(4), 586–591 (1996).

51. K. Shimomura, H. Shimano, N. Sakashita, F. Okuda, T. Oashi, Y. Yamaguchi, T. Eimori, M. Inuishi, K. Arimoto, S. Maegawa, Y. Inoue, S. Komori, and K. Kyuma, "A 1-V 46-ns 16-Mb SOI-DRAM with Body Control Technqiue," *IEEE J. Sol. St. Ckts.*, **32**(11), 1712–1720 (1997).

52. D. Kahng and S. M. Sze, "A Floating Gate and Its Application to Memory Devices," *Bell Sys. Tech. J.*, **46**, 1288–1295 (1967).

53. N. Ohtsuka, S. Tanaka, J. Miyamoto, S. Saito, S. Atsumi, K. Imamiya, K. Yoshikawa, N. Matsukawa, S. Mori, N. Arai, T. Shinagawa, Y. Kaneko, J. Matsunaga, and T. Iizuka, "A 4-Mbit CMOS EPROM," *IEEE J. Sol. St. Ckts.*, **22**(5), 669–675 (1987).

54. M. Kuriyama, S. Atsumi, K. Imamiya, Y. Iyama, N. Matsukawa, H. Araki, K. Narita, K. Masuda, and S. Tanaka, "A 16-ns 1-Mb CMOS EPROM," *IEEE J. Sol. St. Ckts.*, **25**(5), 1141–1152 (1990).

55. A. Scheibe and W. Krauss, "A Two-Transistor SIMOS EAROM Cell," *IEEE J. Sol. St. Ckts.*, **15**(3), 353–357 (1980).

56. M. Lenzlinger and E. H. Snow, "Fowler-Nordheim Tunneling into Thermally Grown SiO_2," *J. Appl. Phys.*, **40**(1), 278–283 (1969).

57. G. Yaron, S. J. Prasad, M. S. Ebel, and B. M. K. Leong, "A 16K E^2PROM Employing New Array Architecture and Designed-In Reliability Features," *IEEE J. Sol. St. Ckts.*, **17**(5), 833–840 (1982).

58. J. Miyamoto, J. Tsujimoto, N. Matsukawa, S. Morita, K. Shinada, H. Nozawa, and T. Iizuka, "An Experimental 5-V-Only 256-kbit CMOS EEPROM with a High-Performance Single-Polysilicon Cell," *IEEE J. Sol. St. Ckts.*, **21**(5), 852–860 (1986).

59. G. Samachisa, C.-S. Su, Y.-S. Kao, G. Smarandoiu, C.-Y. M. Wang, T. Wong, and C. Hu, "A 128K Flash EEPROM Using Double-Polysilicon Technology," *IEEE J. Sol. St. Ckts.*, **22**(5), 676–683 (1987).

60. C. Kuo, M. Weidner, T. Toms, H. Choe, K.-M. Chang, A. Harwood, J. Jelemensky, and P. Smith, "A 512-kb Flash EEPROM Embedded in a 32-b Microcontroller," *IEEE J. Sol. St. Ckts.*, **27**(4), 574–582 (1992).

61. V. N. Kynett, A. Baker, M. L. Fandrich, G. P. Hoekstra, O. Jungroth, J. A. Kreifels, S. Wells, and M. D. Winston, "An In-System Reprogrammable 32K × 8 CMOS Flash Memory," *IEEE J. Sol. St. Ckts.*, **23**(5), 1157–1163 (1988).

62. A. Umezawa, S. Atsumi, M. Kuriyama, H. Banba, K. Imamiya, K. Naruke, S. Yamada, E. Obi, M. Oshikiri, T. Suzuki, and S. Tanaka, "A 5-V-Only Operation 0.6-μm Flash EEPROM with Row Decoder Scheme in Triple-Well Structure," *IEEE J. Sol. St. Ckts.*, **27**(11), 1540–1546 (1992).

63. T. Jinbo, H. Nakata, K. Hashimoto, T. Watanabe, K. Ninomiya, T. Urai, M. Koike, T. Sato, N. Kodama, K. Oyama, and T. Okazawa, "A 5-V-Only 16-Mb Flash Memory with Sector Erase Mode," *IEEE J. Sol. St. Ckts.*, **27**(11), 1547–1554 (1992).

64. S. Kobayashi, H. Nakai, Y. Kunori, T. Nakayama, Y. Miyawaki, Y. Terada, H. Onoda, N. Ajika, M. Hatanaka, H. Miyoshi, and T. Yoshihara, "Memory Array Architecture and Decoding Scheme for 3V Only Sector Erasable DINOR Flash Memory," *IEEE J. Sol. St. Ckts.*, **29**(4), 454–458 (1994).

65. H. Kume, "Flash Memory Technology," *Japan Appl. Physics (Oyobuturi)*, **65**(11), 1114–1124, (in Japanese, 1996).

66. T. Kawahara, T. Kobayashi, Y. Jyouno, S. Saeki, N. Miyamoto, T. Adachi, M. Kato, A. Sato, J. Yugami, H. Kume, and K. Kimura, "Bit-Line Clamped Sensing Multiplex and Accurate High Voltage Generator for Quarter-Micron Flash Memories," *IEEE J. Sol. St. Ckts.*, **31**(11), 1590–1600 (1996).

67. A. P. Brokaw, "A Simple Three-Terminal IC Bandgap Reference," *IEEE J. Sol. St. Ckts.*, **9**(6), 388–393 (1974).

68. R. J. Widlar, "New Developments in IC Voltage Regulators," *IEEE J. Sol. St. Ckts.*, **6**(1), 2–7 (1971).

69. T. Sumi, "Ferroelectric Nonvolatile Memory Technology," *IEICE Trans. Elec.*, **E79-C**(6), 812–818 (1996).

70. J. T. Evans and R. Womack, "An Experimental 512-bit Nonvolatile Memory with Ferroelectric Storage Cell," *IEEE J. Sol. St. Ckts.*, **23**(5), 1171–1175 (1988).

71. T. Kawakubo, K. Abe, S. Komatsu, K. Sano, N. Yanase, and H. Mochizuki, "Novel Ferroelectric Epitaxial (Ba,Sr)TiO$_3$ Capacitor for Deep Sub-Micron Memory Applications," *IEDM Dig.*, 695–698 (1996).

72. S. Onishi, K. Hamada, K. Ishihara, Y. Ito, S. Yokoyama, J. Kudo, and K. Sakiyama, "A Half-Micron Ferroelectric Memory Cell Technology with Stacked Capacitor Structure," *IEDM Dig.*, 843–846 (1994).

73. H. Hirano, T. Honda, N. Moriwaki, T. Nakakuma, A. Inoue, G. Nakane, S. Chaya, and T. Sumi, "2-V/100-ns 1T/1C Nonvolatile Ferroelectric Memory Architecture with Bitline-Driven Read Scheme and Nonrelaxation Reference Cell," *IEEE J. Sol. St. Ckts.*, **32**(5), 649–654 (1997).

74. H. Koike, T. Otsuki, T. Kimura, M. Fukuma, Y. Havashi, Y. Maejima, K. Amanuma, N. Tanade, T. Matsuki, S. Saito, T. Takeuchi, S. Kobayashi, T. Kunio, T. Hase, Y. Miyasaka, N. Shohata, and M. Takada, "A 60ns 1Mb Nonvolatile Ferroelectric Memory with Non-Driven Cell Plate Line Write/Read Scheme," *ISSCC Dig.*, 368–369 (1996).

75. D. Takashima and I. Kunishima, "High-Density Chain Ferroelectric Random Access Memory (Chain FRAM)," *IEEE J. Sol. St. Ckts.*, **33**(5), 787–792 (1998).

Problems

1. Consider the SRAM cell as shown Fig. 5.8 during the read access. Find out the noise margin using SPICE simulation. The equivalent circuit in Fig. 5.8 is neglecting the PMOS pull-up load. When the PMOS pull-up load is included in Fig. 5.8, what will be the equivalent circuit? What is the noise margin?

2. Find out the difference in the performance of the sense amp in Figs. 5.15–5.18 by circuit simulation. Which one has the top sensing speed? Which has the biggest gain? For low-voltage operation, which one is the most appropriate?

3. Consider the DRAM architecture as shown in Fig. 5.48. What is the key point in enhancing its performance? Compared to the synchronous SRAM, what is the difference in the performance? Which one is more appropriate for low-voltage operation?

4. What factors affect the initial voltage difference in the DRAM bit lines during the read cycle? When the supply voltage is lowered, what is the influence in the initial voltage difference in the bit lines during the read cycle?

5. What are the advantages of the SOI DRAM as compared to the bulk CMOS DRAM? What are the disadvantages?

6. What are the programming/erase mechanisms in the flash memory cell? What are the disadvantages of flash memory?

7. What is the difference between the FRAM and other non-volatile memory? What are the advantages?

6
CMOS VLSI Systems

In the previous chapters, CMOS technology, CMOS static and dynamic logic circuits, and CMOS memory circuits have been described. In this chapter, CMOS VLSI systems are described. Starting from the fundamental basic building blocks for VLSI systems, adder and multiplier circuits are described. Then, register files and cache memory are analyzed, followed by programmable logic arrays (PLA), and central processing unit circuits. Finally, VLSI systems realized by other CMOS technologies including BiCMOS and SOI are also described.

6.1 ADDERS

Adders are an important component in a CMOS VLSI system. In this section, adder circuits are described. Starting from the basic principles of the adder circuits, the key component in the adder circuits—the carry look-ahead (CLA) circuit is explained. Several kinds of adder circuits including the Manchester CLA, pass-transistor based CLA, pass-transistor based carry-select CLA, enhanced MODL adder, and carry-skip CLA are described one by one. In the final portion of this section, adders using parallel and pipelined structures are also presented.

6.1.1 Basic Principles

The most straightforward adder is the carry propagate/ripple adder as shown in Fig. 6.1, which is based on cascading of n 1-bit full adder cells. In each full adder, two inputs a_i and b_i and the carry-in signal c_{i-1} are used to produce the sum signal

346   CMOS VLSI SYSTEMS

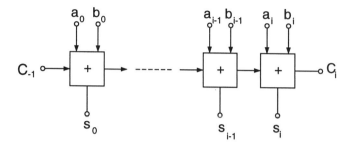

Fig. 6.1 Carry propagate/ripple adder.

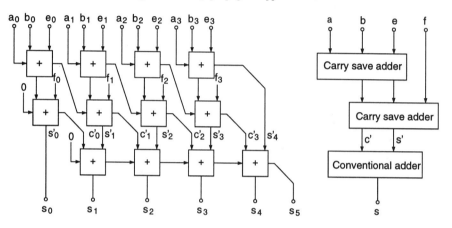

Fig. 6.2 Carry save adder.

s_i and the carry-out signal c_i:

$$s_i = a_i \oplus b_i \oplus c_{i-1},$$
$$c_i = a_i \cdot b_i + b_i \cdot c_{i-1} + a_i \cdot c_{i-1} = a_i \cdot b_i + c_{i-1} \cdot (a_i + b_i), \quad (6.1)$$

where $\oplus$ stands for the exclusive OR operation: $a \oplus b = a \cdot \bar{b} + \bar{a} \cdot b$. The carry-out signal of the current bit is used as the carry-in signal for the next-bit full adder cell. The operation of the full adder of the next bit will not be initiated until the carry-out signal of the current bit is obtained. As a result, speed of the carry propagate/ripple adder is determined by the propagation delay of the carry signals. If there are more than two numbers, for example, $a + b + c$ to be added, two carry propagate/ripple adders are required. Therefore, due to the propagation delay of the carry signals, the speed is slow. While handling the operation of an adder with more than two additions, the 1-bit full-adder cell can be utilized to for the carry save adder for avoiding a long delay in the carry propagation ripple. Fig. 6.2 shows the carry save adder for handling the addition of four numbers ($a + b + e + f$). As shown in the figure, there are three adders in three levels to do the operation of four numbers. In each carry save adder, the carry-out signal of the current bit at a level is not transferred to the next-bit adder of the same level as the carry-in signal. Instead, the carry-out signal is transferred to

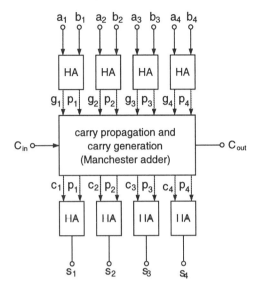

Fig. 6.3 Block diagram of a 4-bit carry look-ahead adder.

the next-bit adder in the lower level as the carry-in signal. In the top layer adder, three numbers can be added simultaneously. In addition to two numbers, the third number can be used as the carry-in signal. In each of the lower levels, one extra number can be added at a time. The adder in the bottom level is a conventional carry propagate adder, which is used to produce the final sum. Therefore, the propagation delay of the whole adder is equal to the delay of the two full-adder cells and the delay of the carry propagate adder. Compared to the carry propagate/ripple adder, the delay is much smaller, especially when many numbers are to be added. The carry save adder is especially effective for handling addition of many numbers. Therefore, the carry save adder is suitable for building multipliers and digital filters, where complicated additions are required.

6.1.2 Carry Look-Ahead Adder

From Eq. (6.1), the carry-out signal can be re-expressed as:

$$c_i = g_i + c_{i-1} \cdot p_i, \tag{6.2}$$

where $g_i = a_i \cdot b_i$ and $p_i = a_i + b_i$ are the generate and the propagate signals, respectively. Fig. 6.3 shows the block diagram of a 4-bit carry look-ahead adder, which is usually implemented by the pass-transistor logic or the multiple output domino logic (MODL) circuits. While using the carry look-ahead adder, there is a problem in the pass-transistor logic and the MODL circuits, which is caused by the situation when $g_i = 1$, $p_i = 1$ ($a_i = b_i = 1$), and $c_{i-1} = 0$. From a logic point of view, there is no problem. However, when realized using the pass-transistor logic or MODL gate using the pass transistors controlled by p_i, there is a sneak path problem—logic-1 and logic-0 blocks are short-circuited as described before may occur. Therefore, many designers redefine the propagate signal as $p_i = a_i \oplus b_i$ such that $a_i \oplus b_i$ and $a_i \cdot b_i$ are not 1 simultaneously. In addition, using the new definition for p_i, $c_i = g_i + p_i \cdot c_{i-1}$ is still valid. Besides, $s_i = p_i \oplus c_{i-1}$. Using the propagate and the generate signals, the carry signals do not need to be propagated bit by bit. The property of carry look-ahead can be obtained. Let's define an operator o as:

$$(g_i, p_i) \circ (g_j, p_j) = (G_i = g_i + p_i \cdot g_j, P_i = p_i \cdot p_j),$$
$$(g_i, p_i) \circ g_j = g_i + p_i \cdot g_j. \quad (6.3)$$

From the above equation, if c_{i-1} is used for g_j, one obtains $c_i = (p_i, g_i) \circ c_{i-1}$. The operator o has the following two properties: First,

$$(G_i, P_i) = \begin{cases} (g_1, p_1) & \text{if } i = 1, \\ (g_i, p_i) \circ (G_{i-1}, P_{i-1}) & \text{if } 2 \leq i \leq n. \end{cases} \quad (6.4)$$

Second, the operator o is referred to be associative as:

$$[(g_3, p_3) \circ (g_2, p_2)] \circ (g_1, p_1) = (g_3, p_3) \circ [(g_2, p_2) \circ (g_1, p_1)]. \quad (6.5)$$

From the above equations, the carry-out signal can be expressed as:

$$c_i = (G_i, P_i) \circ c_0 = (g_i, p_i) \circ (g_{i-1}, p_{i-1}) \circ \ldots \circ (g_1, p_1) \circ c_0, \quad (6.6)$$
$$= [(g_i, p_i) \circ (g_{i-1}, p_{i-1}) \circ \ldots \circ (g_j, p_j)] \circ (g_{j-1}, p_{j-1}) \circ \ldots \circ (g_1, p_1) \circ c_0.$$

Thus, we may define the group generate and the group propagate signals from jth bit to ith bit ($i > j$) as:

$$(g_{i,j}, p_{i,j}) = (g_i, p_i) \circ (g_{i-1}, p_{i-1}) \circ \ldots \circ (g_j, p_j). \quad (6.7)$$

The group generate and the group propagate signals become:

$$(g_{i,j}, p_{i,j}) = [(g_i, p_i) \circ \ldots \circ (g_{m+1}, p_{m+1})] \circ [(g_m, p_m) \circ \ldots \circ (g_j, p_j)],$$
$$= (g_{i,m+1}, p_{i,m+1}) \circ (g_{m,j}, p_{m,j}). \quad (6.8)$$

In addition,

$$c_i = (g_{i,1}, p_{i,1}) \circ c_0,$$
$$= (g_{i,j+1}, p_{i,j+1}) \circ (g_{j,1}, p_{j,1}) \circ c_0,$$
$$= (g_{i,j+1}, p_{i,j+1}) \circ c_j. \quad (6.9)$$

Thus, by generating $g_i = a_i \cdot b_i$ and $p_i = a_i \oplus b_i$ for each bit half adder, and with the carry-in signal, the carry-out signal can be obtained. In addition, the sum signal is $s_i = c_{i-1} \oplus p_i$. When the carry-in signal to the adder is $c_0 = 0$, and if $c_0 = g_0$,

$$g_{1,0} = (g_1, p_1) \circ g_0 = g_1 + p_1 \cdot g_0 = g_1 + p_1 \cdot c_0 = c_1. \quad (6.10)$$

Similarly,

$$g_{i,0} = (g_{i,1}, p_{i,1}) \circ g_0 = (g_{i,1}, p_{i,1}) \circ c_0 = c_i. \quad (6.11)$$

Therefore, if the carry-in signal is 0, from the above equation, the group generate $g_{i,0}$ is equal to carry-out c_i.

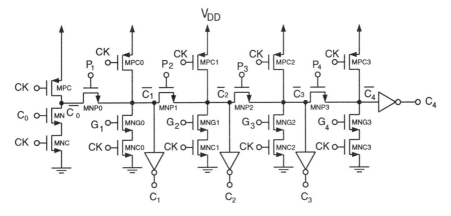

Fig. 6.4 Manchester carry look-ahead circuit.

6.1.3 Manchester CLA Adder

In the Manchester carry look-ahead (CLA) adder, the key component is the Manchester carry look-ahead circuit. Among all CLA circuits, Manchester CLA (carry chain) circuit as shown in Fig. 6.4 is based on the dynamic logic circuit and the pass-transistor logic circuit techniques[1]. The Manchester carry chain circuit is used to process the propagate and generate signals produced by the half adders to have the carry signals. With the pass-transistor configuration, Manchester carry chain circuit has the smallest transistor count among all carry look-ahead circuits including domino and other static techniques. The function of the Manchester carry chain circuit is: $C_i = G_i + C_{i-1} \cdot P_i$, for $i = 1 \sim n$, where n is the bit number, G_i and P_i are the generate and propagate signals ($G_i = a_i \cdot b_i$, $P_i = a_i \oplus b_i$) produced from two inputs (a_i, b_i) to the half adder. In the Manchester carry chain circuit, each bit carry signal ($\overline{C_i}$) is low if the generate signal (G_i) is high or if the propagate signal (P_i) is high and the carry signal of the previous bit ($\overline{C_{i-1}}$) is low. Pass transistors have been used to control the operation of the Manchester carry chain circuit. However, when the carry chain is long, as in a 64-bit adder, the ripple-carry propagation delay becomes unacceptable due to the RC delay of the pass transistor. In other words, the density advantage of the Manchester carry chain circuit is offset by the drawback in the speed. This is especially serious when the power supply voltage is scaled down, which is a trend for deep-submicron VLSI technology. In addition, since Manchester CLA circuit is a dynamic circuit, in order to accommodate the precharge period, pull-high logic has been inhibited. Instead, pull-high logic has been replaced by the precharge scheme. In contrast, for static logic circuits, both pull-high and pull-low logics need to be implemented. Due to the precharge scheme, the carry signal ($\overline{C_i}$) at each internal output node can have only two transition states—(1) from high to high, and (2) from high to low. If for some reasons such as race problems as described before, the carry signal at an internal output node switches to the wrong state during the logic evaluation period, it will never recover. On the other hand, static logic

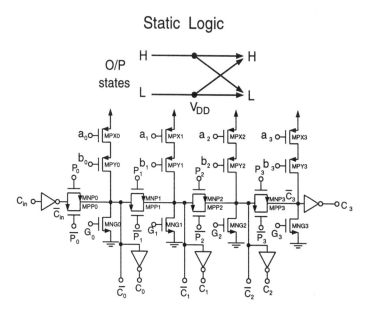

Fig. 6.5 Static Manchester-like CLA circuit. (Adapted from Hung et al. [2].)

circuits can have four output transition states—(1) from high to high, (2) from high to low, (3) from low to low, (4) from low to high. No signal race problems exist in the static logic circuits. In addition, due to the inclusion of the precharge period, the outputs of the dynamic logic circuits are available only during the logic evaluation period—during the precharge period no inputs are usable and hence no outputs are available. In contrast, in static logic circuits, inputs do not need to be restricted in a certain phase, hence outputs are available any time. As for the dynamic logic circuits, Manchester CLA circuit cannot operate at a low clock frequency because of the leakage current problems.

Fig. 6.5 shows a static Manchester-like CLA circuit[2]. As for the Manchester CLA circuit, the static Manchester-like CLA circuit as shown in figure still has a low transistor count. Different from the dynamic Manchester CLA circuit, the precharge PMOS devices have been removed. Instead, two PMOS devices controlled by two input signals (a_i,b_i), which are connected in series, have been used to replace the precharge devices. By using this circuit structure, both pull-high logic and pull-low logic as for the static logic circuits can be implemented. Depending on the two input signals (a_i, b_i), the carry signal $\overline{C_i}$ at each internal output node may be pulled low to ground when the NMOS device controlled by the generate signal (G_i) is on or when the transmission gate controlled by the propagate signal (P_i) is on and the carry signal at the internal output node of the previous bit ($\overline{C_{i-1}}$) is low. The carry signal $\overline{C_i}$ at the internal output node may be pulled high when both PMOS devices controlled by the input signals (a_i,b_i) are on or when the transmission gate controlled by the propagate signal (P_i) is on and the carry signal at the internal output node of the previous bit ($\overline{C_{i-1}}$) is high. By using this circuit arrangement, a concise static Manchester-like CLA circuit, which has the properties of static logic circuits, has been created. Owing to the structure without the precharge scheme, the static Manchester-like CLA circuit can provide valid outputs more frequently as compared to the conventional Manchester CLA circuit. Since this static Manchester-like CLA circuit is a static logic circuit, there are no race, charge-sharing and leakage current problems, which may exist in the dynamic logic circuits. One other advantage of the static Manchester-like CLA circuit is its conciseness. Since it is a static logic circuit, the carry signal ($\overline{C_i}$) from each internal output node can be used to drive other circuits. Therefore, by adding inverters both complementary carry signals ($C_i, \overline{C_i}$) can be obtained from the circuit. In contrast, for the Manchester dynamic CMOS CLA circuit as shown in Fig. 6.4, in order to avoid the race problem, the carry signal ($\overline{C_i}$) from each internal output node can not be used to drive other circuits directly. Instead, an inverter is required to relay the carry signal ($\overline{C_i}$) from each internal output node as in the domino logic circuits. When both complementary carry signals ($C_i, \overline{C_i}$) are needed in a system, a set of two complementary dynamic Manchester CMOS CLA circuits are necessary. In contrast, the static Manchester-like CLA circuit provides the carry signals ($C_i, \overline{C_i}$) simultaneously. Therefore, the conciseness of the static Manchester-like CLA circuit can be seen.

6.1.4 PT-Based CLA Adder

The strength of the Manchester CLA circuit is its conciseness. However, due to the pass transistor structure, when the size is large, the associated propagation delay is long—speed is sacrificed. This is especially serious while operating at low voltage. Fig. 6.6 shows the 1.5V bootstrapped pass-transistor-based carry-look-ahead circuit[3]. As shown in the figure, the bootstrapper circuit[4], which functions as an inverter with its output high level boosted to over V_{DD}, is used to boost the input signal to the gate of the pass transistors. As shown in the figure, when the input to a bootstrapper circuit $\overline{P_i}$ switches from high to low, the output of it (P_i) changes from low to surpass the supply voltage V_{DD}. The operation of the circuit is described here. When CK is low, it is the precharge period, where the bootstrapper input $\overline{P_i}$ is precharged to V_{DD}. MPB is off and the NMOS device MN is on. Therefore, the output of P_i is pulled low to ground. At the same time, the inverter output V_I is low, hence the PMOS device MP is on and the internal node voltage V_b is pulled high to the supply voltage V_{DD}. At this time, the bootstrap transistor MPC, which is made of a PMOS device with its source and drain tied together, stores an amount of charge—$(V_{DD} - |V_{TP}|)C_{ox}WL$, where C_{ox} is the unit area gate oxide capacitance ($C_{ox} = \epsilon_{ox}/t_{ox}$), and V_{TP} is the threshold voltage of the PMOS device. When CK is high, it is the logic evaluation period—when one of a_i and b_i is 1 and the other is 0 ($a_i \oplus b_i = 1$), the bootstrapper input ($\overline{P_i}$) is discharged to low. The NMOS device MN turns off and the PMOS device MPB turns on. Meanwhile, the gate voltage V_I of the bootstrap transistor MPC changes to high and the PMOS device MP turns off. Since the bootstrap transistor MPC turns off, the evacuated holes from the channel of the PMOS device make the capacitor bottom-plate voltage V_b go

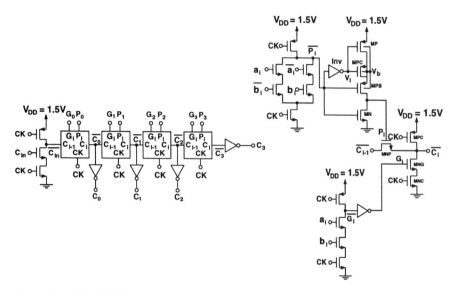

Fig. 6.6 1.5V CMOS bootstrapped PT-based carry-look-ahead circuit. (Adapted from Lou and Kuo[3].)

up to exceed the supply voltage V_{DD}—this is the internal voltage overshoot. Since MPB also turns on ($\overline{P_i}$ is low), the output P_i also goes up to surpass the supply voltage V_{DD}. With the gate overdrive voltage, the pass transistor can turn on earlier and have a larger current driving capability. Owing to the critical path formed by the serial-connected pass transistors, which are controlled by $p_0,...,p_3$, the voltage overshoot at the output of the bootstrapper circuit enhances the speed performance of the pass-transistor-based Manchester carry chain circuit. The advantages of the bootstrapper circuit are especially noticeable for the low supply voltage applications.

6.1.5 PT-Based Conditional Carry-Select CLA Adder

Fig. 6.7 shows 4-bit CLA circuits using the conventional AND-OR CLA circuit and the conditional carry select (CCS) circuit based on multiplexers[5]. The 4-bit carry signal can be produced using the expressions:

$$\begin{align} C_3 &= G_3 + P_3 \cdot [G_2 + P_2 \cdot (G_1 + P_1 \cdot G_0)] + P_3 \cdot P_2 \cdot P_1 \cdot P_0 \cdot C_{-1}, \\ &= GG + GP \cdot C_{-1}, \\ GG &= G_3 + P_3 \cdot [G_2 + P_2 \cdot (G_1 + P_1 \cdot G_0)], \\ GP &= P_3 \cdot P_2 \cdot P_1 \cdot P_0. \end{align} \tag{6.12}$$

Therefore, the critical path of a 4-bit carry look-ahead circuit contains three AND-OR circuits. If implemented by pass transistors, the circuit can be concise. However, serial connected paths as shown in the figure are formed. Therefore, it's not suitable for low-voltage operation. As shown in the figure, the operation principle of the

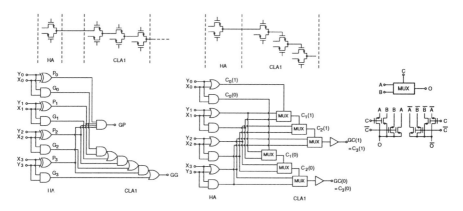

Fig. 6.7 Four-bit CLA circuits using the conventional AND-OR CLA circuit and the conditional carry select (CCS) circuit based on multiplexers. (Adapted from Suzuki et al. [5].)

conditional carry select (CCS) circuit is described here:

$$C_i(k) = G_i + P_i \cdot C_{i-1}(k),$$
$$= \begin{cases} G_i + P_i = a_i + b_i & \text{when } C_{i-1}(k) = 1, \\ G_i = a_i \cdot b_i & \text{when } C_{i-1}(k) = 0, k = 0 \text{ or } 1. \end{cases} \quad (6.13)$$

Therefore, the previous carry bit can be used to control the multiplexer. If C_{i-1} is high, $C_i = a_i + b_i$. If C_{i-1} is low, $C_i = a_i \cdot b_i$. Using this principle, the multiplexer controlled by C_{i-1} has been used to integrate the CLA circuit. Using this multiplexer, the operation of the circuit becomes simple. In addition, two sets of circuits are required. One set assumes that $C_{in} = 0$ and its outputs are $C_0(0)$, $C_1(0)$, $C_2(0)$, and $C_3(0)$. The other set assumes that $C_{in} = 1$ and its outputs are $C_0(1)$, $C_1(1)$, $C_2(1)$, and $C_3(1)$. From $(C_i(0), C_i(1))$, the sum-related signals $(S_i(0), S_i(1))$ are used to form the final sum signal, which is selected by another multiplexer based on the value of C_{in}. Using this algorithm, the block carry look-ahead circuit (CLA2) can be redesigned. Via a multiplexer controlled by C_{in}, C_3 is determined from $(C_3(0), C_3(1))$. From C_3, C_7 is chosen from $(C_7(0), C_7(1))$. Similar results can be obtained for other carry signals (C_{11}, C_{15}). Using this multiplexer-based structure, the problems with the long serial critical path are avoided—when the output of each multiplexer is connected to the gate of the next multiplexer instead of the source/drain. Therefore, there is no long serial path involved, which is especially useful for low-voltage operation. Note that in the half adders only AND and OR gates are required. Compared to the conventional structure, it is simpler. As shown in the figure, in the conditional carry select (CCS) circuit, the serial connected path disappears. At the beginning of the CCS circuit, instead of C_{-1}, $C_0(1) = a_0 + b_0$, and $C_0(0) = a_0 \cdot b_0$ are used. The sum is selected from $S_i(0)$ and $S_i(1)$ by C_{-1}. Fig. 6.8 shows the block diagram of the 32-bit adder using the conditional carry select (CCS) and the multiplexer-based circuits[5]. CCS techniques have been applied to realize the 4-bit CLA circuit (CLA1) and the block CLA circuit (CLA2), using the appropriate signals from the four CLA1s ($GC_i(0)$, $GC_i(1)$, $i = 0$ to 3) based on the

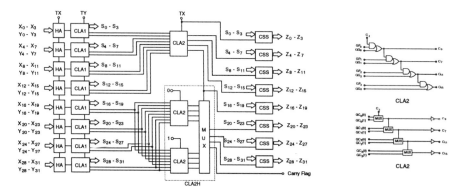

Fig. 6.8 Block diagram of the 32-bit adder using the conditional carry select (CCS) and the multiplexer-based circuits. (From Suzuki et al. [5], ©1993 IEEE.)

carry-in signal. As shown in the figure, the block CLA (CLA2) circuit is similar to the conventional 4-bit CLA circuit as shown in Fig. 6.7—it also has AND-OR circuits in the critical path. Therefore, similar problems also exist in CLA2. Therefore, CLA2 is also based on the multiplexer structure. Hence, CLA2 has strong points as for CCS-based CLA circuit. Therefore, the principle of conditional carry select (CCS) is similar to the technique used in the conventional carry select adder—the carry signal is used as the control signal to the multiplexer. Using CCS, the complexity in the design of the conditional carry select adder can be substantially reduced.

Fig. 6.9 shows the pass-transistor-based carry select adder [6]. As shown in the figure, instead of P_i, C_{i-1} is used as the control signal. In addition, pass-transistor logic has been used to replace the multiplexer structure. The operation principle is described below. When $C_{i-1} = 0$, the carry signal ($C_i = G_i + C_{i-1} \cdot P_i$) becomes $C_i = OR_i = G_i + P_i$, hence OR_i is used as the input to a transmission gate controlled by C_{i-1}. When $C_{i-1} = 0$, the transmission gate is off. Under this situation, the circuit is similar to a $\overline{G_i}$-controlled inverter (when $\overline{C_{i-1}}$ is high, hence the NMOS device $MNCi$ is on). Therefore, $C_i = G_i$. Similarly, $\overline{C_i} = \overline{G_i} \cdot (\overline{C_{i-1}} + \overline{P_i}) = \overline{C_{i-1}} \cdot \overline{G_i} + \overline{G_i} \cdot \overline{P_i}$. Since $\overline{G_i} \cdot \overline{P_i} = (\overline{a_i} + \overline{b_i})(a_i b_i + \overline{a_i} \cdot \overline{b_i}) = \overline{a_i} \cdot \overline{b_i} = \overline{OR_i}$, therefore, $\overline{C_i} = \overline{OR_i} + \overline{C_{i-1}} \cdot \overline{G_i}$. Similarly, $\overline{C_{i-1}}$ is used as the control signal. When $\overline{C_{i-1}}$ is high, $\overline{C_i} = \overline{OR_i} + \overline{G_i} = \overline{a_i b_i} + \overline{a_i} + \overline{b_i} = \overline{a_i} + \overline{b_i} = \overline{G_i}$. Therefore, $\overline{G_i}$ can be used as an input to the transmission gate controlled by $\overline{C_{i-1}}$. When $\overline{C_{i-1}}$ is low, the transmission gate is off. Under this situation, the above circuit becomes an $\overline{OR_i}$-controlled inverter (when C_{i-1} is high, the NMOS device $MNCi$ turns on), $\overline{C_i} = \overline{OR_i}$. Using the pass-transistor logic in the carry selector adder, there are following advantages: As for the multiplexer-based design, the output of one stage is connected to the gate of the next stage instead of the source/drain in the conventional Manchester CLA circuit. Therefore, no long serial critical path is involved in the circuit, which is especially useful for low-voltage operation. In the conditional carry select (CCS) adder using the multiplexer structures described before, G_i and OR_i are connected as inputs to multiplexers. Since complementary

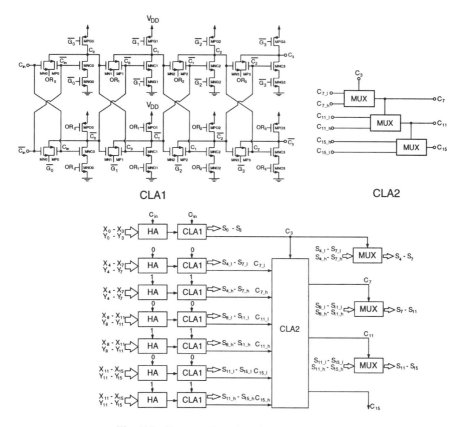

Fig. 6.9 Pass-transistor-based carry select adder.

inputs to MUX are required, therefore two sets of complementary signals $(G_i, \overline{G_i})$ and $(OR_i, \overline{OR_i})$ are required. Using the pass-transistor-based structures, only $(\overline{G_i}, OR_i)$ are required. Therefore, conciseness of the circuit can be obtained. In addition, $(\overline{G_i}, OR_i)$ are NAND circuits with (a_i, b_i) as inputs to G_i and with $(\overline{a_i}, \overline{b_i})$ as inputs to OR_i. In comparison, the layout of the NOR circuit for $(G_i, \overline{OR_i})$ is larger. In addition, the output of the C_i circuit in the pass-transistor-based circuit described above is connected to two NMOS devices and one PMOS device. In contrast, in the multiplexer-based CSS structure, it is connected to two PMOS devices and two NMOS devices. A smaller fan-out can be identified for the pass-transistor-based structure.

6.1.6 Enhanced MODL Adder

Fig. 6.10 shows a 32-bit carry look-ahead adder implemented by the enhanced MODL[7]. As shown in the figure, the B1–B4 subcells are used to produce the group generate, the group propagate signals and the $a_i \oplus b_i$ (x_i) signal for s_i. B1

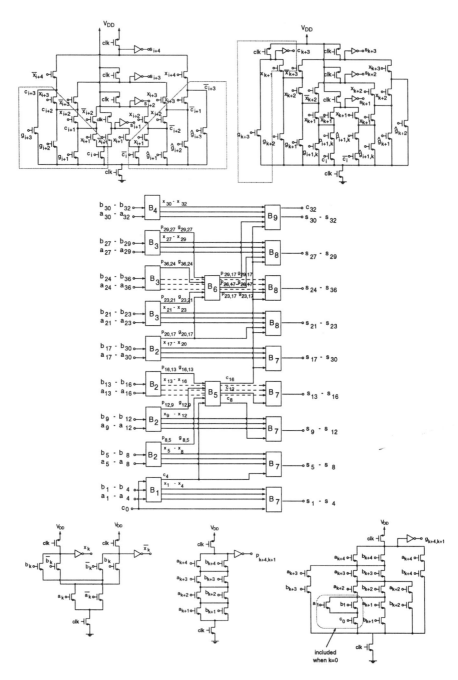

Fig. 6.10 32-bit carry look-ahead adder implemented by the enhanced MODL. (From Wang et al. [7], ©1997 IEEE.)

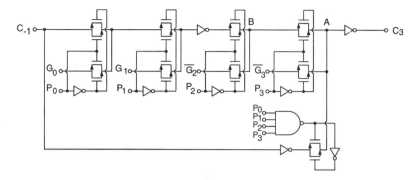

Fig. 6.11 Manchester CLA chain with the carry skip path.

and B2 are 4-bit. B3 and B4 are 3-bit. In total, there are $5 \times 4 + 4 \times 3 = 32$ adders. The group generate output of B1 is c_4. Therefore, no group propagate circuit is required in B1. On the contrary, for B2, both group generate and group propagate circuits are needed. No group generate and group propagate circuits are needed for B4. In B5 and B6, the group generate and the group propagate signals produced by the first stage are further processed. In B5, carry-out signals c_8, c_{12}, c_{16} are generated directly. In B6, $(g_{i,17}, p_{i,17})$, $i = 23, 26, 29$ are generated for further processing in the third sum generation circuit. The third stage is the sum generation stage. B7 is 4-bit. B8 and B9 are 3-bit. Both B8 and B9 use c_{16} as c_{in}. As shown in the figure, in B7, $s_{i+1},...s_{i+4}$ are produced by c_{i+1}, c_{i+2}, c_{i+3}, c_{i+4}, exclusive-ORed with $x_{i+1}...x_{i+4}(a_i \oplus b_i)$. Note that $\overline{c_i}$ is realized by the following equation:

$$\overline{c_i} = \overline{g_i + p_i \cdot c_{i-1}} = \overline{p_i \cdot g_i} + \overline{g_i} \cdot \overline{c_{i-1}} = \overline{a_i} \cdot \overline{b_i} + \overline{(\overline{a_i} + \overline{b_i})} \cdot \overline{c_{i-1}} = \hat{g}_i + \hat{p}_i \cdot \overline{c_{i-1}}. \tag{6.14}$$

In order to make $\hat{g}_i$ and $\hat{p}_i$ not high simultaneously, $\hat{p}_i = \overline{a_i} \oplus \overline{b_i} = a_i \oplus b_i = x_i$. Therefore, $\hat{g}_i$ is similar to g_i except that the inputs are complementary.

6.1.7 Carry-Skip CLA Circuit

Fig. 6.11 shows the Manchester CLA chain with the carry skip path based on the CMOS pass transistor logic. As shown in the figure, the carry skip path is used to reduce the propagation delay associated with the critical path when $P_0 \cdot P_1 \cdot P_2 \cdot P_3 = 1$ and $G_0 = G_1 = G_2 = G_3 = 0$. As shown in the figure, when P_0, P_1, P_2, and P_3 are 1, via a NAND-controlled transmission gate, the carry input C_{-1} becomes the output carry signal. Owing to the reduction in the transmission-gate RC delay, with carry skip the propagation delay of C_3 can be substantially improved. However, this approach still has problems. In this carry-skip CLA circuit, when $P_0 \cdot P_1 \cdot P_2 \cdot P_3 = 1$ and the input carry signal C_{-1} switches from low to high, the carry signal has not propagated to B—B is still high. With the transmission gate turned on by P_3, the high-state of B is passed to A, where it is being pulled low by the carry-skip transmission gate—a conflicting situation occurs. The speed in discharging node A is

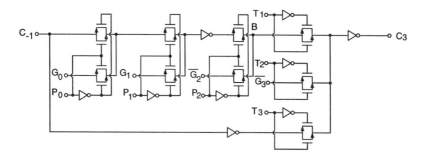

Fig. 6.12 Conflict-free carry-skip circuit. (Adapted from Sato et al.[9].)

affected. Thus the effect of the carry skip circuit is reduced.

Fig. 6.12 shows a conflict-free carry-skip circuit[9]. As shown in the figure, in order to avoid the conflict in the carry-skip circuit as described in Fig. 6.11, a selector, which is composed of three transmission gates, has been used to generate the C_3 signal. Three transmission gates are controlled by the following three signals, T_1, T_2, and T_3:

$$\begin{aligned} T_1 &= \overline{P_0 \cdot P_1 \cdot P_2} \cdot P_3, \\ T_2 &= \overline{P_3}, \\ T_3 &= P_0 \cdot P_1 \cdot P_2 \cdot P_3. \end{aligned} \quad (6.15)$$

Only one of these three signals is activated to form the conduction path. When T3 is activated, T1 and T2 are blocked. Using this selector, the signal conflict in the carry-skip circuit can be avoided.

Fig. 6.13 shows the circuit diagram of a 112-bit transmission gate adder[9]. As shown in the figure, there are seven 16-bit adder blocks (MS16) and a block carry generator (BCG). The 16-bit adder block (MS16) is composed of a carry propagate signal generating circuit (MLA), a block carry generate and propagate signal generating circuit (MB16), and a sum signal generating circuit (ML16). As shown in the figure, the operation of the 16-bit adder block (MS16) accommodates the following functions:

$$\begin{aligned} BG_n &= G_n + P_n \cdot BG_{n-1} = P_n \cdot BG_{n-1} + \overline{P_n} \cdot A_n \cdot B_n, \quad (6.16) \\ &= P_n \cdot BG_{n-1} + \overline{P_n} \cdot A_n, \\ BP_n &= P_n \cdot BP_{n-1} + \overline{P_n} \cdot 0, \\ SUM_n &= \overline{P_n} \cdot (BP_{n-1} \cdot BC_m) + (P_n \oplus BG_{n-1}) \cdot (\overline{BP_{n-1} \cdot BC_m}). \end{aligned}$$

As shown in the figure, in the 16-bit adder multi-carry-skip network, in order to avoid signal conflict, transmission gates controlled by $\overline{T_2} \cdot P_{15}(i)$, $\overline{T_1} \cdot T_2 \cdot P_{15}(ii)$, and $T_1 \cdot T_2 \cdot P_{15}(iii)$, where $T_1 = P_6 \cdot P_7 \cdot P_8 \cdot P_9 \cdot P_{10}$ and $T_2 = P_{11} \cdot P_{12} \cdot P_{13} \cdot P_{14}$, are used. In addition, a block carry generator circuit based on $C_i = BG_{i,j} + BP_{i,j} \cdot C_j$ has been used to obtain the block carry signal.

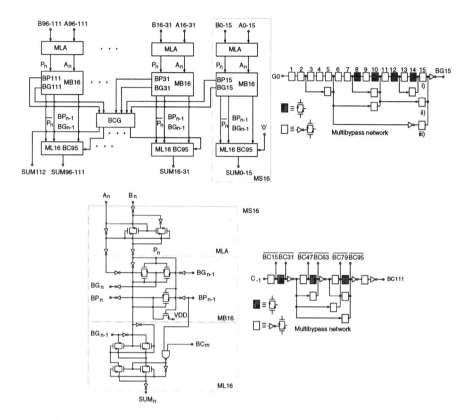

Fig. 6.13 112-bit transmission gate adder. (From Sato et al.[9], ©1992 IEEE.)

6.1.8 Parallel and Pipelined Adders

From Eqs. (6.3)–(6.11), the generation of the carry signal of the present bit does not need to wait for the carry signal of the previous bit. Carry look-ahead can be achieved via circuit techniques—the parallel processing capability can be obtained. When two large numbers are added, further improvement of the adder circuit is required. Fig. 6.14 shows the block diagram of a parallel carry chain structure[10]. As shown in the figure, at the first stage at top, there is a set of generate and propagate signals referred to a bit (g_i, p_i, $i = 1, 2, ...16$). At the second stage, group generate and propagate signals referred to two bits ($g_{i,i-1}$, $p_{i,i-1}$, $i = 2, 4, 8, ...16$) are involved via $(g_i, p_i) \circ (g_{i-1}, p_{i-1}) = (g_i + p_i \cdot g_{i-1}, p_i \cdot p_{i-1}) = (g_{i,i-1}, p_{i,i-1})$. At the next stage, group generate and propagate signals referred to ($g_{i,i-3}$, $p_{i,i-3}$, $i = 4, 8, 12, 16$) are involved via $(g_{i,i-1}, p_{i,i-1}) \circ (g_{i-2,i-3}, p_{i-2,i-3}) = (g_{i,i-1} + p_{i,i-1} \cdot g_{i-2,i-3}, p_{i,i-1} \cdot p_{i-2,i-3}) = (g_{i,i-3}, p_{i,i-3})$. At the fourth stage, group generate and propagate signals referred to ($g_{i,i-7}$, $p_{i,i-7}$, $i = 8, 16$) are involved. At the fifth stage, group generate and propagate generate circuits referred to ($g_{16,1}, p_{16,1}$) are included. Therefore, for 2^n bit add operation, n stages of group generate/propagate signals are involved. Using the group generate/ propagate signals, the carry prop-

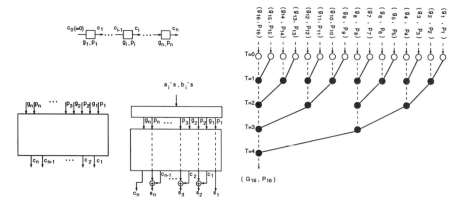

Fig. 6.14 Block diagram of a parallel carry chain structure. (Adapted from Brent and Kung[10].)

agation time can be shortened substantially as described before. Fig. 6.15 shows the computation of all carriers for $n = 16$[10]. As shown in the figure, at $T = 4$, $g_{16} = c_{16}$ is produced. As for $c_9 - c_{15}$, it is produced by $g_8 = c_8$. From the figure, the critical path of this adder is not on c_{16}. Instead, it on c_{15}. Since c_{15} is produced from c_8, three more delays are involved. In order to optimize the delay balance in the parallel structure, in the circuit there are several buffers without any logic functions—they are used to make an effective delay such that delay balance can be obtained.

The parallel adder described above is suitable to be integrated in the wave-pipelined circuit[11]. Wave pipelining is a timing methodology, which is also called max-rate pipelining. In a conventional pipelined system, a new data cannot enter the combinational logic until the logic output data is stabilized. In the wave pipelined system, before the logic output data are stabilized, the input data are entered. As long as the new input data do not affect the undergoing logic operation in producing the output data, they are fine. Using the wave pipelined system, the operating speed does not have to increase, but the throughput has been enhanced, which is similar to the conventional pipelined system except that it is a combinational logic effectively pipelined with multiple coherent data waves. (Note that various data waves represent different stages.) As shown in Fig. 6.16, in the wave pipelined system, data enter the combinational logic array via the input register. The output register provides the output data from the combinational logic, which is controlled by the clock for the input register with a delay. The clock period is bounded by the maximum and the minimum of the delay difference:

$$T_{clk} \geq (t_{max} - t_{min}) + \text{unintentional skew} + \text{setup/hold time}. \qquad (6.17)$$

Therefore, as shown in the figure, if the delay difference between t_{max} and t_{min} is smaller, the uncertain area is smaller—the clock period (T_{clk}) can be shrunk further to obtain a higher clock frequency. It implies a higher throughput although the logic

ADDERS   361

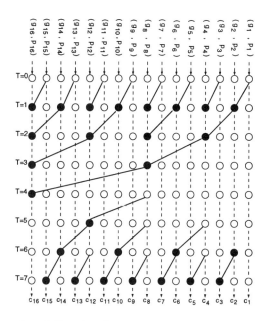

Fig. 6.15 Computation of all carriers for $n = 16$. (Adapted from Brent and Kung[10].)

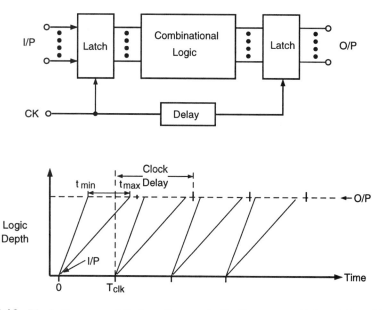

Fig. 6.16 Block diagram and timing of wave pipelined systems. (Adapted from Liu et al.[11].)

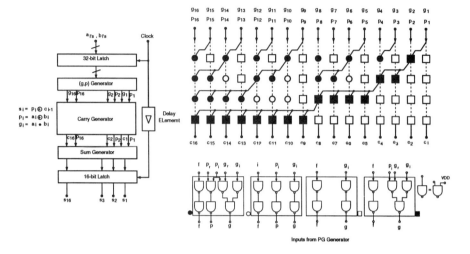

Fig. 6.17 Block diagram of the parallel adder. (Adapted from Liu et al.[11].)

delay is not shrunk. This is different from the conventional pipelined system. In the conventional pipelined system, in order to raise the clock frequency, the logic delay should be shrunk accordingly. In the wave pipeline, the logic delay does not need to be scaled down accordingly. Instead, the logic delay is required to be balanced. Therefore, the parallel adder structure is suitable to be integrated in the wave pipelined system. As long as the path delay difference can be minimized, the effective clock rate can be optimized. Fig. 6.17 shows the block diagram of the parallel adder[11]. As shown in the figure, the parallel adder is compatible to any common adder, only with input/output registers and a delay element for the clock for the wave pipeline operation. As shown in the figure, in the carry generator, since $C_{in} = 0$, $C_i = G_i$, where

$$(G_i, P_i) = \begin{cases} (g_1, p_1) & \text{for } i = 1, \\ (g_i, p_i) \circ (G_{i-1}, P_{i-1}) & \text{for } i > 1. \end{cases} \quad (6.18)$$

In the figure, the solid black square represents that only G_i is generated. The solid black circle represents group generate/propagate generating circuit, whose function is characterized by the ∘ operator: $(g_i, p_i) \circ (g_r, p_r) = (g_i + p_i \cdot g_r, p_i \cdot p_r)$. From the figure, at a delay associated with four blocks, which is equal to about eight NAND2 delays, all carry signals can be obtained. This is much more efficient than other adders. In order to reduce the difference in the delays between the buffer and the gate such that the effective bandwidth of the wave pipelined system can be maximized, all buffers use NAND2 gates with one end connected to V_{DD}. Basically, the delay of all basic cells in the circuit is associated with NAND2 gate. Thus the delay difference between t_{min} and t_{max} can be reduced to increase the throughput of the wave pipelined system.

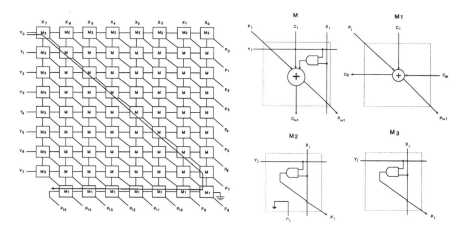

Fig. 6.18 Block diagram of an 8-bit×8-bit array-type multiplier

6.2 MULTIPLIERS

Multipliers are important subsystem circuits frequently used. The speed performance of the multipliers often affects the overall speed performance of a VLSI system. Fig. 6.18 shows the block diagram of an 8-bit×8-bit array-type multiplier. As shown in the figure, its longest critical path occurs as the marked direction in the figure. There are four types of cells in the multiplier circuit—M, M1, M2, M3. In the M cell, two inputs X_i, Y_i, the carry-in signal C_i, the sum of the partial product of the previous bit (P_i) and $X_i \cdot Y_i$ are added together to generate the P_{i+1} and C_{i+1} for the following bit. When the bit number of the multiplicand and the multiplier is large, the propagation delay associated with the longest critical path is long. Therefore, the speed is slow. Its delay is linearly proportional to the bit number.

For the partial products x_0y_7, x_1y_6, ..., x_6y_1, x_7y_0 in the array multiplier mentioned before, it is based on the carry ripple/propagate adder. When the bit number is increased, its delay is increased proportionally. Using the carry save adder techniques to handle the addition of the partial products, the addition time can be reduced effectively. This is the principle of Wallace tree reduction architecture. Fig. 6.19 shows the block diagram of an 8-bit×8-bit multiplier using Wallace tree reduction architecture via the carry save adders. As shown in the figure, in the 8-bit×8-bit multiplier, several add operations are required. In each add operation referred to an output bit, up to eight partial products (x_0y_7, x_1y_6,...,x_7y_0) in a column are to be added. Using the carry save adders, three partial product terms can be added at a time to form the carry and the sum. The sum signal is used in the next-level carry save adder referred to the same bit. The carry signal is used in the next-level carry save adder of the next bit. Therefore, using Wallace tree reduction architecture there are $log_{\frac{3}{2}} n$ (n is the bit number) carry save additions in the multiplier. At the bottom stage, a carry look-ahead adder is used to add up all the results. Using the carry save

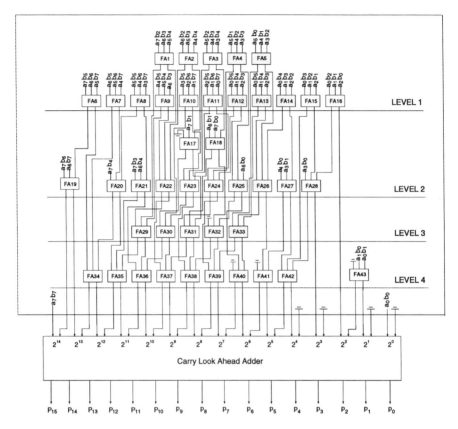

Fig. 6.19 Block diagram of an 8-bit × 8-bit multiplier using carry save adders.

adders, the delay of the addition of the partial products can be shrunk to within $\log_{\frac{3}{2}} n$ stages.

The carry save adders used in the multiplier described above function to compress 3 partial products at the input into 2 outputs—carry and sum signals at the output. So, the 1-bit full adder is called the 3-to-2 compressor in the multiplier. Using two carry save adders, a 4-to-2 compressor can be obtained—four partial products are added to produce sum and carry signals. Fig. 6.20 shows a 4-to-2 compressor, which is made of two 1-bit full adders. From the figure, the 4-to-2 compressor should be the 5-to-3 compressor since C_{in} and C_{out} are also included. The sum signal is $S = X_1 \oplus X_2 \oplus X_3 \oplus X_4 \oplus C_{in}$. C is related to the sum of the first adder, X_4, and C_{in}. C_{out} is the carry signal of the addition of X_1, X_2, and X_3. By this arrangement, the sum can be obtained via four XOR gate delays ($S = [[(X_1 \oplus X_2) \oplus X_3] \oplus X_4] \oplus C_{in}$), which is identical to the result in the Wallace tree structure using 2-layer carry save adders. Therefore, it should be rearranged to be $S = [(X_1 \oplus X_2) \oplus (X_4 \oplus C_{in})] \oplus C_{in}$. Under this arrangement, three XOR gate delays are involved—speed advantage can

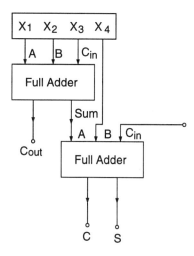

Fig. 6.20 4-to-2 compressor.

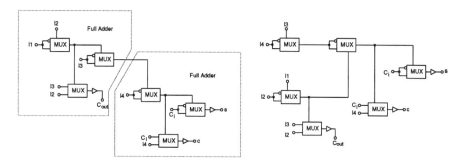

Fig. 6.21 4-to-2 compressor circuits using pass-transistor multiplexers. (Adapted from Ohkubo et al. [13].)

be obtained.

Fig. 6.21 shows the 4-to-2 compressor circuits using pass-transistor multiplexers[13]. As shown in the figure, in the conventional approach, two full-adder cells are used. In an improved approach, $S = (((I_1 \oplus I_2) \oplus I_3) \oplus I_4) \oplus C_i$ has been changed to $S = ((I_1 \oplus I_2) \oplus (I_3 \oplus I_4)) \oplus C_i$. Compared to the previous approach, in the improved approach, the gate delay has been reduced. Thus a higher speed of the 4-to-2 compressor can be expected.

Fig. 6.22 shows the logic diagram of another 4-to-2 compressor (4W) unit[16]. As shown in the circuit,

$$\begin{aligned} C_o &= I_1 \cdot I_2 + I_3 \cdot I_4, \\ S &= C_i \oplus [(I_1 \oplus I_2) \oplus (I_3 \oplus I_4)], \end{aligned} \qquad (6.19)$$

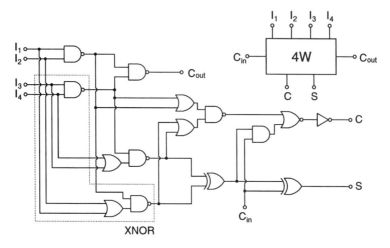

Fig. 6.22 Logic diagram of the 4-to-2 compressor (4W) unit. (Adapted from Goto et al.[16].)

$$C = I_1 \cdot I_2 \cdot I_3 \cdot I_4 + (I_3 \oplus I_2) \cdot (I_3 \oplus I_4) + [(I_1 \oplus I_2) \oplus (I_3 \oplus I_4)] \cdot C_i.$$

Although the logic functions are different from those in the 4-to-2 compressor described before, their effects are equivalent.

One thing worth pointing out is that while using the 4-to-2 compressor in the Wallace tree reduction architecture, the carry signal (C_{out}) is used for propagation to the next bit in the same level. The sum and the other carry signals are to be used in the next level. Fig. 6.23 shows the block diagram of an 8-bit × 8-bit multiplier using 4-to-2 compressors. As shown in the figure, in the 8-bit×8-bit multiplier using the 4-to-2 compressors, the logic depth can be further reduced by an order of $log_2 n$. As a result, the speed can be enhanced further.

6.2.1 Modified Booth Algorithm

Consider two n-bit numbers A and B to be multiplied. B can be expressed as:

$$\begin{aligned} B &= -B_{n-1}2^{n-1} + B_{n-2}2^{n-2} + \ldots + B_0 2^0, \\ &= -(2B_{n-1} + B_{n-2} + B_{n-3})2^{n-2} + (-2B_{n-3} + B_{n-4} + B_{n-5})2^{n-4} + \ldots \\ &\quad + (-2B_1 + B_0 + B_{-1})2^0, \end{aligned}$$

(6.20)

where $B_{-1} = 0$, and $B_{n-3}2^{n-2} - 2B_{n-3}2^{n-4} = B_{n-3}2^{n-3}$ have been used in the expression. Therefore, Eq. (6.20) can be represented by

$$B = \sum_{i=0}^{i=\frac{n}{2}-1} (-2B_{2i+1} + B_{2i} + B_{2i-1}) \cdot 2^{2i} = \sum_{i=0}^{i=\frac{n}{2}-1} Y_i \cdot 2^{2i}. \qquad (6.21)$$

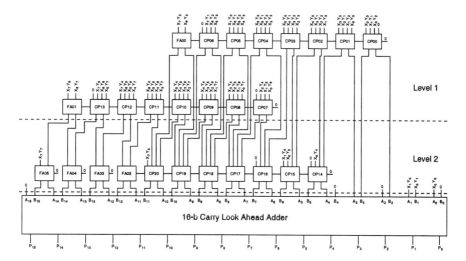

Fig. 6.23 Block diagram of an 8-bit × 8-bit multiplier using 4-to-2 compressors.

Thus

$$A \times B = \left(-A_{n-1}2^{n-1} + \sum_{i=0}^{i=n-2} A_i \cdot 2^i\right)\left(-B_{n-1}2^{n-1} + \sum_{j=0}^{j=n-2} B_j \cdot 2^j\right),$$

$$= \left(-A_{n-1}2^{n-1} + \sum_{i=0}^{i=n-2} A_i \cdot 2^i\right)\left(\sum_{j=0}^{j=\frac{n}{2}-1} Y_j \cdot 2^{2j}\right). \quad (6.22)$$

From the sequence of B_{2i+1}, B_{2i}, and B_{2i-1}, Y_i, which can be −2, −1, 0, 1, 2, can be known. Therefore, the partial products of n-bit × n-bit multiplication can be simplified as the effective partial products of n-bit × $\frac{n}{2}$-bit multiplication—the multiplication time can be reduced. The relation between Y_i and B_{2i+1}, B_{2i}, and B_{2i-1} is summarized as:

B_{2i+1}	B_{2i}	B_{2i-1}	Y_i	operation with A	X_i	$2X_i$	M_i
0	0	0	0	$0 \cdot A$	0	0	0
0	0	1	+1	$1 \cdot A$	1	0	0
0	1	0	+1	$1 \cdot A$	1	0	0
0	1	1	+2	$2 \cdot A$	0	1	0
1	0	0	−2	$-2 \cdot A$	0	1	1
1	0	1	−1	$-1 \cdot A$	1	0	1
1	1	0	−1	$-1 \cdot A$	1	0	1
1	1	1	0	$0 \cdot A$	0	0	0

From the above table, from B_{2i+1}, B_{2i}, and B_{2i-1}, three signals X, 2X, and M (stands for minus) can be encoded. For example, for $(B_{2i+1}, B_{2i}, B_{2i-1})$=(1,0,0),

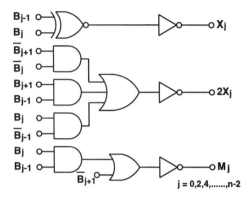

Fig. 6.24 Booth encoder circuit.

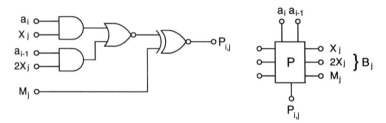

Fig. 6.25 Reduced partial products based on the modified Booth algorithm.

the corresponding (X, 2X, M)=(0,1,1) is obtained. For $(B_{2i+1}, B_{2i}, B_{2i-1})$=(1,0,1), one obtains (X,2X,M)=(1,0,1). Therefore, based on X, 2X, and M, the partial product becomes $P_{i,j} = (A_i \cdot X_j + A_{i-1} \cdot 2X_j) \oplus M_j$, where M_j is referred to M of jth's bit.

Fig. 6.24 shows the Booth encoder circuit[15], which is implemented in logic circuits. In $(X_j, 2X_j, M_j)$, X_i represents that the multiplicand to be multiplied by 1. $2X_j$ stands for the multiplicand to be multiplied by 2, which implies shift one bit. M_j means multiplication by -1. Fig. 6.25 shows the reduced partial products based on the modified Booth algorithm. As shown in the figure, the P unit is used to accommodate:

$$P_{i,j} = (a_i x_j + a_{i-1} 2x_j) \oplus M_j, \ i = 0, 1, 2, ..., n-1, \ j = 0, 2, 4, ..., n-4, n-2. \tag{6.23}$$

6.2.2 Advanced Structure

In a large-size multiplier such as 54-bit×54-bit, the modified Booth algorithm can be used to reduce the partial product depth to one half (from 54 to 27). In addition, the Wallace tree reduction technique based on the 4-to-2 compressor can be used to further simplify the addition of partial products in a column. As shown in Fig. 6.26, via the Booth algorithm and the Wallace tree reduction technique, the addition of partial

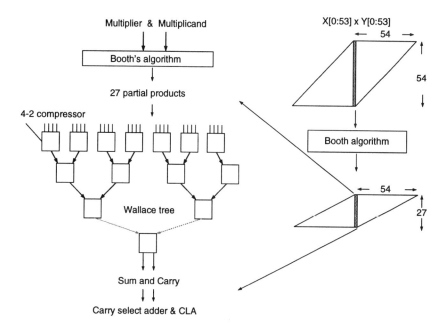

Fig. 6.26 54-bit×54-bit parallel structured full array multiplier. (From Mori et al.[14], ©1991 IEEE.)

products in a column contains seven 4-to-2 compressors and 27 additions (instead of 54 additions) of partial products, which can be realized in four layers of 4-to-2 compressors. Based on modified Booth algorithm, the bit sequence of the multiplier B is encoded into $(X_j, 2X_j, M_j)$. According to $(X_j, 2X_j, M_j)$ and the multiplicand A, in the Booth selector, the function: $P_{i,j} = (a_i \cdot X_j + a_{i-1} \cdot 2X_j) \oplus M_j$ is executed. If M_i is 1, an inversion is added. Then, via the carry select adder array (Wallace tree), partial products are compressed. Finally, via the carry propagation adder (CPA), the final product is obtained.

Fig. 6.27 shows the interconnection of 4-to-2 compressors[14]. As shown in the figure, only sum (S) is transferred to the next level. Carry-out1 is transferred to the next-bit 4-to-2 compressor of the next level. Carry-out2 is transferred to the 4-to-2 compressor of the next bit at the same level. Fig. 6.28 shows the method for the Wallace tree handling the function of a 28-to-2 compressor[15]. At the first stage, it is the 7-to-2 compressor, which is made of a 3-to-2 compressor, a 4-to-2 compressor, and another 4-to-2 compressor. Therefore, using four-layer compression, the function of the 28-to-2 compressor can be accomplished. The circled S refers to the Booth selector output (reduced partial product). 4W is the 4-to-2 compressor. 3W is the 3-to-2 compressor (1-bit full adder).

Fig. 6.29 shows 7D (7-to-2 compressor), 4D, and 3D compressors in a 54-bit×54-bit regularly structured tree multiplier[16]. In the 7-to-2 compressor, in order to have

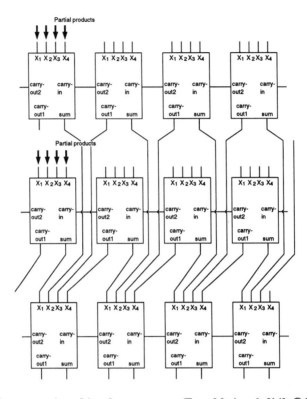

Fig. 6.27 Interconnection of 4-to-2 compressors. (From Mori et al. [14], ©1991 IEEE.)

a regular structured tree, the outputs of the 4D propagate toward right. The outputs of the 3D compressors propagate toward left. Therefore, the 4D compressors at the lowest layer accept the output carry/sum signals from the right 3D compressor blocks and the output carry/sum signals from the left 4D compressor block. 4D contains the 4-to-2 compressor (4W) and the reduced partial product generators based on Booth algorithm as shown in Fig. 6.25. 3D contains the 3-to-2 compressor (3W) and the reduced partial product generators.

6.3 REGISTER FILE AND CACHE MEMORY

Register files in a CPU are an important storage element. Register files are used to provide read and write data access for the CPU in a clock cycle. Therefore, its access speed is the fastest in the whole system. Due to the access speed requirement, register files usually cannot be large. Along with the evolution of the CPU design, in a superscalar structure with many functional units, simultaneous access to the registers for data read/write may be required. For the purpose of the concurrent access to the ALU, register files need to have at least two read ports in addition to the write port. Therefore, register files need to use multi-port memory cells. Fig. 6.30 shows the

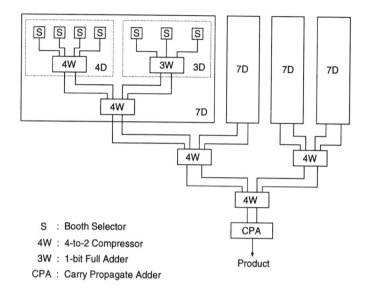

Fig. 6.28 Sequence of Booth selectors and carry save adder tree in the multiplier. (Adapted from Goto et al.[15].)

cell structures in the basic register file: (a) three-port memory cells with 2 read and 1 write ports (2R1W), (b) memory cells with read/write port (1RW). The memory cell contains a cross-coupled inverter pair and a buffer inverter for the read port to avoid the influence of the read operation in the cell data. The write operation occurs when both column write enable (CWE) and write word line (WL_W) are high. As shown in the figure, the write bit line (BL_W) of the adjacent bits can be shared. As long as one of CWE_a and CWE_b is high, the cell to write can be determined. As shown in Fig. 6.30(b), for a pair of memory cells with read/write port (1RW), when WL_RW is high, read and write can be carried out simultaneously.

Fig. 6.31 shows the data I/O circuit for the register file[17]. During the read operation, when the read word line is high, the cell data are accessed via INV1 and N1 to the read bit line. When the voltage drop at the read bit line is sufficient for readout, RE is high, and TG, which is controlled by the read enable (RE) and the read column decoder output, turns on. As a result, the read bit line is connected to $\overline{RD}$. Then via P2 and N2, the data output is available at RD. Due to the body effect, the pass transistor TG cannot effectively provide a logic-1 output. At the 0 → 1 transition of $\overline{RD}$, special attention should be taken. When RE is high, N4 is off. Therefore, the discharge path of the feedback inverter, which is composed of P3, N3, and N4, is off—it does not affect the 0 → 1 transition on $\overline{RD}$. When the inverter, which is made of P2 and N2, switches low, P3 turns on—$\overline{RD}$ is pulled high to V_{DD}. When $\overline{RD}$ has the transition from high to low, a ratioed condition between INV1, N1, TG and P3 exists. Therefore, the aspect ratio of P3 should not be too large. When RE is low, N4 turns on. Therefore, P3, N3, and N4 act as an inverter—the latch formed by P2,

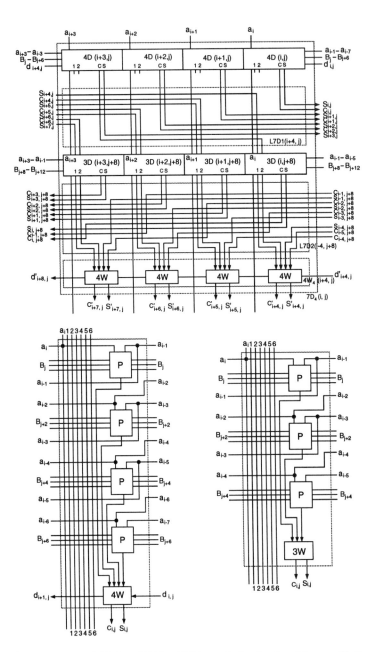

Fig. 6.29 7D, 4D, and 3D in a 54-bit×54-bit regularly structured tree multiplier. (From Goto et al.[16], ©1992 IEEE.)

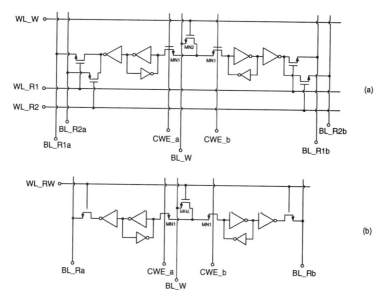

Fig. 6.30 Basic register file cell structures: (a) three-port memory cells with 2 read and 1 write ports (2R1W); (b) memory cells with read/write port (1RW). (From Shinohara et al.[17], ©1991 IEEE.)

N2, P3, N3, and N4 maintains the stability of RD and DOUT. The write operation is controlled by the write word line and column write enable (CWE), which is controlled by write enable (WE) and write column decoder output. In the memory cell of the register file, minimum width devices should be arranged for the feedback inverter. Considering the write operation, the gate length of the device in the feedback inverter should be increased to reduce the ratioed logic problem during the write operation. In addition, the logic threshold voltage of the feedforward inverter should be small to avoid the problems during the operation of write 'logic-1'. Therefore, a β ratio of 5 to 1 between the NMOS and the PMOS devices should be kept to make sure a low gate threshold. At the same time, the β ratio between MN2 and the NMOS device in the feedback inverter should be larger than 10 to ensure the correct $0 \rightarrow 1$ write operation. Fig. 6.32 shows the floor plan of a register files in M-word by N-bit[17]. In order to facilitate layout, a module-oriented layout floor plan is used. A memory cell subarray for each data bit consists of M/CPB rows and CPB columns. (CPB is the column per bit in the memory cell array.) The memory cell subarray and data I/O modules for each port are stacked to form a 1-bit module unit. The data I/O module of the WRITE port contains a D_{in} buffer, a write column decoder, and CWE drivers. The data I/O module of the READ port contains a read column decoder, a bit-line selector, and a sense latch. The differences between the READ/WRITE port (1RW) and a pair of the READ and WRITE ports (1R1W) are in the number of the row decoders, the number of word lines, and the dimensions of the word line drivers.

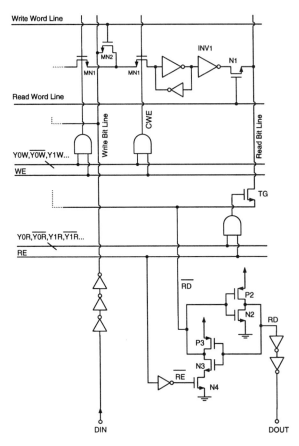

Fig. 6.31 Data I/O circuit for the register file. (From Shinohara et al. [17], ©1991 IEEE.)

While the processor technology evolves, computing power has been enhanced by the parallel architecture techniques such as superscalar. Therefore, the demand on the concurrent read/write capability of the register files becomes higher. Fig. 6.33 shows three types of on-chip register files— (a) shared memory across a bus, (b)local, private, or semi-private memory with each functional unit, and (c) multi-access common storage implemented with a multi-ported register file[18]. For type (a), with a shared bus, load is large—its bandwidth is limited. For type (b), duplicate register files increase the area/complexity. For type (c), the problems with types (a) and (b) have been avoided. However, in type (c), multi-ported register files are required. For a typical ALU system, two three-port (2R1W) register files should be equipped. When the register files are simultaneously accessed by more units, the number of ports should be larger. Fig. 6.34 shows the memory cell and its associated READ and WRITE circuits in a 9-port register file[18]. As shown in the figure, only the read/write circuits referred to three-ports (2R1W) are shown. The data on bus A are for read and write. Bus B is dedicated for read only. The differential write operation

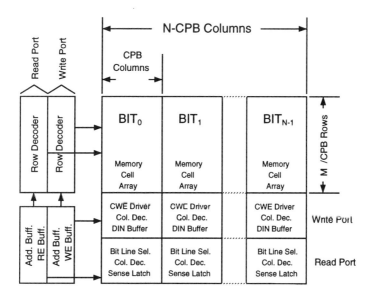

Fig. 6.32 Floor plan of an M-word N-bit register files with two-ports (1R1W). (Adapted from Shinohara et al. [17].)

has been adopted for the write port for higher reliability and robustness. When the WRITE signal is high, the transistor wired-ORed with the port B address decoder turns on. As a result, the word line of port B is pulled high— the data can be written differentially. The READ operation is via single-ended read. The read word line, which is controlled by the address decoder of Port A and B, provides the data from the cell to the bit line of A or B port, followed by the amplification by the sense amp. READ operation can be carried out at several ports at the same time. Therefore, the stored data corruption problems during read should be avoided—during READ, the initial charge sharing between the precharge-to-1 bit line and the cell node storing logic-0 may occur. Thus the access transistor and the memory pull-down transistor should be designed with an appropriate β ratio. Please refer to the description in the SRAM sections in Chapter 5 for details.

Fig. 6.35 shows the translation lookaside buffer (TLB) and the data cache implementing the process from a virtual address to a data item. TLB itself is also a cache memory, which stores the physical page number of the virtual address. Therefore, its tag field stores the virtual page number information. The address sent by the CPU is a virtual address. The more significant bits in the virtual address contain the virtual page number (in the figure [12:31]). Less significant bits are referred to page offset without transformation. The virtual page number is compared with the tag field in the TLB. If matched, the corresponding physical page number information in TLB is taken out and combined with the page offset to form the physical address. The more significant bits of the physical address ([16:31]) are compared with the tag field in the cache memory. If matched, it implies a cache hit—its data are provided to

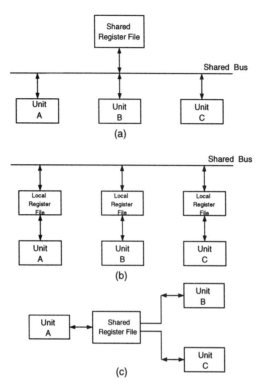

Fig. 6.33 Three types of on-chip register files. (Adapted from Jolly[18].)

the CPU for further processing. If not matched, in the main memory the data at the corresponding location will be read and provided to the CPU for further processing. In addition, it will be written back in the cache memory for future direct access without touching the main memory for fetching data.

Content addressable memory (CAM) is often used to implement the tag field memory in the translation lookaside buffer (TLB) for a CPU. Via TLB, the virtual memory page address can be transformed into the physical page address for accessing the data in the cache. Fig. 6.36 shows the memory cell in the content addressable memory (CAM)[19]. As shown in the figure, the content addressable memory stores the address tag. One key feature of the content addressable memory cell is the transistors $M1$–$M4$ used for tag compare. When doing tag compare, the input at the bit line is the virtual page address bit (Note that at this time word line is low.). Assume that '1' is stored at node a. When the address bit is identical to the data stored in CAM, $M3$ is on and $M4$ is off. In addition, $M1$ is off and $M2$ is on. Therefore, there is no conducting path from the sense line to ground via $M1$, $M3$, $M2$, and $M4$. At the beginning, the sense line is precharged to high. If the address bit is not identical to the tag, consider the case with the stored logic-0 at node a in the memory cell. Thus $M1$ and $M3$ turn on—the sense line is discharged to ground—address

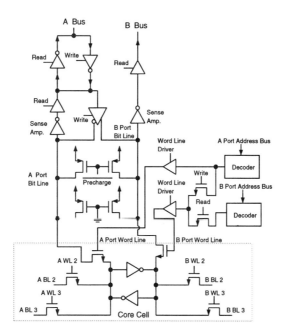

Fig. 6.34 Memory cell and its associated READ and WRITE circuits in a 9-port register file. (From Jolly[18], ©1991 IEEE.)

mismatch. Therefore, using the wired-OR sense line, as long as a bit is not identical, the sense line will be discharged to ground. Only when it is totally matched (tag hit) will the sense line still be high.

Fig. 6.37 shows CAM and SRAM in the TLB[20]. In the CAM, the virtual address tag is stored. In SRAM, the transformed physical address is stored. When a hit is determined by the address tag compare, the sense line is high and the corresponding data in the SRAM can be used for obtaining the physical address. The end of sense (EOS) signal generator is used to simulate the critical path of the sense line discharge when there is only one bit mismatch such that the self-timed signal can be obtained. When the EOS line is discharged to low, if the sense line is high (match), the hitting word driver (HWD) will pull up the SRAM word line to access the data in the SRAM. If the sense line is low (mismatch), the HWD output maintains low, thus SRAM remains inactive. During tag miss, based on the least recently used (LRU) replacement algorithm, the address tag of an LRU entry is changed to the new virtual address tag. At the same time, SRAM is written with the new physical address data. After TLB transforms the virtual address into the physical address, its address is compared with the tag field in the cache memory to determine if it is a cache hit. Fig. 6.38 shows the block diagram of an integrated cache memory, which includes a 32-kbyte data memory, 34-kbit tag memory, 8-kbit valid flags, a 2-kbit LRU flag, a tag compare circuit, and system-level interface circuits[22]. As shown in the figure, the input

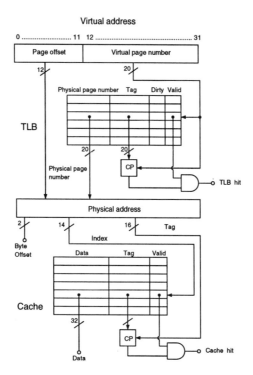

Fig. 6.35 Translation lookaside buffer (TLB) and the data cache implementing the process from a virtual address to a data item. (After Hennessy et al. [38].)

address bits A31–A15 are used for the tag field, which is to be compared with the tag field in the cache memory. If matched, read is initiated. A14–A4 are used to look for the corresponding set. A2–A3 are used for the byte offset. The data in the tag memory are detected by the tag sense amp, and compared with the tag data. As long as the tag RAM data are identical to the compared tag data, two NMOS gates in the C^2MOS-type compare circuit are high. Thus the output is discharged to low. In the following portion, it is a NOR gate. Only when two tag compares are identical, its output will be high. Otherwise, its output will maintain low. After this, it is an AND operation. When all tag bit compares are matched, its output will be high. This tag compare circuit is static—no standby power is consumed.

Fig. 6.39 shows a separated bit-line memory hierarchy architecture (SBMHA) suitable for low-voltage and low-power operation, which combines the caches of Level 1 and Level 2 separated by transmission gates. When accessing the Level 1 cache, there is a smaller bit-line load. When the Level-2 cache is accessed via the hierarchy switch (HSW1), the same bit line and read/write amplifiers are used. Compared to the conventional Level 1-Level 2 hierarchy, SBMHA is simpler. Thus smaller power consumption is expected. For operation at a small voltage (1V), the multi-threshold technique described before has been used. In the memory cell, high

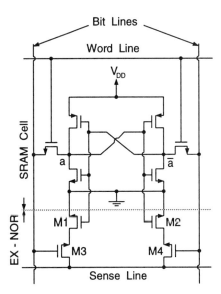

Fig. 6.36 Memory cell in content addressable memory (CAM). (From Kadota et al.[19], ©1985 IEEE.)

V_T devices are used to reduce power consumption. In the peripheral circuits, low V_T devices are used to obtain high speed. As shown in the figure, it is a four-way set associative memory structure. If the request data are in the Level-1 cache, the Level-1 cache data on $BL1$ and $\overline{BL1}$ will propagate via the way switch (WSW) to the read/write amplifier. At this time SA1 switches from low to high (SA2 maintains low) to activate the read amplifier. When the read amplifier is active, SA2 switches high. Thus the data on $BL1$ and $\overline{BL1}$ propagate to $BL3$ and $\overline{BL3}$. After amplification, they are available at the output node. If Level-2 cache contains the data to access, then HSW1 is high. Thus the information in the Level-2 cache is read. In addition, the corresponding information in the Level-1 cache is updated. At this time, the data in Level-2 cache are read first, then are written into Level-1 cache via the same read/write amplifier. Therefore, the structure is concise. When the Level-2 cache is accessed, there is an extra load—the bit-line load of the Level-1 cache.

Fig. 6.40 shows the schematic diagram of the domino tag comparator (DTC) in the cache tag array[23]. As shown in the figure, the tag data in the cache are compared with the physical address tag. If they are identical, it is a cache hit. Different from the CAM described before, the tag comparators are not implemented in the memory cells. Instead, in the domino tag comparator (DTC), a butterfly comparator is added to the bit line. If the address tag bit is high, both S1 and S4 are on. Thus $BL1$ and $BL2$ are connected. In addition, $\overline{BL1}$ and $\overline{BL2}$ are connected. If the address bit is low, $BL1$ and $\overline{BL2}$ are connected; $\overline{BL1}$ and $BL2$ are connected. If the data in the tag memory are identical to the address tag, $BL2$ maintains high and $\overline{BL2}$ is discharged to low. At the beginning, $SA1$ is high, which activates the bit-0 sense

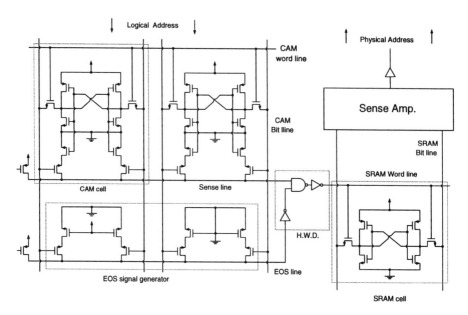

Fig. 6.37 CAM and SRAM in the TLB. (From Kadota et al. [20], ©1987 IEEE.)

amp. If both the address tag and the tag memory data are 1, $\overline{BL1}$ is discharged. In addition, $\overline{BL2}$ is discharged. If both the address tag and the tag memory data are 0, both $BL1$ and $\overline{BL2}$ are discharged ($BL1$ and $\overline{BL2}$ are connected). As long as a match occurs at tag comparisons, $\overline{BL2}$ will discharge to low, and output $H0$ will switch to high, which activates the next-bit comparison (K1). If matches exist for subsequent tag comparisons, $H0$, $H1$, $H2$,.... will switch from low to high one by one as for the domino logic. As long as a mismatch occurs at the tag comparisons, for example, K1, then $H1$ maintains 0. Instead, M1 turns on (since its driving gate signal switches to high), thus the mismatch line is pulled down. The output miss signal is switched high. In addition, $H1$ is 0 and the following K2–K22 sense amps will not be activated. Therefore, power consumption can be effectively reduced. If all miss signals in Way Selector 1–3 ($Miss1$–$Miss3$) are high, in the way selector, $WAY0$ becomes high—its related Way 0 bit lines are connected to the sense amp as shown in Fig. 6.39. $SA2$ is used to speed up domino tag comparison. Under most situations, the variation of the physical address is within several less-significant-bits (LSB) (e.g., 5 LSBs). Thus, when K0–K4 tag comparisons are done, $SA2$ is high—the following K5–K22 sense amps turn on to reduce the comparison time. Using this approach, higher power dissipation has been used to obtain a higher comparison speed.

6.4 PROGRAMMABLE LOGIC ARRAY

Programmable logic arrays (PLA) are frequently used to realize the control logic circuits used in CPU and DSP VLSI circuits. Conventional PLAs are based on

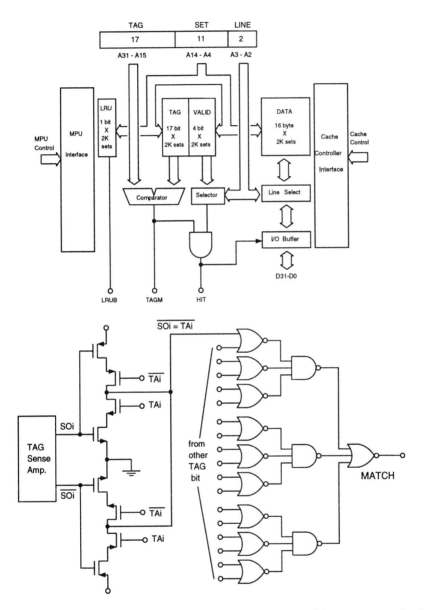

Fig. 6.38 Block diagram of integrated 32-kbyte cache memory and the tag compare circuit. (From Sawada et al. [22], ©1989 IEEE.)

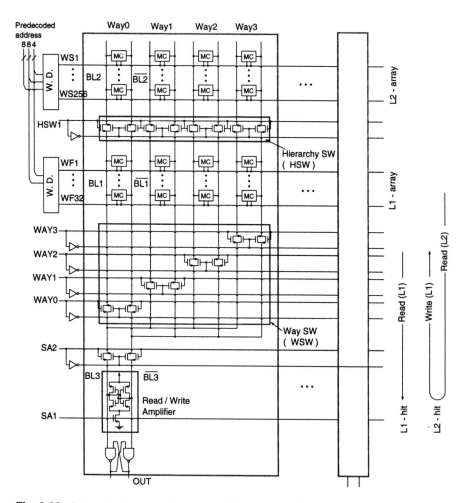

Fig. 6.39 Schematic diagram of separated bit line memory hierarchy architecture (SBMHA) in the cache data array. (From Mizuno et al. [23], ©1996 IEEE.)

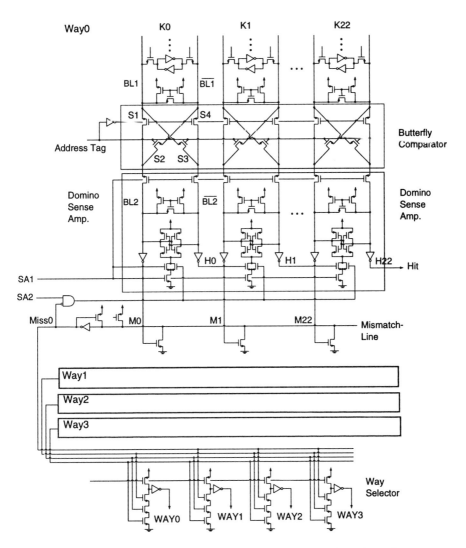

Fig. 6.40 Schematic diagram of the domino tag comparators (DTC) in the cache tag array. (From Mizuno et al. [23], ©1996 IEEE.)

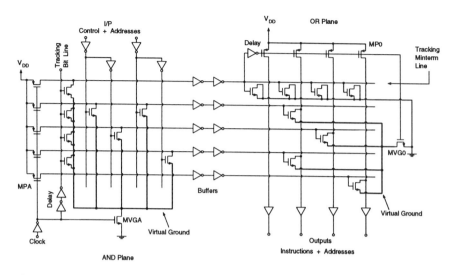

Fig. 6.41 NOR-NOR programmable logic array (PLA). (Adapted from Linz[24].)

pseudo-NMOS logic, which has a ratioed pulled-down structure. As a result, static power consumption is large. Fig. 6.41 shows a NOR-NOR PLA based on the dynamic logic circuits in the OR plane[24]. In order to avoid the signal race problems associated with the high-to-low transition at the input of the dynamic input, the second-stage clock signal is replaced by the self-timed signal produced by the first-stage reference tracking bit line. When the OR-plane enters the logic evaluation period, the output data from the AND plane have already been obtained. Although the speed performance is good, this PLA circuit still has power consumption problems—only the selected minterms are high while others are discharged from high to low. Thus most of the precharged data are discharged to ground. As a result, unnecessary switching power waste exists. In addition, in a large-size PLA, two inverters are needed to serve as buffers. Thus the power waste problem appears in the two inverters. Most inverters have a transition in a clock cycle. The parasitic capacitance of the virtual ground line of the AND plane, which is made of the diffusion line to reduce the layout area, is large. In order to speed up the discharge speed, the discharge NMOS MVGA needs to be large. Note that in the pseudo-NMOS logic, no such problem exists since it is static. When the size of the PLA is not large, these two inverters can be removed to save power and to enhance speed. In order to avoid the power waste problem, Fig. 6.42 shows a NAND-NOT-NOR PLA structure[24]. This is the frequently used 'dynamic PLA'. As shown in the figure, the AND plane works as a domino logic circuit. The OR plane works similarly. Compared with Fig. 6.41, the parasitic capacitance of the virtual ground line in the AND plane is much smaller. Due to its long serial path, the propagation delay increases substantially when the size increases. In the circuit, only the selected minterm will be discharged. Thus the switching power is much smaller.

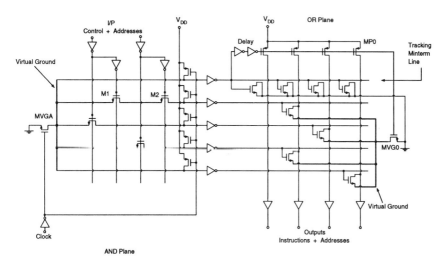

Fig. 6.42 NAND-NOT-NOR PLA. (Adapted from Linz[24].)

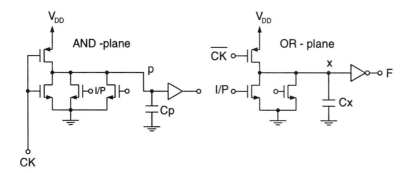

Fig. 6.43 Single-phased dynamic CMOS PLA. (Adapted from Blair[25].)

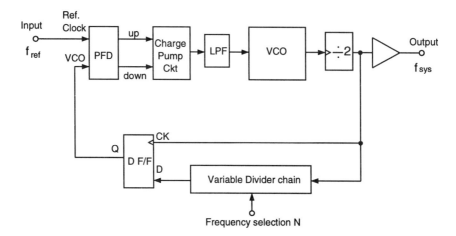

Fig. 6.44 PLL architecture. (Adapted from von Kaenel et al.[26].)

Fig. 6.43 shows a single-phased dynamic CMOS PLA based on pseudo-NMOS and dynamic logic circuits[25]. As shown in the figure, pseudo-NMOS technique has been used to realize the AND plane. When CK is high, it's the predischarge period (precharge period for the OR plane). Node p is predischarged to ground, and node x is precharged to V_{DD}. When CK is low, it is the evaluation period. At this time, the CK-controlled PMOS device in the AND plane turns on—the whole circuit performs as a pseudo-NMOS logic circuit. The output of the AND-plane is transferred to the dynamic OR-plane for OR operation. Thus a NOR-NOR structure can be obtained. In addition, the power consumption problem in the pseudo-NMOS circuit can be reduced. During the evaluation period, only the AND plane is pseudo-NMOS—the OR plane does not have any static power consumption. Compared to the dynamic NOR-NOR PLA, this single-phase dynamic CMOS PLA has smaller switching power consumption. Compared with the dynamic NAND-NOT-NOR PLA, there is no extra delay due to the long serial path. However, in the AND plane, the pseudo-NMOS is ratioed logic, which has high power consumption.

6.5 PHASE-LOCKED LOOP

Nowadays, high-speed VLSI chips operate at their maximum frequency which cannot be directly supplied from an external clock. A further increase in the frequency of external clock is difficult. In order to further increase the system clock frequency to enhance speed, phase-locked loop (PLL) has been used to raise the internal clock frequency in a VLSI system. Fig. 6.44 shows the phase-locked loop (PLL) architecture[26]. As shown in the figure, via the frequency divider, the frequency of the internally generated clock is lowered. After comparing the phase and the frequency errors with the input reference clock at the phase frequency detector (PFD), its phase and frequency differences are used as an instruction for the charge pump

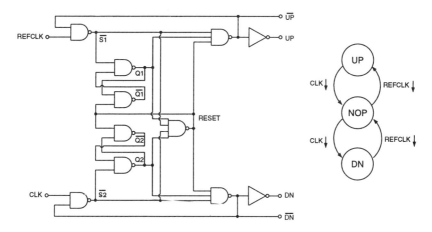

Fig. 6.45 Phase-frequency detector. (Adapted from Young et al.[27].)

(CP) circuit to pump up/down its output. After a low-pass filter to remove the high frequency terms, it has been used as a control voltage for the voltage-controlled oscillator (VCO). The output of VCO is used to produce a system clock with a 50% duty cycle by the ÷2 frequency divider. The frequency of the system clock is determined by the divisor (÷N) of the variable frequency divider. Thus the system clock frequency is: $f_{sys} = N \cdot f_{ref}$. By changing N, the frequency of the system clock can be changed.

Fig. 6.45 shows a phase-frequency detector (PFD) and its state diagram[27]. It is based on the S-R flip-flop and NAND gate. The outputs (UP, DN) of PFD are used to control the charge pump. As shown in the figure, two cross-coupled NAND gates form static S-R flip-flops. State UP represents that a high is presented at the output UP and a low at output DN. State DN represents that a high is presented at the output DN and a low at UP. State NOP represents both outputs UP and DN are low. Assume PFD is at the DN state. If the negative-edge transition of the CLK arrives earlier than REFCLK, the NOP state is changed to the DN state. After the negative edge of REFCLK arrives, it is returned to the NOP state.

Fig. 6.46 shows the charge pump circuit with the loop filter[27]. According to UP and DN, the pump up/down of $V_{control}$ by the current sources I_{up} and I_{dn} can be determined. If $UP = 1$ and $DN = 0$, switches S1 and S4 turn on. $V_{control}$ is connected to I_{up}. In addition, N2 is connected to the output of the unity gain buffer to equalize the voltages at N2 and N1 such that $V_{control}$ is not affected by N2. As shown in the figure, the low-pass filter consists of two MOS capacitors C_1 and C_3 and a resistor R_2. $C_1 R_2$ forms a first-order low-pass filter. Combined with C_3, a second-order low-pass filter is formed to increase the stability of the system.

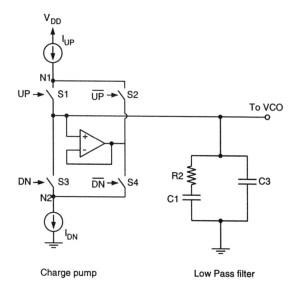

Fig. 6.46 Charge pump circuit and low-pass filter. (Adapted from Young et al.[27].)

The voltage-controlled oscillator (VCO) is made of a ring oscillator consisting of $2n+1$ cascading voltage controlled delay elements such that its output frequency can be adjusted by the control voltage. Fig. 6.47 shows the delay elements for the ring oscillator used in the VCO[27]. As shown in the figure, there are two conventional delay elements: the current starved inverter and the variable capacitance load. The current starved inverter is used to control the current of the inverter by controlling the voltage such that its propagation delay time can be changed. The variable capacitance load is used to adjust the load capacitance of the inverter. The conventional delay elements have drawbacks in the high sensitivity to the power supply noise. As shown in Fig. 6.47(c), in the differential delay element, its delay cell is based on a PMOS source coupled pair with voltage-controlled resistor (VCR) load element. A cascoded tail current source is used to obtain a high impedance to V_{CC} and thus a higher power noise rejection. Its delay is a function of the tail current I_{source}. Via $R_{control}$, the VCR load is controlled while the voltage swing is held constant and independent of the supply voltage. The delay time is proportional to $\tau_d \propto \frac{C \cdot V_{sig}}{I_{source}}$. Thus the frequency is proportional to the current I_{source}. The high power-supply noise rejection is accomplished by the differential signal reference to V_{SS} and the cascoded variable current source from V_{DD}. Fig. 6.48 shows a differential delay element VCO[27]. As shown in the figure, a replica biasing block is used to stabilize the property of VCR. An op amp and a delay cell are used to generate the appropriate bias for the VCR. All currents flow via the differential pair in the left path for emulating the maximum signal condition. A high power-supply noise rejection for the tail current source in the delay stage has been obtained by the voltage-to-current converter and the current mirror approach. AMP is used to amplify the small signal swing to the CMOS level.

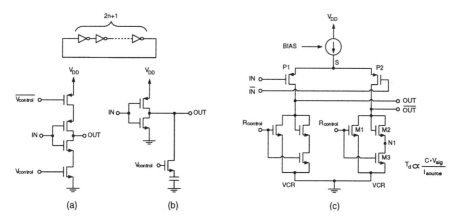

Fig. 6.47 Conventional delay elements and the differential delay elements for the ring oscillator used in VCO. (From Young et al. [27], ©1992 IEEE.)

In order to obtain a high PLL frequency, instead of voltage-controlled, current-controlled oscillator as shown in Fig. 6.49 has been used [28]. It has an input buffer and a loop made of a mixer, a low-pass filter (LPF), a current amplifier, and a current-controlled oscillator (CCO). The CCO output drives an NMOS device with an open drain to provide current for the external 50Ω load. In this PLL, current-mode control signals have been used to lower the supply and substrate noise sensitivity for high-speed operation. As shown in the figure, the CCO is a 3-stage ring oscillator with its PMOS load as the controlled current source. CCO has high speed in a wide tuning range. In addition, its property is affected by the PMOS device characteristics. Fig. 6.50 shows the mixer-LPF-current amplifier[28]. As shown in the figure, M1–M6 comprise the exclusive NOR (XNOR) gate. M7 and C_L form a low-pass filter. M8–M10 organize the current amplifier. V_{b1} is set to $\frac{V_{DD}}{2}$. I_1 is used to accomplish the coarse frequency control.

Fig. 6.51 shows the block diagram of a digital phase-locked loop (PLL) [29]. The digital approach to realize the phase-locked loop (PLL) provides advantages in better noise immunity and less sensitivity to process variations. In addition, the digital PLL is more suitable to operate at a low supply voltage, as compared to the analog PLL. The structure of the digital PLL is similar to the analog PLL except that the analog filter is replaced by a digital loop filter. In addition, the conventional voltage controlled oscillator (VCO) is replaced by the digital controlled oscillator (DCO). The operation of the digital PLL is described here. The reference clock edge and the divided-down DCO edge are detected by the phase detector. Depending on the lead or the lag in the phase difference, the frequency of the DCO can be increased or decreased by the phase detector output. When the difference of the two edges is within the tolerance of the phase detector, no correction pulse is generated. The loop filter re-checks the output from the phase detector to determine how to modify the frequency of the DCO. The loop filter is based on a first-order low-pass filter. When

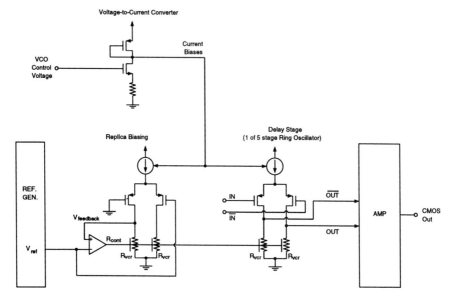

Fig. 6.48 Differential delay element VCO. (Adapted from Young et al.[27].)

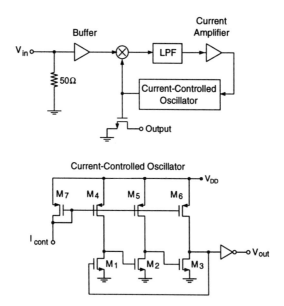

Fig. 6.49 Current-controlled oscillator in a high-speed phase-locked loop. (Adapted from Razavi et al. [28].)

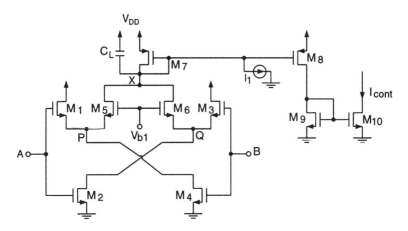

Fig. 6.50 Mixer-LPF-current amplifier. (From Razavi et al. [28], ©1995 IEEE.)

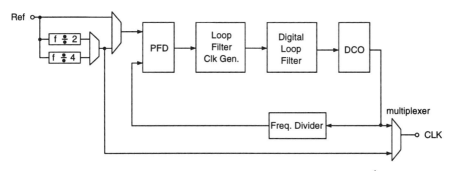

Fig. 6.51 High-level architecture of digital phase-locked loop (PLL). (Adapted from Lee et al.[29].)

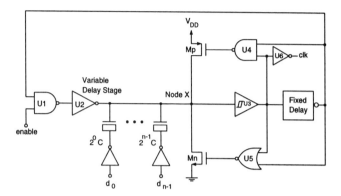

Fig. 6.52 Schematic of digitally controlled oscillator (DCO) with switched-capacitor frequency control. (From Lee et al. [29], ©1997 IEEE.)

the DCO frequency is modified, after the frequency divider, the phase difference is further compared with the reference clock. As a result, a lock-in of the phase is obtained. Fig. 6.52 shows the schematic of digitally controlled oscillator (DCO) with switched-capacitor frequency control[29]. As shown in the figure, this DCO uses a binary-weighted switched capacitor array to control the oscillation frequency. The inverter $U2$ drives a variable load, which is composed of the binary weighted capacitor array. The capacitor is made of NMOS devices with source and drain tied together. When d_{in} is high, the source/drain end of the capacitors is grounded. Therefore, inversion charge exists in the channel of the capacitor device. The load capacitance is the gate capacitance C of the MOS device. When d_{in} is low, the MOS device turns off—the load capacitance is equal to the gate overlap capacitance, $C_{off} = C/k$, $k \gg 1$. Therefore, the load capacitance at node x is approximately equal to:

$$C_x = C_p + d \cdot C + (2^n - 1 - d)C_{off} = C'_p + \frac{k-1}{k} d \cdot C, \qquad (6.24)$$

where $d = d_0 2^0 + d_1 2^1 + ... + d_{n-1} 2^{n-1}$ is the value of the digital control word, $C'_p = C_p + ((2^n - 1)/k)C$. The function of the switched capacitor array is to produce a delay at the output of the inverter $U2$, which is roughly linearly proportional to the digital control word. Schmitt trigger $U3$ is used to detect if x surpasses high or low threshold such that a sharp output can be generated. After a fixed delay element, which is used to adjust an appropriate frequency offset, the signal is connected to $U4$ and $U5$. If x is higher than the $U3$ high threshold, the output of $U3$ is high. Thus, at the output of the fixed delay element, it is high within the fixed delay and low exceeding the delay. Thus, within the delay, both the $U3$ output and the delay element output are high. Therefore, MP is on and MN is off—node x is pulled up to V_{DD}. Until the output of the delay element becomes low, both MP and MN are off. $U2$ begins to discharge x. Thus, MP, MN, $U4$, and $U5$ are used to make sure that at the beginning of the charge/discharge, x is at the rail voltage.

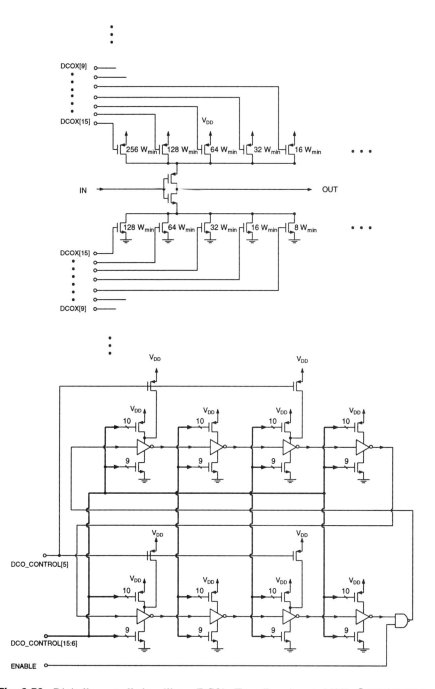

Fig. 6.53 Digitally controlled oscillator (DCO). (From Dunning et al.[30], ©1995 IEEE.)

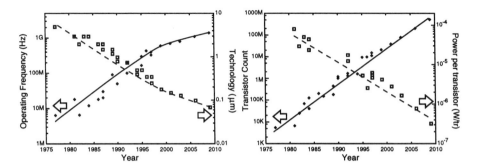

Fig. 6.54 Evolution of the VLSI CMOS technology and CPU. (Adapted from 1997 National Technology Roadmap for Semiconductors[32].)

Fig. 6.53 shows the circuit diagram of another digitally controlled oscillator (DCO) [30]. It has a 16-bit resolution via two techniques. First, by adjusting the W/L aspect ratios of the PMOS and the NMOS devices, a 9-bit resolution has been obtained for the DCO cell. Basically, DCO cell is the current starved inverter mentioned in Fig. 6.47(a). Second, it has cascading 8 DCO cells with the NAND gate and the control bit—$ENABLE$ as shown in the figure. In the 8 DCO cells, there are extra minimum-size PMOS devices. In contrast, there are no extra NMOS devices. Therefore, the pull-up speed of the inverter can be increased. On the other hand, the pull-down speed does not change. From the delay averaging effect of the eight DCO inverters, an extra bit resolution has been reached. Similar approaches can be used in four, two, or one inverters among 8 inverters in the DCO. After the delay averaging of the 8 DCO inverters, from the average of the inverter delays, an extra 3-bit resolution has been obtained. By adding 1/2, 1/4, 1/8 the aspect ratio of the minimum-size PMOS device (by increasing its channel length) to one of the 8 DCO inverters, another 3-bit resolution can be enhanced. Consequently, a total of 16-bit resolution of the DCO has been obtained.

6.6 PROCESSING UNIT

Processing units are very important components in a VLSI system. In this section, three kinds of processing units—central processing unit, floating point processing unit, and digital signal processing unit, are described.

6.6.1 Central Processing Unit (CPU)

The overwhelmingly rapid progress in the computer technology is partly owing to the continuous evolution of the CPU technology. Fig. 6.54 shows the evolution of the VLSI CMOS technology and CPU from 1997 National Technology Roadmap for Semiconductors[32] and ISSCC as described in Chapter 1. From this figure, along with the progress in VLSI CMOS technology and CPU, the number of transistors

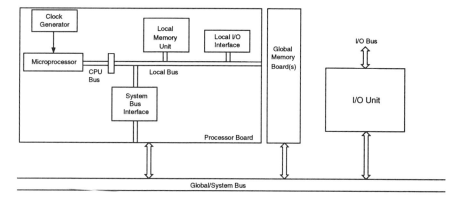

Fig. 6.55 A microprocessor-based system.

in a CPU chip and its operating frequency are exponentially growing. In contrast, the channel length of the CMOS devices used to integrate the CPU chips and its per-transistor power dissipation are decaying exponentially. From this figure, the rapid progress of the CPU is owing to the advancement in the CMOS processing technology and circuit techniques.

Fig. 6.55 shows a typical microprocessor-based system often used in nowadays PCs and workstations[33]. The microprocessor-based system is composed of several important parts— (1) clock generator, which is used to provide the system clock for the microprocessor and other components; (2) memory units, which include main memory made of DRAM chips to store program codes and data and SRAM chips for the level-2 cache memory; (3) I/O units, which provide communication between the microprocessor and the external devices such as keyboard, mouse, printer, and modem; (4) microprocessor (CPU), which is the most important part. The following paragraphs will focus on microprocessors (CPU).

Fig. 6.56 shows the internal components of a microprocessor[33]. Marked by the dashed lines are the functional units contained in an advanced microprocessor. Marked by the solid lines are the standard units in the microprocessor—the bus interface unit, the control unit, and the integer execution unit. Owing to the rapid progress in processing technology, in an advanced microprocessor nowadays, more and more functional blocks can be included in a single chip. In addition to the functional units, in a modern CPU, multimedia functional blocks such as multimedia extension (MMX unit), graphics, and networking are also added such that a better cost/performance (CP) value can be obtained.

For a microprocessor, its operations contain the following five steps—(1) instruction fetch (IF), (2) instruction decode (ID), (3) arithmetic logic unit (ALU) operations, (4) memory access (MEM), and (5) write back (WB).

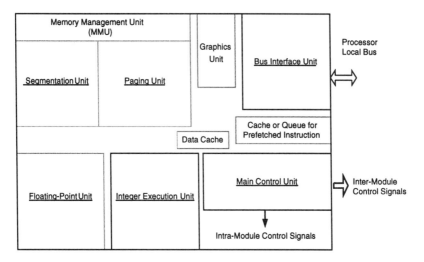

Fig. 6.56 Internal components of a microprocessor.

(1) Instruction Fetch (IF): An instruction cache (I-cache) address is produced by the CPU, which is delivered to the I-cache for determining if the I-cache contains this address information (cache hit).

(2) Instruction Decode (ID): The instruction bytes in the I-cache are read out and delivered to the decoder in the control unit for decoding. Then, the information in the CPU registers, which may be needed during operation, is accessed to become operands.

(3) Arithmetic Logic Unit (ALU) Operations: Depending on the format of the instruction, the instruction execution of the ALU is determined.

(4) Memory Access (MEM): Basically, load/store instructions for accessing D-cache or main memory for data load/store are executed.

(5) Write Back (WB): The data read from memory/D-cache and the results after execution of ALU are written back to the CPU registers.

In a microprocessor, its five-stage instruction sequence and simplified physical representation are shown in Fig. 6.57[33]. In order to raise the efficiency of the CPU instructions, parallelism techniques including pipelining (temporal parallelism) and superscalar (spatial instruction-level parallelism) techniques have been applied widely in microprocessor designs nowadays. If stages as shown in Fig. 6.57 are divided by latches, the pipeline structure can be obtained—the original serial instruction sequence can be transformed into the pipelined instruction sequence. In the original serial instruction sequence, the operation of each stage occurs once for five clock cycles. Using the pipelined structure, the operation of each stage occurs at least once for each clock cycle. Therefore, performance can be enhanced—CPI (cycle per instruction) value becomes close to one. Since the operation of each stage

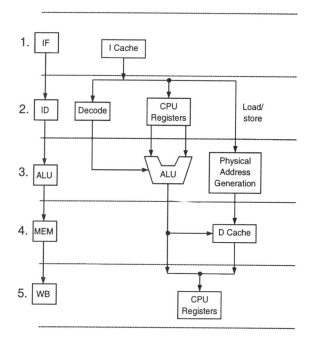

Fig. 6.57 Five-stage instruction sequence. (Adapted from Alexandridis[33].)

occurs during every clock cycle, power dissipation becomes n times higher. As for the superscalar structure, n units are in operation simultaneously. Within a clock cycle, instructions ($CPI = n$) can be carried out in a superscalar system. Therefore, system performance can be enhanced at power dissipation five times higher. For every 20ns two instructions are dispatched simultaneously and executed concurrently in a clock cycle at a clock frequency of 50MHz. When the degree of parallelism becomes larger, the system throughput becomes higher. However, power dissipation also becomes higher.

Fig. 6.58 shows the block diagram of the micro-architecture of the 266-MHz MMX (multi-media extensions)-enabled processor[34]. It contains a 32-kbyte 2-way set associative level-1 instruction cache, a 32-kbyte 2-way set associate level-1 dual-port data cache, and a 20-kbyte predecode cache. The data cache can simultaneously support a read operation and a write operation in a clock cycle. In each cycle, up to 16 instruction bytes are fetched from the instruction cache to the instruction buffer. In each cycle, from the instruction buffer, up to two instructions are decoded. In addition, branches are identified immediately for branch prediction, and up to four RISC86 operations are generated. In each cycle, up to six RISC86 instructions issued to seven parallel execution units are speculatively executed based on the branch prediction result. It includes one memory read, one memory write, two register or one multimedia register operation, one floating point operation, and one branch condition evaluation. In a cycle, up to four RISC86 operations can be retired in order. In some

398 CMOS VLSI SYSTEMS

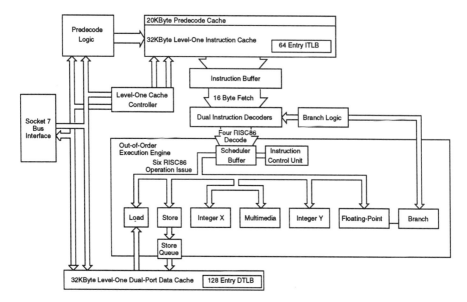

Fig. 6.58 Block diagram of the microarchitecture of the 266-MHz MMX (multi-media extensions)-enabled processor. (From Draper et al. [34], ©1997 IEEE.)

processors, it also contains multi-media extensions (MMX) unit to the X86 instruction set. MMX unit adopts the single instruction multiple data (SIMD) technique to handle the 8, 16, 32, 64-bit multiple operands in the 64-bit data path for performing highly parallel applications. In addition, eight new 64-bit registers are provided as a part of the SIMD pipeline structure.

Programmable logic arrays (PLA) are often used to realize the control logic in a microprocessor. The requirements for PLA to be used in the microprocessor are high speed and immunity against coupling between bit lines. Fig. 6.59 shows the twisted differential bit line structure for a PLA[34], which is based on twisted fully differential bit lines and AND/OR arrays to improve speed and to reduce influences from the bit-line coupling. For a standard PLA, the single-ended bit line structure is often used, which has drawbacks in the coupling problems. (Since at the bit line of a PLA, there exist multiple pull-down structures.) The needed minterms are not pulled down. Only the unnecessary minterms are pulled down. Due to the NOR/OR structure, multiple pull-downs may exist, hence the coupling problems may be serious. The mechanism of reducing coupling is similar to the folded bit line structure in a DRAM as described in Chapter 5. As shown in the figure, both true bit line and the complementary bit line are present. The complementary bit line is used for reference, where PLACAP devices are connected to mimic the identical load as the true bit line. Therefore, at the complementary bit line, all devices are dummy devices—their source end is not programmed. Only at the read bit has a pull-down device with an appropriate aspect ratio been designed to mimic the pull-down waveform for the

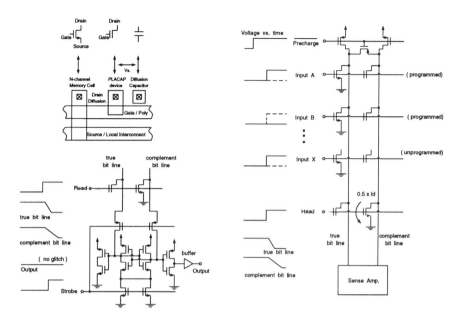

Fig. 6.59 PLACAP matching load used in PLA. (From Draper et al. [34], ©1997 IEEE.)

device with half of the drain current. Therefore, if there is a discharge operation at the true bit line, its discharge speed is at least two times that at the complementary bit line. When not at discharge, a noticeable difference between the true and the complementary bit lines can be identified. Also shown in the figure is the strobed sense amp for the true and the complementary bit lines. Before the strobe signal is active (high), the sense amp is inactive. When the voltage difference between the true and the complementary bit lines is sufficiently large, the strobe signal changes to high to activate the sense amp, which is made of a latch with back-to-back connected cross-coupled inverters.

Fig. 6.60 shows the edge-triggered latch used in the microprocessor[34]. For standard latches, they are level sensitive—when CK is high, the latch output varies with the input data. When CK is high, any change in the input may lead to a corresponding change in the output. In order to have a latch with output only at the CK transition edge (edge-triggered), for example, for the 0 to 1 transition edge, cascading of the master and the slave latches is required. (Note that when $CK = 0$, the master latch is in the transparent mode—the output varies with the input. When $CK = 1$, the master latch in the latch/hold mode—the output does not change with the input. When $CK = 0$, the slave latch is in the latch/hold mode. When $CK = 1$, the slave latch is in the transparent mode.) As shown in the figure, the edge-triggered latch is implemented by only one latch. When CK is low, node X is precharged to high, N1 is off, and N3 is on ($CKDB$ is high). Thus, P4 turns off (X is high), and N4 is off (CK is low). Therefore, the output Q is in the latch/hold mode. When CK changes

400 CMOS VLSI SYSTEMS

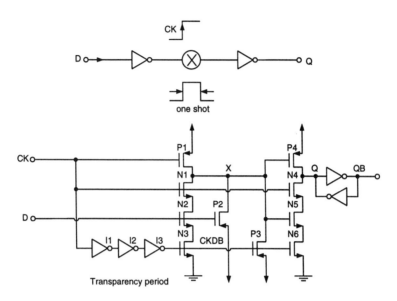

Fig. 6.60 Edge-triggered latch used in the microprocessor. (Adapted from Draper et al. [34].)

from 0 to 1, P1 is off, N1 is on, and N4 is on. At this time, $CKDB$ is still high. Hence, N3 is still on, and N6 is still on. D determines if X stays high. If $D = 0$, N2 is off and P2 is on. Thus, X stays high and Q is low. If $D = 1$, N2 is on and P2 is off. X is pulled down—Q is high. After three inverter delays, $CKDB$ changes to low—N3 turns off, and N6 turns off. P3 turns on—node X is charged to high to turn off P4. Thus, Q holds the data. This latch is different from the conventional level sensitive latch—when $CKDB$ changes from high to low, in spite of $CK = high$, a change in D will not affect the output Q. Therefore, it implies that between the $low \rightarrow high$ CK transition and the $high \rightarrow low$ $CKDB$ transition, a one-shot is formed. Only when both CK and $CKDB$ are high does the output Q change with D—an effective edge triggered latch(ETL).

Fig. 6.61 shows the dual-rail pulsed edge trigger latch (ETL) for self-resetting logic[34]. By including the self-resetting circuit, the output resets to low after the output data have been used by the next logic stage. In the quiescent state (CK is low), $rst1$ is high, both q and qb are low, and QP/QBP are low. When CLK changes from low to high, $CKDB$ is high initially, then changes to low at a delay of three inverters. Thus, when both CK and $CKDB$ are high, one of q and q_b is discharged (depending on D—if D is high, it is qb; if D is low it is q.) Thus one of QP or QBP switches from low to high. As a result, the output of the NOR gate changes from high to low. Thus, after two inverter delays, $rst1$ changes from high to low. One of q and q_b is charged from low to high again. When CK changes from low to high, one of QP and QPB switch from low to high. After some gate delays, it is reset back to the quiescent value 0. Therefore, after the output data being used in the next stage,

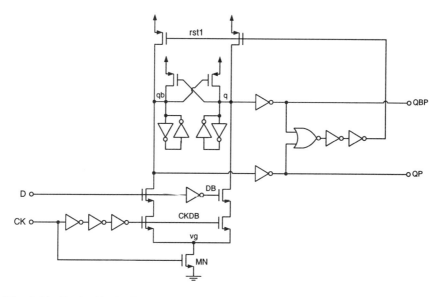

Fig. 6.61 Dual-rail pulsed edge triggered latch (ETL) for self-resetting logic. (Adapted rom Draper et al. [34], ©1997 IEEE.)

for some gate delays, the output is reset to 0.

In a high-speed system, dynamic logic circuit techniques are frequently used. The drawback of the dynamic logic circuits is its small noise immunity. Consider a dynamic domino logic gate used in a CPU as shown in Fig. 6.62. During the evaluation period, if there is no conducting path in the NMOS logic, the internal node is floating. Under this situation, the circuit is susceptible to noise. If the floating node voltage changes to a value smaller than the gate threshold voltage of the following inverter, a false output occurs. Therefore, while doing the circuit design, techniques to prevent the floating dynamic output from noise are used. As shown in the figure, three pull-up PMOS devices have been used to prevent noise. In type A, a PMOS load with its gate connected to ground has been used to reduce the interference from noise. However, while using the PMOS load, the circuit should be properly designed such that the speed of the NMOS logic pull-down (it is ratioed logic) is not affected. In type B, a PMOS device controlled by an output inverter feedback is used to reduce noise effects. During the pull-down of the NMOS logic, it is still ratioed logic. Then, the inverter feedback turns off the PMOS device such that power dissipation can be reduced. Using the type B concept, type C is for driving a large load. In order to prevent the feedback signal from being too late, a driving inverter and a feedback inverter are used separately. Fig. 6.63 shows the noise-tolerant precharge (NTP) circuit[35]. As shown in the circuit, the key point is that, except the precharge PMOS device, another noise-tolerant PMOS logic is added to prevent the dynamic node from floating when there is no conducting path in the NMOS logic. The most straightforward way is to design the noise-tolerant PMOS logic using the static CMOS method. As shown in

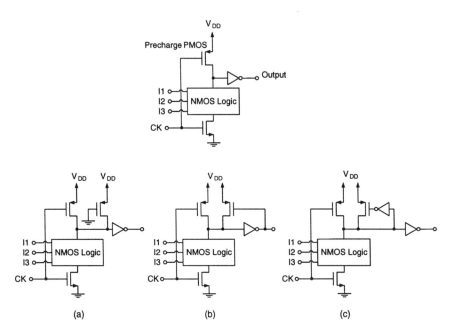

Fig. 6.62 Noise problems of a dynamic domino logic gate in a CPU at low voltage. (Adapted from Murabayashi et al.[35].)

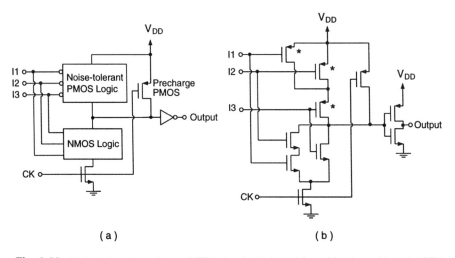

Fig. 6.63 Noise tolerant precharge (NTP) circuit. (Adapted from Murabayashi et al. [35].)

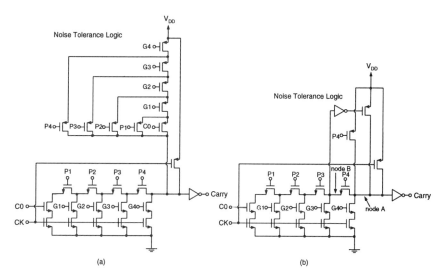

Fig. 6.64 Four-bit carry propagation noise-tolerant-precharge (NTP) circuits—(a) fully functional noise-tolerant design; and (b) simplified noise-tolerant design. (From Murabayashi et al. [35], ©1996 IEEE.)

the figure, for a logic function of $In1 \cdot In2 + In3$, the noise-tolerant PMOS logic is the p-block of the static logic except that minimum sizes are used for minimizing the parasitic capacitances of the p-block. The purpose of this circuit is to prevent dynamic nodes from the influence of the noise. Therefore, its objective is different from that of the static logic. It has advantages in the noise tolerance and the property of static logic circuits. In addition, its parasitic capacitance is smaller than that of the static logic circuits since minimum-size PMOS devices are used. On the other hand, its complexity (and thus load) are higher as compared to the dynamic logic circuits. Fig. 6.64 shows a 4-bit carry propagation noise-tolerant-precharge (NTP) circuits—(a)fully functional noise-tolerant design and (b) simplified noise-tolerant design[35]. As shown in the figure, for the fully functional noise-tolerant design, it is based on the static logic circuit. In the simplified noise-tolerant design, $\overline{C_3}$ at node B is used to prevent node A ($\overline{C_4}$) from floating. The conditions for node A being high (floating) are (1) P_4 is low and G_4 is low, (2) P_4 is high and node B is high. In the simplified noise-tolerant design, the above problem is avoided via controlling a pull-up PMOS by the inverted signal of the node B, thus the gate can be simplified. However, node B is still susceptible to the noise effects—the noise-tolerant PMOS logic may not be applicable. In addition, when node B is high or P_4 is low and G_4 is high, there is a problem—NMOS and PMOS devices turn on simultaneously. This problem can be resolved by adding a PMOS device, which is controlled by G4, cascading with the two pull-up PMOS devices.

Fig. 6.65 shows the memory cell in an 8-port register file with 4 read-ports and 4 write-ports[35]. In the memory cell, a feedforward inverter is added for the read port.

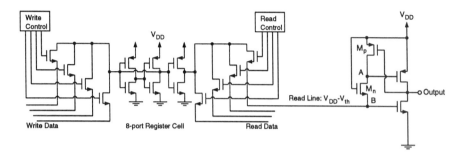

Fig. 6.65 The memory cell in an 8-port register file. (Adapted from Murabayashi [35].)

The read data are passed via NMOS devices. Due to body effect, during reading logic-1, the voltage of the readline is $V_{DD} - V_{th}$. At a supply voltage of 2.5V, power dissipation of following buffers may increase. Also shown in Fig. 6.65 is the leakless buffer circuit[35]. In the inverter, the PMOS gate terminal (A) is separated from the NMOS gate terminal (B). Consider the read logic-1 case on the read line. Initially, B is low, the output is high. Therefore, the pull-up PMOS device is off. Since the gate of the separate NMOS device M_n is connected to V_{DD}, node A is grounded. When the voltage of the readline rises, M_n is on and functions as a pass transistor—node A rises with node B. When the output switches to low, the pull-up PMOS device M_p turns on. As a result, node A is pulled up to V_{DD}. Therefore, NMOS and PMOS devices in the output inverter do not turn on simultaneously—no static power. This buffer is different from the conventional output feedback-controlled pull-up PMOS device, which is used to pull up the read line to V_{DD}. The maximum voltage of the read line is still $V_{DD} - V_{th}$, but the gate voltage of the inverter PMOS gate can be V_{DD} via the pull-up PMOS device. With this buffer, node A can be pulled up to V_{DD} and node B can be pulled up to $V_{DD} - V_T$. Therefore, a small read-line swing enhances the switching speed. When the read line at B has a transition from logic 1 to logic 0, during the discharge process, its ratio logic-related problem is similar.

In implementing a high-speed circuit for a VLSI CPU circuit operating at a high clock rate, there is an important circuit technique—the self-timed pipeline technique. In the domino logic, after all logic gates finish evaluation, precharge of all gates, which is controlled by clock, is initiated. In the self-timed pipeline technique, the precharge logic is not driven by the clock. Instead, when the next-stage logic gate accomplishes its logic operation, the current-stage logic gate enters logic precharge. Therefore, it is called post-charge logic or self-reset logic. Using the self-reset logic, the clock load can be substantially decreased. In addition, the selective discharge/precharge also decreases power consumption—if there is no change at the output, no precharge is required. Therefore, the self-reset logic is suitable to be used in the RAM structure—it is also called delayed reset logic, as shown in Fig. 6.66[36]. The operation of delayed reset logic is like a dynamic logic circuit. Therefore, the n-p-n-p arrangement is used to ensure that at the beginning of the evaluation period the inputs make all logic transistors turn off. For the 2-input NAND and inverter

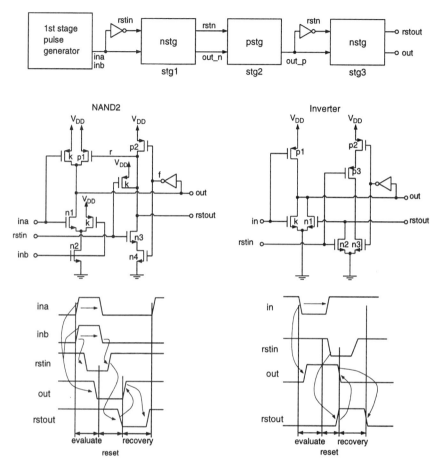

Fig. 6.66 Delayed reset logic used in a microprocessor. (From Lev et al. [36], ©1995 IEEE.)

as shown in the figure, in addition to the common logic inputs, there is the rst_in input, which is used to control the output reset (rst_out). For example, in the NAND2 circuit, the device marked k is used for precharge. If in_a is low at the beginning, the output is precharged to V_{DD}. When both in_a and in_b are high, since both NMOS n1 and n2 turn on, the output is discharged to low. The operation of the circuit is similar to a dynamic logic gate except that the precharge transistor is controlled by the input signal, which is similar to the CDPD dynamic logic described in Chapter 4. After the output is discharged to 0, the reset is accomplished via the reset circuit, which is composed of p1, p2, n3, n4, and the feedback inverter. When rst_in switches from low to high, it represents that output must be reset. At this time, n3 is on. If the output is discharged to 0, f is high, thus n4 turns on. As a result, rst_out is discharged to low— the reset is initiated. Thus, p1 is turned on and the *out* is reset to high. After *out* is reset to high, f switches to low—p2 turns on. Therefore, rst_out is charged

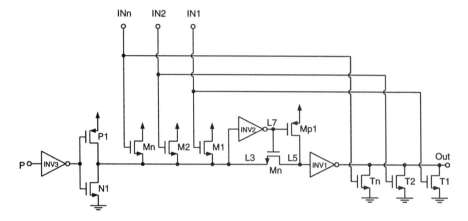

Fig. 6.67 Cascode NOR gate. (Adapted from Lev et al. [36].)

to high— this implies the end of reset. Then, after completing the reset operation, p1 turns off. rst_out becomes rst_in at the next stage. The operation of the p-logic is similar to that of the n-logic except that a transition from high to low at the rst_in results in the reset of the output. Note that the rst_in for driving the n-logic is obtained by taking the inversion of the input. In contrast, the rst_in for driving the p-logic is obtained by the rst_out of the previous-stage n-logic.

Fig. 6.67 shows the cascode NOR gate used to process wide OR[36]. As shown in the figure, P is used to control the operation of the gate. When P is high, it is the shut-down operation: PMOS P1 turns on, thus node L3 is high and output is low. At this time, no matter what the input value is, the output is not affected. Under this situation, the gate does not consume any dynamic power. When P is low, the gate can operate. NMOS N1 is on and node L3 at quiescent is low. When any input is high—assuming IN1 is high—M1 turns on and node L3 is pulled up. At the same time, T1 turns on and Out is pulled down. When the voltage at L3 exceeds the Inv2 gate threshold, node L7 is pulled down and M_{P1} is turned on. Thus node L5 is pulled up to V_{DD}. Therefore, Inv1 helps Out pull down to low. When all inputs are low, $M1,...Mn, T1,...Tn$ are all off. At this time, N1 will pull down node L3. As a result, node L7 is pulled high—NMOS MN is turned on. Consequently, node L5 is discharged with node L3. Then via the inverter, Inv1 pulls Out up to high.

Until now in this section, design issues regarding CPU designs were focused on high performance and computing power. Power consumption is not the only major concern. Low power dissipation is for the purpose of achieving improved reliability and reduced electromigration problems. Nowadays, the demand for mobile computing becomes higher and higher. How to lower the consumed power under a condition that the computing power is not sacrificed has been the general trend in designing next-generation CPU. Table. 6.1 shows the strategies of reducing power dissipation for a 32-bit RISC microprocessor[37]. Power consumption of a VLSI CPU chip can

Alpha 21064: 26W @ 200MHz and 3.45V

Methods	Power Reduction	Expected Power
V_{DD} reduction $3.45V \rightarrow 1.5V$	5.3×	4.9W
Reduce function	3×	1.6W
Scale process	2×	0.8W
Reduce clock load	1.3×	0.6W
Reduce clock rate	1.25×	0.5W

Table 6.1 Strategies of reducing power dissipation for a 32-bit RISC microprocessor. (Adapted from Montanaro et al. [37].)

be reduced via several techniques. Via reducing V_{DD}, reducing functions, scaling the processing technology, reducing the clock load, and reducing clock rates, power dissipation can be lowered. Among all techniques, the V_{DD} reduction approach is the most effective in reducing power consumption. When the supply voltage is reduced from 3.45V to 1.5V, power consumption is shrunk by 5.3×.

In order to have a stable operation at a low power-supply voltage, in the CPU, many circuits are designed to be static or pseudo-static. In addition, no circuits are adopted if they are not suitable for low-voltage operation. Under some situations where wide OR operation is required, pseudo-static techniques, which are based on weak static feedback circuits or self-timed circuits to latch the output data and to restore the dynamic node to the precharged state, have been used. Fig. 6.68 shows a self-timed register file (RF) precharge technique[37]. As shown in the figure, a dummy bit cell added for self-timing is used to mimic the read data delay time at the output. The register file read is done by the READ_WORDLINE controlled NMOS device and the bit line-controlled NMOS device. Only when the data bit is high and the READ_WORDLINE is high, is the DATA_BITLINE discharged. The dummy cell is used to mimic this situation. Therefore, in every cycle the self-bit line is discharged. When CLOCK_L is low, it is the precharge period. When CLOCK_L is high, READ_WORDLINE is high. Thus SELF_BITLINE is discharged. In addition, the SELF_ENABLE signal is also discharged to low. When the SELF_ENABLE signal is low, it implies that the data in the register file have been read out. Therefore, READ_WORDLINE and PRECHARGE_L are pulled down to low. Thus register file read is disabled and the DATA_BITLINE is precharged to high. From the above analysis, only at the very beginning of the read cycle, may DATA_BITLINE be floating. Then it is back to the precharged state.

Fig. 6.69 shows a differential edge-triggered latch circuit, which is designed to operate as a static circuit as much as possible[37]. Only at the positive clock edge

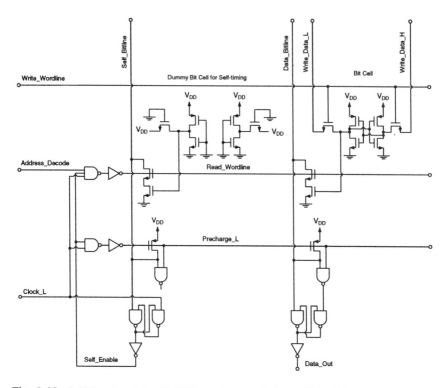

Fig. 6.68 Self-timed register file (RF) precharge technique. (From Montanaro et al. [37], ©1996 IEEE.)

will the output change in accordance with the change in the input. The operation of this latch is described here. When CLK is low, the output is held. L1 and L2 are precharged to V_{DD} individually. When CLK is high, the differential sense amp starts to work. If IN is high and $\overline{IN}$ is low, MNH turns on and MNL is off. Thus nodes L3 and L2 are discharged to low. Node L1 is pulled high since $MP2$ turns on. If without MN, when IN switches to low and $\overline{IN}$ switches to high, MNH is off and MNL is on. Node L4 is low and both nodes L3 and L2 are floating. At this time, the leakage of the PMOS device, which is off, may make nodes L3 and L4 drift upward. At a low power-supply voltage, with a low threshold-voltage device, this leakage problem becomes worse. With MN, nodes L2 and L3 can be stabilized at low (since L4 is low; MN and $MN1$ are on). Thus a static operation can be maintained. The inclusion of MN does not consume static power. With MN, the leakage current effect can be removed. In addition, the leakage static power consumption is small, which does not cause substantial influence in the overall power dissipation.

6.6.2 Floating-Point Processing Unit

The floating-point processing unit is important in current computer systems. Fig. 6.70

PROCESSING UNIT **409**

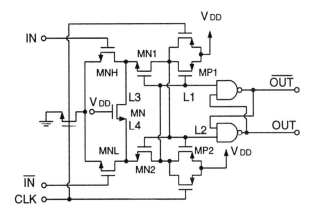

Fig. 6.69 Differential edge-triggered latch circuit designed to operate as a static circuit as much as possible. (Adapted from Montanaro et al. [37].)

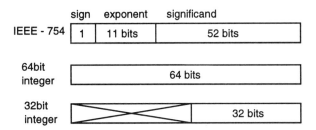

Fig. 6.70 Floating-point and integer formats.

shows the floating-point and integer formats. As shown in the figure, IEEE-754 standard is the frequently used floating-point format, which contains a sign bit (s), 11 exponent bits, and 52 significand (mantissa) bits in the CPU of a 64-bit computer system including workstations and PC:

$$A = (-1)^s \cdot (1 + significand) \times 2^{(exponent-1023)}. \tag{6.25}$$

For example, $(\frac{3}{2})_{10} = 1 + \frac{1}{2} = (1.1)_2$ is expressed as: $A = (-1)^0 \cdot (1 + 0.1) \times 2^{(1023-1023)}$, where the sign bit is 0, the exponent bits are 01111111111, the significand bits are 1000...000 (51 0's). As for the integer format, there are two modes: 32-bit and 64-bit.

In the procedure of the floating-point addition[38], addition of two floating numbers is started with comparing the exponents of the two numbers. The number with the smaller exponent is shifted right until two exponents are identical. Then, two significands are added. The sum is normalized by either shifting right or/and incrementing the exponent or shifting left and decrementing the exponent. The next step is to check if there is an overflow or an underflow. If normal, the result is normalized again such that the result can meet the designated bit resolution. Thus the addition of two floating numbers is done.

For multiplication of two floating numbers, the biased exponents of the two numbers are added, followed by subtracting the bias from the sum to get a new biased exponent. Then, the significands are multiplied. The product is normalized by shifting it right and incrementing the exponent if necessary. Then, the result is checked for overflow or underflow. If normal, the significand is rounded to the appropriate number of bits. Finally, the sign of the product is determined. If the signs of the original operands are the same, the sign of the product is positive. If they are different, the sign of the product is made to be negative.

For a floating-point multiplier designed based on the IEEE 754 standard, the bit numbers of its multiplicand and multiplier are important. From Eq. (6.25), the '1' in the (1+significand) is a hidden bit, which should be considered in multiplication and addition. Including the sign bit, the floating-point multiplier is designed to be 54-bit×54-bit (52 bits+hidden bit+sign bit) in general.

Fig. 6.71 shows the block diagram of a 320-MFLOPS CMOS floating-point processing unit for superscalar processors[39]. As shown in the figure, it contains an arithmetic logic unit (ALU), a multiply/divide unit (MDU), and a register file. In order to meet more and more rigorous requirements in computing power, the floating-point processing unit (FPU) needs to do ALU and MDU operations simultaneously in each cycle (1-cycle/2-operations). Via the pipeline and 4W6R register file structures, except floating-point division, the throughput of all other operations as shown in Fig. 6.71 is 1-cycle/1-operation. Fig. 6.72 shows the block diagram of ALU[39]. In addition to the floating-point add/subtract, the ALU can be used to perform other operations. The ALU contains a three-stage pipeline structure. The first stage performs

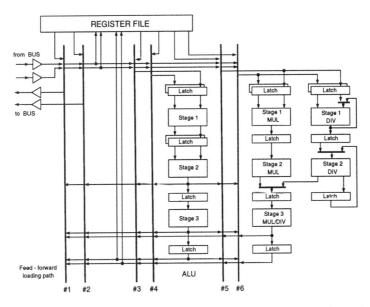

Fig. 6.71 Block diagram of a 320-MFLOPS CMOS floating-point processing unit for superscalar processors. (From Ide et al. [39], ©1993 IEEE.)

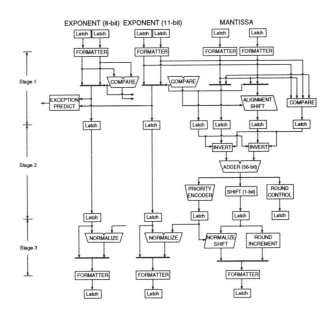

Fig. 6.72 Block diagram of ALU used in the floating-point processing unit. (From Ide et al. [39], ©1993 IEEE.)

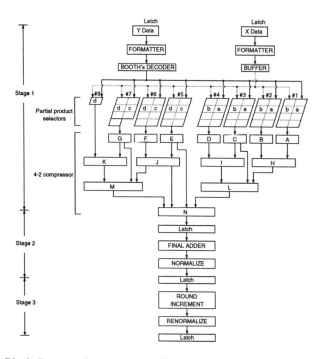

Fig. 6.73 Block diagram of multiplier (MUL) unit in the floating-point processor. (From Ide et al. [39], ©1993 IEEE.)

exponent comparison and significands (mantissa) alignment. The second stage performs significands addition/subtraction and normalization shift step scanning. The third stage performs rounding or normalization. Fig. 6.73 shows the block diagram of the multiplier (MUL)[39]. As shown in the figure, it contains similarly a three-stage pipeline structure. The first stage performs partial product generation, carry save addition of significands, and approximate exponent computing. The modified Booth algorithm and Wallace tree with 4-to-2 compressors, as described before, are adopted. The second stage performs final carry propagation addition and normalization of the significands. The third stage performs rounding and bypassing. Fig. 6.74 shows the block diagram of the divider (DIV)[39]. As shown in the figure, it adopts the radix-4 SRT-division[40] and 5 symbols—2, 1, 0, −1, −2. In each cycle, the 4-bit quotient can be computed. Via the two-level cascaded carry save adder, in the same cycle, the next partial remainder can be computed.

6.6.3 Digital Signal Processing Unit

The main functions of digital signal processors (DSP) are used to replace conventional analog circuits such as amplifiers, modulators, and filters. Fig. 6.75 shows a typical digital signal processing system, where processing of the signals is handled by the digital approach. The advantages of the digital signal processing is precision

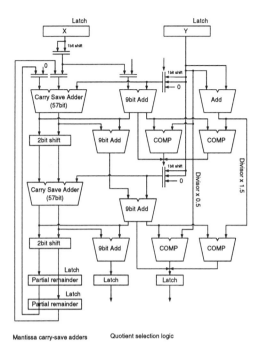

Fig. 6.74 Block diagram of the divider (DIV) unit used in the floating-point processor. (From Ide et al. [39], ©1993 IEEE.)

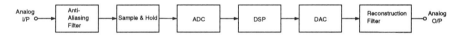

Fig. 6.75 A digital signal processing system.

and repeatability. In addition, designs can be easily implemented, simulated, and modified in software. Using the digital signal processors, complexity of the signal processing becomes virtually independent of the hardward system. Many operations, which are difficult to be handled by the analog circuits, can be easily handled by the digital signal processing approach. As shown in Fig. 6.75, in a digital signal processor-based system, the input analog waveforms are filtered, sampled, then transformed into digital data for digital signal processor, where further processing is implemented in software. The digital output signals of the digital signal processor are converted back to the analog signals and passed through a reconstruction filter to transform the discrete level signals into continuous analog signals for communication with the outside world. In a wireless communication system, digital signal processor (DSP) chips are used to perform speech coding, channel coding, demodulation, and equalization. High computational throughput at small power consumption is important for DSP chips. A frequently used approach is to design the DSP chip for operation at a power supply voltage of 1V and using dual V_T devices to maintain high performance. Fig. 6.76 shows a DSP used in a cellular phone application[29].

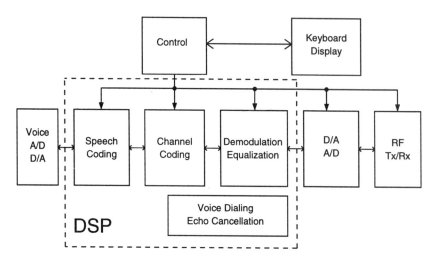

Fig. 6.76 Use of DSP in a cellular phone application. (Adapted from Lee et al. [29].)

As shown in the figure, via analog-to-digital (A/D) conversion and digital-to-analog (D/A) conversion, the analog voice signal from the outside environment is transformed into digital signals for further processing. Thus the digital signal processing capability is important. Fig. 6.77 shows the block diagram of a DSP architecture[29]. As shown in the figure, it adopts pipelining and parallelism to enhance the computing capability. A six-stage instruction pipeline structure, three data buses, a program bus, two data address generators, and a program address generator are included. Therefore, in a clock cycle, two read and one write operations can be carried out. Using the concurrent processing technique, a concise instruction set can be used to enhance the energy efficiency of the processor. Fig. 6.78 shows a gated slave logic to reduce the unnecessary switching activity[29]. As shown in the figure, the gated clock techniques can be used to increase power saving. Latches are clocked only when input data are useful. This arrangement is carried out via a locally master latch clock with a data ready signal, which is produced by the local control logic with the output from the instruction decoder. In the gated slave logic, when the master clock is imposed on the master latch, the enable signal is generated such that the slave latch can receive the gated clock. Thus the functional block can work only when valid data are available. Therefore, via a power down instruction, some portions of the system can be halted to reduce power consumption.

Reducing clock load is an effective method to lower power consumption. Fig. 6.79 shows the conventional CMOS latch and the lower-power latch for reducing clock loading[29]. As shown in the figure, the NMOS pass transistor is used to control the transfer of D_{in} to the output. Since the NMOS pass transistor cannot effectively pass logic 1, another NMOS device $MN2$ and the D_{in}-controlled device $MN1$ are added. When ϕ_m is high and D_{in} is high, both $MN1$ and $MN2$ turn on. Thus node n is discharged to low—node x can be pulled up to V_{DD} via an inverter. If D_{in} is low, $MN1$

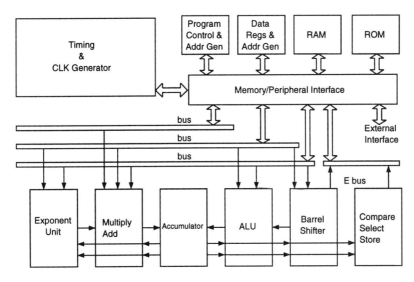

Fig. 6.77 Block diagram of a DSP architecture. (Adapted from Lee et al. [29].)

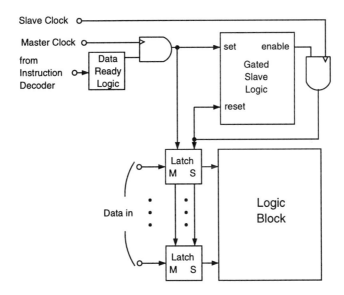

Fig. 6.78 A gated slave logic to reduce the unnecessary switching activity. (Adapted from Lee et al. [29].)

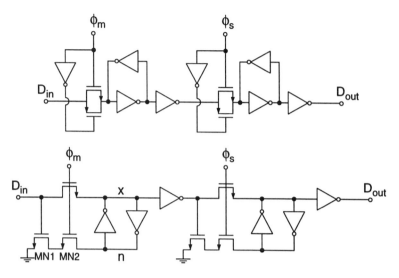

Fig. 6.79 Conventional CMOS latch and the lower-power latch for reducing clock loading. (From Lee et al. [29], ©1997 IEEE.)

is off—the node x discharged by D_{in} is not affected. By this arrangement, the transmission gate becomes NMOS pass transistor, thus clock load is reduced, and area is shrunk. In addition, speed is enhanced, owing to the reduction from an inverter delay.

Fig. 6.80 shows the power-management sequences in the DSP chip for cellular phones[42]. As shown in the figure, when the cellular phone enters the active mode, DSP enters full operation. Thus its active power is large. When the phone is not in use, it is in the waiting state. During the waiting state, the DSP circuit periodically checks if any call comes in. Thus, during the waiting state, there is dynamic power consumption periodically. Even without any operation, due to leakage current, there is some static power consumption. This is especially important for the DSP operating at a low voltage such as 1V. In order to enhance performance, low-threshold voltage devices should be used. The leakage-related power consumption is very small when compared to the case in active mode. On the other hand, the leakage-related power consumption still cannot be overlooked during a long duration of the waiting mode. Thus, as shown in Fig. 6.80, reducing the leakage-related power consumption is important to lengthen the usage time of the battery. Fig. 6.81 shows the connection between the DSP and the embedded power-management processor (PMP) using a multi-threshold technique[42]. As shown in the figure, the low-V_T devices are used in the DSP block such that speed performance can be enhanced. The high-V_T device is used for the power switch transistor. When entering the waiting mode, the power-management processor will output the SL (sleep) signal to turn off the power switch transistor. Since the power switch transistor is a high V_T device, the leakage current is small. The leakage-related power consumption can be reduced substantially, thus the battery can last longer.

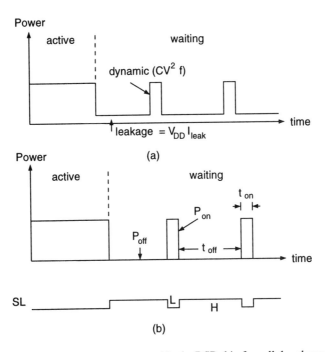

Fig. 6.80 Power-management sequence used in the DSP chip for cellular phones. (Adapted from Mutoh et al.[42].)

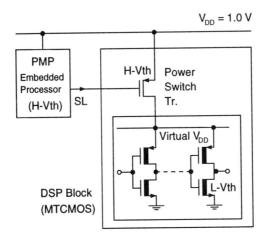

Fig. 6.81 Connection between the DSP and the embedded power-management processor (PMP) using a multi-threshold technique. (Adapted from Mutoh et al.[42].)

6.7 OTHER TECHNOLOGIES

In addition to bulk CMOS technology, other technologies have also been used to integrate VLSI systems. In this section, BiCMOS and SOI systems are described.

6.7.1 BiCMOS Systems

Owing to the progress of CMOS VLSI technology, the advancement in the microprocessor is at a rapid pace. Speed and performance of a microprocessor is improving quickly. Measured by million instructions per second, the performance of a microprocessor has been improving impressively. Nowadays, the speed of a microprocessor exceeds several hundred MIPS. A microprocessor chip may have several million transistors in a die area of several centimeters by several centimeters. In the future, goals of microprocessor chips are a higher speed and a larger integration. When devices become smaller and chips become bigger, the influence of the interconnects becomes more and more important. For a VLSI microprocessor chip, load capacitances associated with the bus may be large, which are particularly suitable for using BiCMOS driver circuits. Therefore, some VLSI microprocessor chips are designed into BiCMOS circuits.

Fig. 6.82 shows the block diagram of the floating-point datapath and the cache memory in a 32-bit BiCMOS microprocessor chip[43]. As shown in the figure, there are four critical paths—(1) adder path, (2) multiplier path, (3) divider path, and (4) register read path. BiCMOS circuits have been used in most circuits except the cache memory cell and the register cell. The BiCMOS circuits, which can drive a large capacitive load, are particularly useful for VLSI microprocessor circuits. The number of bipolar devices in the block exceeds 20% of the total transistor count, including the CMOS devices.

As described before, a multi-port register file is needed in a superscalar microprocessor for parallel execution. To realize a small multiport register file cell, a single-ended bit line is used. The register file with singled-ended common-base BiCMOS sense amp circuit is shown in Fig. 6.83[43]. The BiCMOS sense amp is fundamentally a single-ended common-base circuit with its performance affected by the stability of V_B. Owing to the high conductance of the bipolar transistors, the access time has been shortened. V_B is generated by the biasing circuit with the ground as the reference. When there is a variation in V_{DD}, V_B stays relatively unchanged. Thus a high power supply rejection ratio (PSRR) and a more stable sense amp performance can be obtained. A multi-bit comparator circuit is used to check hits for the translation lookaside buffers (TLB). As shown in Fig. 6.84[43], the 22-bit comparator is made with just two stages using the wired-OR bipolar devices[43]. When all bits match, no transistors turn on—the output $Match$ is high. When any mismatch occurs, the corresponding bipolar transistor turns on— the output $Match$ is low. Due to the speed-up by the bipolar transistors, the delay time of the 22-bit BiCMOS comparator is small. Fig. 6.85 shows the adder circuit for the multiplier

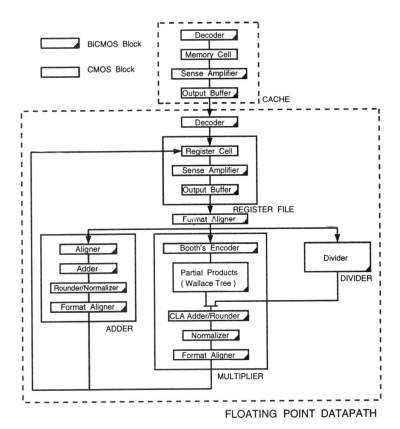

Fig. 6.82 Floating-point datapath and cache memory in a BiCMOS microprocessor. (From Murabayashi et al. [43], ©1994 IEEE.)

used in the BiCMOS microprocessor[43]. As shown in the figure, a BiNMOS buffer is used to drive heavy loads. The logic function is still implemented in the pass-transistor logic. Its output is full-swing, otherwise the speed performance of the next stage is seriously affected. Therefore, via the gate-grounded pull-up PMOS device, the output can be pulled to V_{DD}. Using a 0.5μm BiCMOS technology, this BiCMOS RISC microprocessor includes a 240MFLOPS fully pipelined 64-bit floating-point datapath, a 240-MIPS integer datapath, and 24KB cache, and 2.8 million transistors. Compared to a CMOS microprocessor using 0.5μm, this BiCMOS microprocessor is 11–47% faster.

Fig. 6.86 shows the block diagram of the translation lookaside buffer (TLB)[44]. As shown in the figure, the operation principle is similar to the TLB described before. The virtual page address issued by the CPU is compared with the virtual address tag stored in the CAM of TLB. If matched (tag hit), the physical address stored at the corresponding SRAM is outputed. If not matched (tag miss), the mismatch signal is outputed. Via the least recently used (LRU) algorithm, the new virtual address tag

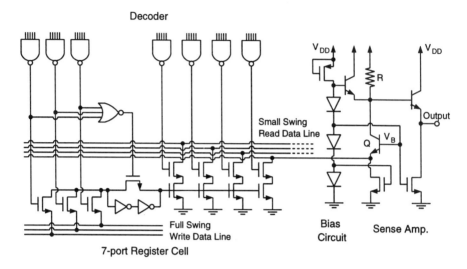

Fig. 6.83 Register file with single-ended common-base BiCMOS sense amp circuit. (Adapted from Murabayashi et al. [43].)

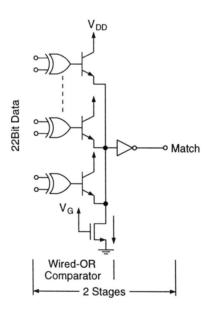

Fig. 6.84 Twenty-two-bit comparator circuits for TLB. (Adapted from Murabayashi et al. [43].)

OTHER TECHNOLOGIES   421

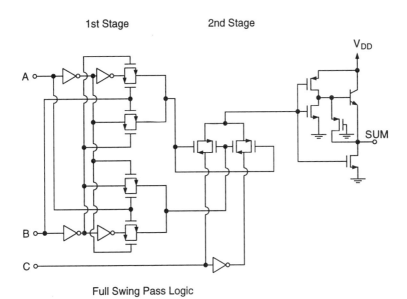

Fig. 6.85 Adder circuit for the multiplier used in the BiCMOS microprocessor. (Adapted from Murabayashi et al. [43].)

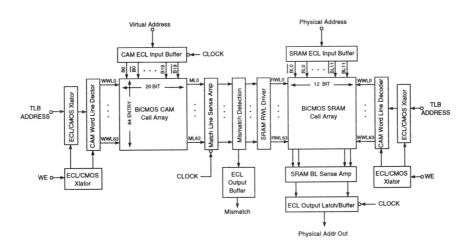

Fig. 6.86 Block diagram of the translation lookaside buffer (TLB). (From Tamura et al. [44], ©1990 IEEE.)

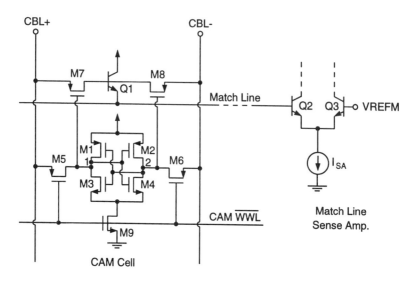

Fig. 6.87 BiCMOS CAM cell. (Adapted from Tamura et al.[44].)

and the corresponding physical address are stored in the CAM and its corresponding SRAM word. Using the BiCMOS technique, the read/write operation can be enhanced.

Fig. 6.87 shows the schematic of the BiCMOS CAM cell[44]. $M1$–$M4$ form the memory cell. $M5$, $M6$, and $M9$ are used for the write operation. $M7$ and $M8$ organize the XOR circuit to compare the content of the memory cell with the tag bit. Assume node 1 stores '1', which represents logic '1' stored in the memory cell. Before the read operation of the CAM, initialization is needed. All CAM bit lines are pulled to low. Since one of $M7$ and $M8$ is turned on, the base voltage of the bipolar device Q_1 is low. Thus match line is low. During read access of CAM, one of the bit lines CBL_- or CBL_+ is pulled up to high. (When tag bit is '1', it is CBL_+; when '0', CBL_-.) If the tag bit is identical to the memory cell data (assuming both are '1'), the source and the gate voltages of $M7$ and $M8$ are identical. The base voltage of Q_1 maintains low. If there is a mismatch, the base voltage of the emitter follower Q_1 is pulled up to high by $M7$ or $M8$. Thus, $Match$ line is pulled up to high. $Match$ line will maintain low only when tag hits occur for all bits. The write operation of CAM is done by lowering CAM $\overline{WWL}$. At this time, $M9$ turns off. The latch formed by $M1$–$M4$ is inactive. Data write can be carried out. When $\overline{WWL}$ becomes high, the regenerative feedback formed by $M1$–$M4$ is constructed to hold the data. Fig. 6.88 shows the BiCMOS SRAM cell used in the TLB[44]. As shown in the figure, the structure formed by $M1$–$M7$ is similar to the one shown in Fig. 6.87. The read operation is via the pull-up of the SRAM RWL, which functions as the supply to a latch formed by $M1$–$M4$. When RWL is high, the data in the memory cell are coupled to the bit lines $SBL+$ and $SBL-$ via the emitter followers Q_1 and Q_2, and amplified by the sense amp. The common base amplifier (Q_3 and Q_4) forms

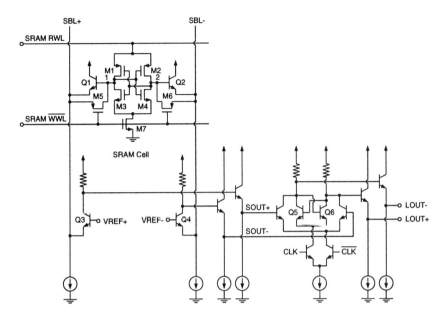

Fig. 6.88 BiCMOS SRAM cell. (Adapted from Tamura et al.[44].)

the sense amp. Via the emitter follower, which serves as the level shifter, its output is connected to the output latch ($SOUT+$, $SOUT-$) for further amplification. The write operation is similar to the case of the CAM cell described before.

Fig. 6.89 shows the circuit diagram of the BiCMOS cache macro[45]. As shown in the figure, the comparator, which is used to do comparison between the TAGIN and the TAG memory data, is implemented by the XOR gate in pass transistor logic. Via a single-ended differential amplifier, followed by a level shifter implemented by an emitter follower, it is connected in the wired-OR configuration. When the result of the TAG comparison is matched, the output of the differential amplifier n is low. When all tag bits are identical, node p maintains a low level. Thus, the output x of the ECL comparator is also low. As along as there is a mismatch in tag bits, node p is pulled up to high. Therefore, the output x of the ECL comparator is also high. The ECL hit logic is also implemented in ECL-like circuit to enhance speed. With a special bit and a valid bit, the output of the ECL comparator is converted to the CMOS level. The hit signal from the ECL hit logic is used to activate way selectors of DATA memory such that the data on the bit lines are transferred to the BiCMOS sense amp for amplification.

6.7.2 SOI Systems

Due to the advantages in radiation hardness, SOI CPUs have been used for special-purpose environments. The trends on SOI CPU are toward low-voltage and low-

424   CMOS VLSI SYSTEMS

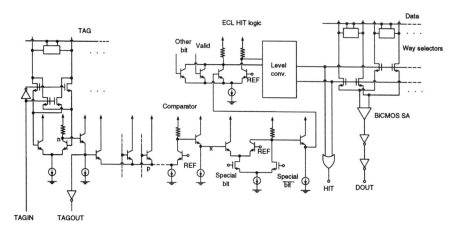

Fig. 6.89 Circuit diagram of the BiCMOS cache macro. (Adapted from Hara et al. [45].)

power. Owing to the superior properties in the subthreshold slope, SOI CMOS devices are especially suitable for low-voltage operation. Based on a 0.5μm SOI CMOS technology, a microprocessor operating at below 1V has been integrated[46]. As shown in Fig. 6.90, at 0.9V, the operation speed is 5.7MHz, which is 1.9 times faster as compared to the bulk one. Fig. 6.91 shows the core frequency performance as a function of the operating temperature over a supply voltage of 0.9V to 2V of the SOI CPU. At a larger V_{DD}, when the operating temperature increases, due to a degraded mobility, the maximum operating frequency is reduced since the driving capability of the devices is decreased. On the other hand, at a smaller V_{DD}, the maximum operating frequency does not decrease when the temperature is increased. Instead, speed increases slightly when the operating temperature is raised. Since the magnitude of the threshold voltages of the CMOS devices decreases when the operating temperature is increased, this offsets the influence from the decreased mobility. When V_{DD} becomes smaller, the variation in the threshold voltage affects the speed more seriously. (Please refer to the problem with the low-voltage in Chapter 3.) At a low supply voltage, when the temperature rises, f_{max} increases.

Fig. 6.92 shows the multi-threshold CMOS/SIMOX circuit scheme[47]. As shown in the figure, this SOI CMOS circuit is composed of two portions—the power switch transistor and the logic block. In logic block, all transistors are fully-depleted SOI CMOS devices with a smaller magnitude in the threshold voltage. The supply to the logic block is called virtual V_{DD}, which is connected to the V_{DD} of 0.5V via the power switch transistor. The power switch transistor is a partially-depleted SOI MOS device with a larger threshold voltage. During the sleep mode, the power switch transistor is turned off to save power. The larger threshold voltage guarantees a smaller leakage current in the turned-off power switch transistor. As shown in the figure, a diode-connected auxiliary fully-depleted SOI device based on the dynamic threshold technique has been used to tie the body of the partially-depleted power switch

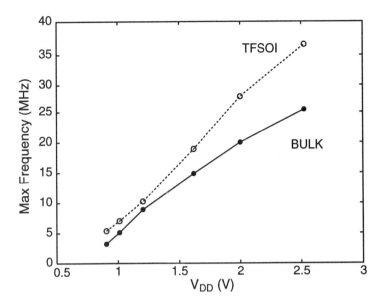

Fig. 6.90 Comparison of microcontroller CPU core frequency performance between SOI and bulk CMOS. (From Huang et al. [46], ©1995 IEEE.)

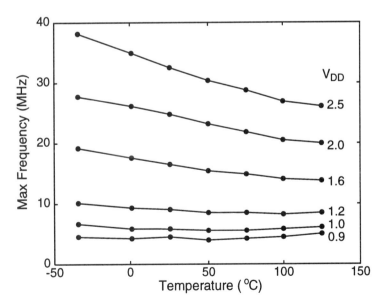

Fig. 6.91 Core frequency performance as a function of the operating temperature over a supply voltage of 0.9V to 2V of the SOI CPU. (From Huang et al. [46], ©1995 IEEE.)

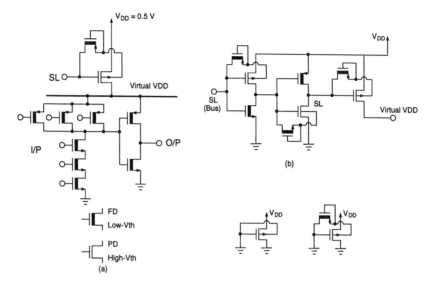

Fig. 6.92 Multi-threshold CMOS/SIMOX circuit scheme. (Adapted from Douseki et al. [47].)

transistor to gate. As a result, when SL switches from high to low, the power switch transistor turns on. Due to the auxiliary PMOS, its body potential drops. Therefore, its threshold voltage $|V_{TP}|$ can be lowered. Hence, when the virtual V_{DD} is connected to V_{DD}, its voltage difference between V_{DD} and virtual V_{DD} can be small. Based on the multi-threshold CMOS/SIMOX circuit scheme, a 16-bit arithmetic logic unit (ALU) based on a 0.25μm SOI CMOS technology has been integrated. At a supply voltage of 0.5V, a speed of 50MHz has been reached. The power consumption is only 0.35mW. During the sleep mode, the power consumption is less than 5nW.

Using the DTMOS technique mentioned in Chapter 3, which is based on the dynamic threshold MOS device with its body tied to the gate, the speed performance of a pass-gate logic can be faster. Fig. 6.93 shows the body-contact (BCSOI) pass-gate logic circuits[48]. One type of the BCSOI circuit is for driving loads with long wires and large fan-out as shown in Fig. 6.93. The other type of BCSOI circuit is for driving local interconnect as shown in Fig. 6.93. The BCSOI pass-gate logic circuit as shown in Fig. 6.93 is the CPL circuit using the DTMOS devices. Fig. 6.93 is the DPTL circuit using the DTMOS devices. Using the BCSOI pass-gate logic circuits based on a 0.3μm SOI technology, a 32-bit ALU has been integrated. At 0.5V, the maximum operating frequency is 200MHz. However, the power consumption is 20mW. Due to the restraints in turn-on of the body/source diode, this BCSOI CPU can only work at a supply voltage smaller than 0.8V.

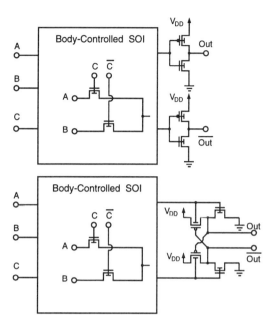

Fig. 6.93 Body-controlled (BCSOI) pass-gate logic circuits. (Adapted from Fuse et al. [48].)

6.8 SUMMARY

In this chapter, fundamental CMOS VLSI system circuits are described. Various CMOS adder circuits have been analyzed. Then, CMOS multiplier circuits have been presented. In a VLSI system, register files and cache memory are important. PLA and PLL circuit techniques have also been described. In addition, register files and cache memory have been described. Processing unit circuits, including the floating-point processing unit, the central processing unit, and the digital signal processing unit circuits, have been introduced. In the final portion of this chapter, VLSI systems integrated by other technologies, including BiCMOS and SOI, have been explained.

REFERENCES

1. N. H. E. Weste and K. Eshraghian, *Principles of CMOS VLSI Design*, Addison-Wesley: New York, 1985.

2. C. C. Hung, J. H. Lou, and J. B. Kuo, "A CMOS Quasi-Static Manchester-Like Carry-Look-Ahead Circuit," *private communication*.

3. J. H. Lou and J. B. Kuo, "A 1.5V Bootstrapped Pass-Transistor-Based Carry Look-Ahead Circuit Suitable for Low-Voltage CMOS VLSI," *IEEE Trans. Ckts. and Sys.* (1998).

4. J. H. Lou and J. B. Kuo, "A 1.5-V Full-Swing Bootstrapped CMOS Large Capacitive-Load Driver Circuit Suitable for Low-Voltage CMOS VLSI," *IEEE J. Sol. St. Ckts.*, **32**(1), 119–121 (1997).

5. M. Suzuki, N. Ohkubo, T. Shinbo, T. Yamanaka, A. Shimizu, K. Sasaki, and Y. Nakagome, "A 1.5-ns 32-b CMOS ALU in Double Pass-Transistor Logic," *IEEE J. Sol. St. Ckts.*, **28**(11), 1145–1151 (1993).

6. J. H. Lou, Y. M. Huang, and J. B. Kuo, "A New Pass-Transistor-Based Carry Select Adder," *private communciation*.

7. Z. Wang, G. A. Jullien, W. C. Miller, J. Wang, and S. S. Bizzan, "Fast Adders Using Enhanced Multiple-Output Domino Logic," *IEEE J. Sol. St. Ckts.*, **32**(2), 206–214 (1997).

8. A. Guyot, B. Hochet, and J.-M. Muller, "A Way to Build Efficient Carry-Skip Adders," *IEEE Trans. Comput.*, **36**(10), 1144–1152 (1987).

9. T. Sato, M. Sakate, H. Okada, T. Sukemura, and G. Goto, "An 8.5-ns 112-b Transmission Gate Adder with a Conflict-Free Bypass Circuit," *IEEE J. Sol. St. Ckts.*, **27**(4), 657–659 (1992).

10. R. P. Brent and H. T. Kung, "A Regular Layout for Parallel Adders," *IEEE Trans. Comp.*, **31**(3), 260–264 (1982).

11. W. Liu, C. T. Gray, D. Fan, W. J. Farlow, T. A. Hughes, and R. K. Cavin, "A 250-MHz Wave Pieplined Adder in 2-μm CMOS," *IEEE J. Sol. St. Ckts.*, **29**(9), 1117–1128 (1994).

12. M. Nagamatsu, S. Tanaka, J. Mori, K. Hirano, T. Noguchi, and K. Hatanaka, "A 15-ns 32×32-b CMOS Multiplier with an Improved Parallel Structure," *IEEE J. Sol. St. Ckts.*, **25**(2), 494–497 (1990).

13. N. Ohkubo, M. Suzuki, T. Shinbo, T. Yamanaka, A. Shimizu, K. Sasaki, and Y. Nakagome, "A 4.4 ns CMOS 54× 54-b Multiplier Using Pass-Transistor Multiplexer," *IEEE J. Sol. St. Ckts.*, **30**(3), 251–257 (1995).

14. J. Mori, M. Nagamatsu, M. Hirano, S. Tanaka, M. Noda, Y. Toyoshima, K. Hashimoto, H. Hayashida, and K. Maeguchi, "A 10-ns 54×54-b Parallel Structured Full Array Multiplier with 0.5-μm CMOS Technology," *IEEE J. Sol. St. Ckts.*, **26**(4), 600–606 (1991).

15. G. Goto, A. Inoue, R. Ohe, S. Kashiwakura, S. Mitarai, T. Tsuru, and T. Izawa, "A 4.1-ns Compact 54×54-b Multiplier Utilizing Sign-Select Booth Encoders," *IEEE J. Sol. St. Ckts.*, **32**(11), 1676–1682 (1997).

16. G. Goto, T. Sato, M. Nakajima, and T. Sukemura, "A 54× 54-b Regularly Structured Tree Multiplier," *IEEE J. Sol. St. Ckts.*, **27**(9), 1229–1236 (1992).

17. H. Shinohara, N. Matsumoto, K. Fujimori, Y. Tsujihashi, H. Nakao, S. Kato, Y. Horiba, and A. Tada, "A Flexible Multiport RAM Compiler for Data Path," *IEEE J. Sol. St. Ckts.*, **26**(3), 343–349 (1991).

18. R. D. Jolly, "A 9-ns, 1.4-Gagabyte/s, 17-Ported CMOS Register File," *IEEE J. Sol. St. Ckts.*, **26**(10), 1407–1412 (1991).

19. H. Kadota, J. Miyake, Y. Nishimichi, H. Kudoh, and K. Kagawa, "An 8-kbit Content-Addressable and Reentrant Memory," *IEEE J. Sol. St. Ckts.*, **20**(5), 951–957 (1985).

20. H. Kadota, J. Miyake, I. Okabayashi, T. Maeda, T. Okamoto, M. Nakajima, and K. Kagawa, "A 32-bit CMOS Microprocessor with On-Chip Cache and TLB," *IEEE J. Sol. St. Ckts.*, **22**(5), 800–807 (1987).

21. A. K. Goksel, R. H. Krambeck, P. P. Thomas, M.-S. Tsay, C. Y. Chen, D. G. Clemons, F. D. LaRocca, and L.-P. Mai, "A Content Addressable Memory Management Unit with On-Chip Data Cache," *IEEE J. Sol. St. Ckts.*, **24**(3), 592–596 (1989).

22. K. Sawada, T. Sakurai, K. Nogami, T. Shirotori, T. Takayanagi, T. Iizuka, T. Maeda, J. Matsunaga, H. Fuji, K. Maeguchi, K. Kobayashi, T. Ando, Y. Hayakashi, A. Miyoshi, and K. Sato, "A 32-kbyte Integrated Cache Memory," *IEEE J. Sol. St. Ckts.*, **24**(4), 881–888 (1989).

23. H. Mizuno, N. Matsuzaki, K. Osada, T. Shinbo, N. Ohki, H. Ishida, K. Ishibashi, and T. Kure, "A 1-V, 100-MHz, 10-mW Cache Using a Separated Bit-Line Memory Hierarchy Architecture and Domino Tag Comparators," *IEEE J. Sol. St. Ckts.*, **31**(11), 1618–1623 (1996).

24. A. R. Linz, "A Low-Power PLA for a Signal Processor," *IEEE J. Sol. St. Ckts.*, **26**(2), 107–115 (1991).

25. G. M. Blair, "PLA Design for Single-Clock CMOS," *IEEE J. Sol. St. Ckts.*, **27**(8), 1211–1213 (1992).

26. V. von Kaenel, D. Aebischer, C. Piguet, and E. Dijkstra, "A 320 MHz, 1.5mW @ 1.35V CMOS PLL for Microprocessor Clock Generation," *IEEE J. Sol. St. Ckts.*, **31**(11), 1715–1722 (1996).

27. I. A. Young, J. K. Greason, and K. L. Wong, "A PLL Clock Generator with 5 to 110 MHz of Lock Range for Microprocessors," *IEEE J. Sol. St. Ckts.*, **27**(11), 1599–1607 (1992).

28. B. Razavi, K. F. Lee, and R. H. Yan, "Design of High-Speed, Low-Power Frequency Dividers and Phase-Locked Loops in Deep Submicron CMOS," *IEEE J. Sol. St. Ckts.*, **30**(2), 101–109 (1995).

29. W. Lee, P. E. Landman, B. Barton, S. Abiko, H. Takahashi, H. Mizuno, S. Muramatsu, K. Tashiro, M. Fusumada, L. Pham, F. Boutaud, E. Ego, G. Gallo, H. Tran, C. Lemonds, A. Shih, M. Nandakumar, R. H. Eklund, and I.-C. Chen, "A 1-V Programmable DSP for Wireless Communications," *IEEE J. Sol. St. Ckts.*, **32**(11), 1766–1776 (1997).

30. J. Dunning, G. Garcia, J. Lundberg, and E. Nuckolls, "An All-Digital Phase-Locked Loop with 50-Cycle Lock Time Suitable for High-Performance Microprocessors," *IEEE J. Sol. St. Ckts.*, **30**(4), 412–422 (1995).

31. Y. Hagihara, S. Inui, F. Okamoto, M. Nishida, T. Nakamura, and H. Yamada, "Floating-Point Datapaths with Online Built-In Self Speed Test," *IEEE J. Sol. St. Ckts.*, **32**(3), 444–449 (1997).

32. 1997 National Technology Roadmap for Semicondcutors, http://notes.sematch.org/97pelec.htm.

33. N. Alexandridis, *Design of Microprocessor-Based Systems*, Prentice-Hall: Englewood Cliffs, NJ, 1993.

34. D. Draper, M. Crowley, J. Holst, G. Favor, A. Schoy, J. Trull, A. Ben-Meir, R. Khanna, D. Wendell, R. Krishna, J. Nolan, D. Mallick, H. Partovi, M. Roberts, M. Johnson, and T. Lee, "Circuit Techniques in a 266-MHz MMX-Enabled Processor," *IEEE J. Sol. St. Ckts.*, **32**(11), 1650–1664 (1997).

35. F. Murabayashi, T. Yamauchi, H. Yamada, T. Nishiyama, K. Shimamura, S. Tanaka, T. Hotta, T. Shimizu, and H. Sawamoto, "2.5V CMOS Circuit Techniques for a 200 MHz Superscalar RISC Processor," *IEEE J. Sol. St. Ckts.*, **31**(7), 972–980 (1996).

36. L. A. Lev, A. Charnas, M. Tremblay, A. R. Dalal, B. A. Frederick, C. R. Srivatsa, D. Greenhill, D. L. Wendell, D. D. Pham, E. Anderson, H. K. Hingarh, I. Razzack, J. M. Kaku, K. Shin, M. E. Levitt, M. Allen, P. A. Ferolito, R. L. Bartolotti, R. K. Yu, R. J. Melanson, S. I. Shah, S. Nguyen, S. S. Mitra, V. Reddy, V. Ganesan, and W. J. de Lange, "A 64-b Microprocessor with Multimedia Support," *IEEE J. Sol. St. Ckts.*, **30**(11), 1227–1238 (1995).

37. J. Montanaro, R. T. Witek, K. Anne, A. J. Black, E. M. Cooper, D. W. Dobberpuhl, P. M. Donahue, J. Eno, G. W. Hoeppner, D. Kruckemyer, T. H. Lee, P. C. M. Lin, L. Madden, D. Murray, M. H. Pearce, S. Santhanam, K. J. Snyder, R. Stephany, and S. C. Thierauf, "A 160-MHz, 32-b, 0.5-W CMOS RISC Microprocessor," *IEEE J. Sol. St. Ckts.*, **31**(11), 1703–1714 (1996).

38. J. L. Hennessy and D. A. Patterson, *Computer Organization and Design: The Hardware/Software Interface*, Morgan Kaufmann Publishers: San Francisco, CA, 1994.

39. N. Ide, H. Fukuhisa, Y. Kondo, T. Yoshida, M. Nagamatsu, J. Mori, I. Yamazaki, and K. Ueno, "A 320-MFLOPS CMOS Floating-Point Processing Unit for Superscalar Processors," *IEEE J. Sol. St. Ckts.*, **28**(3), 352–361 (1993).

40. D. E. Atkins, "Higher-Radix Division Using Estimates of the Divisor and Partial Remainders," *IEEE Trans. Comput.*, **17**(10), 925–934 (1968).

41. A. Bateman and W. Yates, *Digital Signal Processing Design*, Pitman Publishing, London, 1988.

42. S. Mutoh, S. Shigematsu, Y. Matsuya, H. Fukuda, T. Kaneko, and J. Yamada, "A 1-V Multithreshold-Voltage CMOS Digital Signal Processor for Mobile Phone Application," *IEEE J. Sol. St. Ckts.*, **31**(11), 1795–1802 (1996).

43. F. Murabayashi, T. Hotta, S. Tanaka, T. Yamauchi, H. Yamada, T. Nakano, Y. Kobayashi, and T. Bandoh, "3.3-V BiCMOS Circuit Techniques for a 120-MHz RISC Microprocessor," *IEEE J. Sol. St. Ckts.*, **29**(3), 298–302 (1994).

44. L. R. Tamura, T.-S. Yang, D. E. Wingard, M. A. Horowitz, and B. A. Wooley, "A 4-ns BiCMOS Translation-Lookaside Buffer," *IEEE J. Sol. St. Ckts.*, **25**(5), 1093–1103 (1990).

45. H. Hara, T. Sakurai, T. Nagamatsu, K. Seta, H. Momose, Y. Niitsu, H. Miyakawa, K. Matsuda, Y. Watanabe, F. Sano, and A. Chiba, "0.5-μm 3.3-V BiCMOS Standard Cells with 32-kilobyte Cache and Ten-Port Register File," *IEEE J. Sol. St. Ckts.*, **27**(11), 1579–1583 (1992).

46. W. M. Huang, K. Papworth, M. Racanelli, J. P. John, J. Foerstner, H. C. Shin, H. Park, B. Y. Hwang, T. Wetteroth, S. Hong, H. Shin, S. Wilson, and S. Cheng, "TFSOI CMOS Technology for Sub-1V Microcontroller Circuits," *IEDM Dig.*, 59–62 (1995).

47. T.Douseki, S.Shigematsu, J. Yamada, M. Harada, H. Inokawa, and T. Tsuchiya, "A 0.5-V MTCMOS/SIMOX Logic Gate," *IEEE J. Sol. St. Ckts.*, **32**(10), 1604–1609 (1997).

48. T. Fuse, Y. Oowaki, T. Yamada, M. Kamoshida, M. Ohta, T. Shino, S. Kawanaka, M. Terauchi, T. Yoshida, G. Matsubara, S. Yoshioka, S. Watanabe, M. Yoshimi, K. Ohuchi, and S. Manabe, "A 0.5V 200MHz 1-Stage 32b ALU using a Body Bias Controlled SOI Pass-Gate Logic," *ISSCC Dig.*, 286–287 (1997).

Problems

1. Use (a) static CMOS logic circuits, (b) complementary pass-transistor logic (CPL), and (c) double pass-transistor logic (DPL) to design the parallel adder as shown in Fig. 6.17. Which approach has the best speed performance (smallest propagation delay)? For the design with the best speed performance, is its throughput also the highest?

2. Compare the performance of the multipliers using Wallace tree reduction with 3-to-2 and 4-to-2 compressor, modified Booth encoder/decoder, and combining modified Booth encoder/decoder with Wallace tree reduction.

3. Based on Fig. 6.30, design a basic cell for the 3W6R register file. Is there anything worth pointing out while doing the design?

4. Compare the difference between the basic memory cell in the register file in Fig. 6.34 and the one pointed out in Problem 3. When designing the 3W6R register file, what should be taken care of while simultaneous access of the six read ports are being carried on?

5. What is the difference between the CAM in TLB and the tag memory in the cache? What is the difference from a design point of view?

6. What are the methods to reduce noise problems when designing dynamic logic circuits? What are the tradeoffs?

7. For the delayed reset logic in Fig. 6.66, draw the circuit diagram of the nstg and pstg general logic. Design $S = (A \oplus B) \oplus C$ using a 2-stage XOR gate.

Index

α-particle induced critical charge, 266
3D SRAM, 301
4T cell, 239
6T cell, 239
Access transistor, 240, 266, 316, 333
Active pull-down, 89
Adders, 345
Address
  access time, 238
  hold time, 237
  setup time, 237
  tag, 380
  transition detection (ATD), 238
Adiabatic logic, 153
Aging effect, 71–72
All-N-logic TSP BDL circuit, 223
Alpha particle, 171, 266, 302
Analog to digital conversion (A/D), 412
Application-specific SRAM, 257
Arithmetic logic unit (ALU), 132, 396, 409, 425
Asynchronous
  SRAM, 238
  strobe, 274
Avalanche breakdown, 20, 70
Base voltage, 122, 293
BiCMOS
  buffer circuit, 126
  cache macro, 422
  DRAM, 295, 298
  driver, 297
  dynamic carry chain circuit, 206

  dynamic logic circuit, 206, 209, 211, 214, 221
  ECL I/O, 291
  inverter, 116, 120
  latch, 213
  microprocessor, 418
  NAND, 123
  sense amp, 417
  SRAM, 286
  static logic circuit, 115
  systems, 417
  technology, 1, 24–26, 286
Bidirectional read/write shared sense amplifier (BSA), 258
Binary decision diagram (BDD), 114,
BiNMOS, 119
  gate, 294
  pull-down, 119,
BiPMOS, 119
  pull-down, 214
  pull-up, 119
Bird's beak, 44,
Bit line, 313
  load circuit, 246
BJT collector resistance, 297
Block
  carry generator (BCG), 358
  group select (BGS), 251
  select (BS), 249, 251
Body
  charge, 78
  driven equalizer (BDEQ), 307

433

434   INDEX

effect, 41–42, 53, 57, 96, 138
pulsed sense amplifier (BPS), 305
refresh, 303
Body-controlled SOI (BCSOI), 425
Boost voltage generator, 283
Booth selector output, 369
Bootstrap
   capacitor, 218
   technique, 132
   transistor, 351
Bootstrapped
   dynamic logic (BDL) circuit, 218
   MODL (BMODL), 218
Bootstrapper circuit, 351
Buried
   channel, 21, 23–24
   layer, 25
   oxide, 33
Bus
   architecture, 145
   driver circuit, 146
   line, 149
Cache memory, 370, 377
CAD tools, 2, 113
Carry
   in signal, 345, 348
   look-ahead (CLA) adder, 345, 347, 359
   out signal, 346, 348
   propagate/ripple adder, 345
   save adder, 346, 364
   skip, 357
   skip CLA, 345
CE scaling law, 15
Cell bit line, 260
Cellular phone, 412
Central processing unit, 392
Ceramic packaging, 4
Chain FRAM (CFRAM), 334
Channel
   hot-electron (CHE) injection effect, 311, 317
   length, 3, 73
   length modulation, 47, 74
   punchthrough, 30
   stop implant, 44
   stop region, 69
Charge
   pump circuit, 327, 386–387
   recycling bus (CRB) architecture, 150–151, 279
   redistribution, 149
   retention, 329
   sharing, 167, 169
   sharing problem, 181, 209, 214, 333
   sharing problem free, 170
   transfer presensing scheme (CTPS), 284
Chemical mechanical polishing (CMP), 19
Chemical vapor deposition (CVD), 301, 20, 33
Chip select setup time, 237

Chip select, 236
Clock and data precharge dynamic (CDPD), 182
Clock race through problem, 167
Clock skew, 185, 201
Clocked
   CMOS (C^2MOS), 174, 198
   CMOS latch, 175–176
CMOS
   bootstrapper circuit, 218, 222
   differential static logic circuit, 101
   domino logic circuit, 221
   dynamic latch circuit, 168
   dynamic logic circuit, 171, 208, 217
   static logic circuit, 96, 101
   technology, 2–3, 9, 13–14, 29, 31–32, 86, 101
   TSPC dynamic logic circuit, 223
CMOS/SIMOX, 132
Column
   address, 238
   decoder, 373
Complementary pass-transistor logic (CPL), 111
Compressor, 366, 369
Conditional carry select (CCS), 352
Conflict free carry skip, 358
Constant
   electric field (CE) scaling law, 15
   voltage (CV) scaling law, 15, 98
Content addressable memory (CAM), 376
Coordinate transformation, 242
Counter-dope, 40
Coupling capacitance, 31
Critical electric field, 48
Critical path, 250
Crosstalk, 171
Current
   amplification, 247
   continuity equation, 62–63
   controlled oscillator (CCO), 388
   mode logic, 295
   sense amp(CSA), 247
   sensing, 282
CV scaling law, 16
Darlington pair, 119, 296
Data corruption problem, 375
Data throughput, 275
Data
   hold time, 237
   retention time, 5, 278
   setup time, 237
Data-dependent logic swing bus, 147
DC standby leakage, 8
DCVS
   domino logic circuit, 194
   dynamic logic circuit, 191
   with pass gate (DCVSPG) logic, 106–107
Deep submicron, 29
Delta doped MOS device, 30

INDEX 435

Depletion load, 89, 96
Depletion type, 86
Differential
  amplifier, 247
  cascode voltage switch (DCVS) logic, 101, 105, 110, 192, 280
  edge-triggered latch circuit, 407
  pass-transistor logic (DPTL), 110–111
  split-level (DSL) logic circuit, 103
Diffusion self-aligned (DSA), 324
Digital-controlled oscillator (DCO), 389
Digital signal processor (DSP), 392, 411
Digital to analog conversion (D/A), 412
DINOR cell, 324
Divided bit line NOR (DINOR), 324
Divided word line (DWL), 249
Domino
  CMOS logic circuit, 178–180
  logic gate, 187
  tag comparator (DTC), 379
  to static interface circuit, 182
Doping profile, 74
Double gate, 312
Double pass-transistor logic (DPL), 112
Drain body capacitance, 62
Drain gate capacitance, 61
Drain induced barrier lowering (DIBL), 57
  coefficient, 75
Drain transient current, 66,
DRAM, 6, 13–14, 24, 29, 51, 99, 264
  memory cell, 52
Driving source line (DSL) memory cell
    architecture, 261
Dual rail pulsed edge trigger latch (ETL), 400
Dual-gate SOI CMOS, 31
Dummy
  bit line, 258
  cell, 258, 268, 270, 280
  ground line, 147
Dynamic power dissipation, 4
Dynamic sense amp, 260
Dynamic threshold
  MOS (DTMOS), 139
  technique, 129
Dynamic
  differential logic, 190
  logic circuit, 163, 400
E beam lithography, 30,
ECL, 289
  I/O, 290
  I/O SRAM, 287
Edge triggered
  D flip flops (ETDFF), 199
  latch, 399
EEPROM, 314, 318
Effective channel length, 47
Effective electron mobility, 47, 57

Electromagnetic radiation, 171
Electromigration, 257
Electron
  cyclotron resonance (ECR), 34
  energy flux, 54–55
  mobility, 57
  temperature, 53–54, 56, 60
  temperature profile, 53
  traveling velocity, 48
Embedded
  DRAM, 275
  power-management processor (PMP), 415
Enable disabled CMOS differential logic (ECDL), 192
End of sense (EOS), 377
Energy efficient logic (EEL), 154, 156
Energy recycling, 154
Energy
  balance equation, 54
  flux, 54
  relaxation time, 54
  transport, 54
Enhanced MODL adder, 345
Enhancement type, 22, 39–40, 86
EPROM, 311
Equalize, 268
Erase/program verify, 319
Etching, 16, 29, 265
Evaluation period, 151, 164
Exclusive
  NOR, 97
  OR, 97
Extrinsic capacitance, 69
Fan-out, 28
Fast page (FP) DRAM, 274
Feedback inverter, 373
Ferroelectric
  capacitor, 329–330
  RAM (FRAM), 328
First stage sense amp, 255
Flash memory, 317
Flat band voltage, 39
Floating body effect, 36, 299
Floating gate, 311
Floating-point
  addition, 409
  processing unit, 392, 408
FN tunneling, 317, 321
Folded bit line, 269
FRAM cell, 331
Full adder, 347
Full array multiplier, 368
Fully-depleted SOI, 36, 131, 423
GaAs MESFET, 1
Gate
  charge, 66, 78
  current, 72

436 INDEX

Gauss
  box, 59
  law, 38, 59
Geometry dependence effect, 74
Global
  data line, 252
  row decoders, 253
  word line, 243
Gradual channel approximation, 48
Group
  generate, 348, 355
  propagate, 348, 355
HBT, 28
Hierarchical
  power line structure, 277
  word line decoding (HWD), 250
High resistive load (HRL), 287
High-to-low propagation delay time (t_{PHL}), 89
Highest output voltage (V_{OH}), 89
Hitting word driver (HWD), 377
Hot
  carrier effect, 37, 70
  electron effect, 21, 25, 71–73
  hole effect, 25, 73
Hysteresis, 329
I/O line, 272
Impact ionization, 70–71, 311–312
Input noise margin, 179
Input variation race free sequence, 176
Instruction
  cache, 397
  decode (ID), 395
  fetch (IF), 395
Insulator material, 266
Integer format, 408
Integrated injection logic (I^2L), 1
Inter-pair coupling noise, 269
Interconnect, 16, 31, 35, 369
Interface state generation, 71
Internal voltage
  overshoot, 117, 122
  undershoot, 120
Intra-pair coupling noise, 269
Intrinsic capacitance, 61
Inversion layer, 46
Ionic displacement, 329
Junction
  capacitance, 31, 69
  depth, 72
  leakage, 278
  temperature, 278
Kink effect, 36
Latch
  structure, 235
  type sense amp, 261
Latched
  domino logic, 182
  sense amp, 246
Latchup, 86, 239, 296
Latency, 275
Lateral electric field, 3, 21, 47, 56
Lattice temperature, 55–56
Leakage
  charge, 265
  current, 24, 52, 139, 142, 179
  current problem, 181
Least recently used (LRU), 377–378, 418
Less significant bit (LSB), 380
Lightly doped drain (LDD), 9, 20–21, 25, 28–29
Linear load, 89
Load line, 246
Local word line, 250
LOCOS isolation, 19, 33–34, 44, 52
Logic evaluation, 175
Loop filter, 387
Low k dielectric, 31
Low pressure chemical vapor deposition (LPCVD), 34
Low temperature oxide (LTO), 20, 35
Low-power CMOS circuit, 140
Low-voltage dynamic logic, 217
Lowest output voltage (V_{OL}), 89
Main amp, 274, 295
Manchester
  carry chain circuit, 206
  CLA, 345, 349
Manchester-like, 350
Maximum refresh cycle time, 265
Memory
  bit line, 280
  cell, 239, 280, 299, 377, 379
  cell array, 243
  IC, 235
Metal line parasitics, 255
Microarchitecture, 397,
Microprocessor, 15, 99, 395
  processing unit (MPU), 258
Million instructions per second (MIPS), 15, 29, 417
Minimum
  supply voltage, 7
  threshold voltage, 7
Mobile telephone set, 135
Mobility, 37, 44
Modified Booth algorithm, 366, 368
Modulation doped FET (MODFET), 1
MOS capacitances, 61
Multi access, 374
Multi divided
  array, 273
  bit line, 272
Multi media extensions (MMX) enabled processor, 397,
Multi threshold, 276, 305, 425

INDEX 437

CMOS logic circuit, 131, 135
  standby/active technique, 135
Multi-port memory cell, 371
Multiple output domino logic (MODL), 188, 218, 347
Multiplexer, 354, 365
Multiplicand, 368
Multiplication, 367
Multiplier, 363
Multiply/divide unit (MDU), 409
NAND circuit, 97
NAND-NOT-NOR PLA, 386
Narrow channel
  effect, 44
  threshold voltage formula, 44
Newton's approximation method, 51
Node voltage vector, 78
Noise coupling effect, 269
Noise-tolerant precharge (NTP), 401
Noise
  immunity, 228, 400
  margin, 88, 165, 241
  problem, 171
Non-volatile, 235
NOR-NOR structure, 385
NORA, 172–173, 182, 217
Normal read cycle time, 265
Open bit line structure, 268
Oscillation frequency, 389
Output
  buffer, 295
  column line, 310
  conductance, 60
  swing, 293
Overflow, 409
Overlapped capacitance, 93
Oxide capacitance, 57
Oxide-nitride-oxide dielectric, 312
Packing density, 303
Parallel adder, 359–360, 362
Parallelism, 274, 397
Parasitic BJT, 303
Parasitic capacitance, 95, 252, 268, 273, 310
Partial product, 364, 368
Partially-depleted SOI, 36, 131
Pass-transistor based
  carry select CLA, 345
  CLA, 345,
Pass transistor, 166, 300
  logic circuit, 108–109
Passive pull-up, 89
Phase frequency detector (PFD), 386,
Phase locked loop (PLL), 386, 389
  jitter, 185
Photolithography, 16–17, 29–30, 35, 265
Pinchoff point, 47, 53
Pipeline, 157, 174, 274

Pipelined adder, 359
Planarization, 35
Pocket implanted MOS, 30
Polyemitter, 28
Polysilicon
  resistor, 239
  thin-film transistor (TFT), 1, 257, 289, 301
  word line, 243
Post charge logic, 404,
Post-saturation, 58–59
  region, 59
Power consumption, 3, 86, 272
Power-delay product, 98
Power management, 415
Power supply
  line, 276
  rejection ratio (PSRR), 314, 417
  voltage, 6, 98
Power switch transistor, 276
Power utilization, 279
Pre-decoder, 253
Pre-saturation region, 58–59
Precharge, 164, 268
Predecode cache, 397
Prefetch SDRAM, 274
Prepinchoff velocity saturation, 49
Program disturb problem, 321
Program time, 312
Programmable logic arrays (PLA), 384
Propagation delay, 95, 98, 100, 255, 346, 363
Pseudo NMOS, 164, 243, 384–385
Pseudo static, 406
PT-based
  CLA adder, 351
  conditional carry-select CLA adder, 352
Pull-up transient, 119, 122
Punchthrough, 18, 29
Push pull cascode logic (PPCL), 107
Push pull load, 103
Quasi-complementary BiCMOS, 125
Race free, 175
Race problem, 172
Rapid thermal anneal (RTA), 18, 35
RC delay, 243, 273
Reactive ion etching (RIE), 35
Read address access time, 246, 250, 310
Read operation, 237, 261, 264
Reciprocity, 62, 68
Refresh
  cycle, 264
  time, 5
Regenerative push-pull differential logic (RPPDL), 195
Register file, 370–371
Reliability, 73, 257
Resistive load, 89
Restore amp, 268

Restore operation, 305
ROM, 308
Row
  address, 238
  decoder, 243
Sample-set differential logic (SSDL), 192
Saturated velocity, 47–48
Saturation load, 89
Saturation region, 58
Scattering, 56
Schmitt trigger, 213
Self-reset logic, 404
Self-timed
  circuit, 191, 406
  pipeline technique, 404
  pulsed word line, 258
  register file (RF), 406
Semi-dynamic DCVSPG domino logic circuit, 225
Semi-static latch circuit, 169
Sense amp, 246, 313
Sense line, 280, 284
Sensitivity, 75
Separated bit-line memory hierarchy architecture (SBMHA), 378
Separation by implantation of oxygen (SIMOX), 33
Sequential machine, 166
Shallow trench isolation (STI), 9, 19, 44, 52
Shift register, 166
Short channel
  effect, 30, 42–43
  threshold voltage formula, 43
Sidewall
  capacitance, 69
  spacer, 21, 29
Signal processing, 145
Significand (mantissa), 409
Silicon on insulator (SOI), 1, 32
Simplified Brokaw's circuit, 327
Single
  bit line memory cell, 257
  instruction multiple data (SIMD), 398
Single phase
  finite state machine structure, 201
  positive edge triggered flip flop, 201
Single transistor clocked TSPC dynamic latch, 199
Skew tolerant domino logic, 185
Slow recovery problem, 277
Sneak path, 348
Soft error immunity, 239, 302
SOI
  CMOS, 36–37
  CMOS buffer, 130
  CMOS inverter, 130
  CMOS static logic, 128
  CMOS technology, 13, 128, 299
  CPU, 422

  DRAM, 302
  SRAM, 299
  system, 422
Source body capacitance, 62
Source charge, 65, 78
Source coupled split gate (SCSG), 318
Source gate capacitance, 62
Source resistance, 57, 59–60
Space charge, 44, 62
SPICE
  parameter, 93
  simulation, 94
Sputtering, 35
SR flip-flop, 260
SRAM, 24, 99
Stabilized feedback current sense amp (SFCA), 248
Stacked emitter polysilicon (STEP), 290
Standby (sleep) mode, 135
Standby subthreshold current, 277
Standby/active mode logic, 138
Static feedback, 406
Static random access memory (SRAM), 236
Static to domino interface circuit, 182
Storage capacitor, 265, 280
Sub-data line, 271
Sub-global word line, 250,
Substrate current, 72
  reduction technique, 297
Subthreshold
  conduction, 17, 74
  current hump, 52
  current, 49
  leakage, 52, 266, 277
  slope, 8, 22, 24, 51, 77, 101
Surface
  mobility, 56
  potential, 38
  scattering effect, 24
Swing reduction architecture, 146
Swing restored pass-transistor logic (SRPL), 111
Switched output differential structure (SODS) dynamic logic, 193
Switched source impedance (SSI), 138, 276
Synchronous DRAM (SDRAM), 274
Tag
  compare circuit, 376–377
  data, 379
  hit, 418
  memory, 378
  memory data, 380
Thermal conducting coefficient, 55
Thermal relaxation time, 57,
Threshold voltage, 6–7, 38–41, 52, 57, 71, 75, 130, 267, 281
  formula, 38, 42,
Time increment method, 78

INDEX **439**

Transient saturation technique, 210
Transistor triggered logic (TTL), 1
Translation lookaside buffer (TLB), 375–376, 418
Transmission gate, 166, 260, 354, 357–358
Transport current, 64
Triode region, 56, 60,
True-single-phase clocking (TSPC), 196, 198, 222
   latch, 198,
TTL, 289
   I/O SRAM, 289
Tunneling, 30
Twisted bit line (TBL), 270
Two-stage pipeline extended data-out (EDO), 274
Two-step word (TSW) voltage method, 261
Ultra-violet light, 312
Underflow, 409
Unity gain frequency, 24, 28
UV erase, 317
Velocity overshoot, 37, 57, 61
Vertical electric field, 45
Virtual power line, 276
Virtual
   address tag, 418
   page address, 418
   page number, 375

VLSI
   circuits, 17
   CMOS technology, 16
Volatile, 235
Voltage-controlled
   oscillator (VCO), 386–387, 389
   resistor (VCR), 387
Wallace tree reduction technique, 363, 368
Wave-pipelined circuit, 360
Weak inversion, 50–51
Wireless communication, 412
Word line
   boost technique, 261
   drivers, 374
   load, 293
   noise, 269
Work function difference, 38
Write
   access, 237, 249
   back (WB), 396
   enable (WE), 236
   operation, 237, 261, 264, 373,
Write recovery (WR), 246
   time, 237
XOR circuit, 102
Zipper CMOS dynamic logic, 177, 217